Mastication

Proceedings of a Symposium on the Clinical and Physiological Aspects of Mastication held at the Medical School, University of Bristol on 14–16 April 1975

Edited by
D. J. Anderson and B. Matthews

John Wright and Sons Limited Bristol
1976

ISBN 0 7236 0438 X

Printed in Great Britain by Henry Ling Ltd., at the Dorset Press, Dorchester, Dorset.

Preface

The purpose of this symposium was to bring together research workers, clinical investigators and interested dental clinicians, whose common aim is an understanding of mastication. We believe that while the high levels of specialization and sophistication in this, as in other fields, have brought many advantages, the opportunities for exchange of ideas and cross-fertilization are too few. The authors of papers were invited: the other participants came because they wished to learn. Our main aim and most difficult ask in editing the Proceedings has therefore been to enable as many people as possible to understand work going on in other specialties, without detracting from the scientific merit of the contributions. The discussions have presented a particularly difficult problem. We felt that questions asked and comments made should as far as possible remain unedited, revealing thereby gaps in knowledge and differences of thought and approach to biological problems between scientists and clinicians. This has often called for a tight hold on the editorial blue pencil. It is necessary for us formally to state that the papers presented in this volume have not been refereed in the sense that a scientific journal would do this.

There are a few technical points of editorial detail. We have not conformed with any particular anatomical terminology, but have accepted all terms which we understand. We have allowed those presenting papers and taking part in the discussions to make small modifications to what they said before it appeared in print.

All the references appear at the end of the volume because we think that this is a sensible plan and it was convenient to list those cited in the discussions separately under the heading 'Additional References'. No reprints of individual papers are available. Our publishers have been most co-operative and speedy; however, we must record our regret that a British publisher has forced upon us an alien spelling of the word neurone and compound forms thereof.

We are very happy to express our sincere gratitude to the sponsors, whose names appear on p. vii. Needless to say the symposium and this volume would not have been possible without their help. Although no trace remains in the printed proceedings, we still remember with gratitude the social side of our meeting for which we received valuable support financially and in kind.

Finally, we should like to thank all our helpers in Bristol who contributed in very many ways to the preparations and the minute-to-minute running of the meeting. Our thanks go especially to Mrs Wendy Coote who at the typewriter, on the telephone and in many other ways worked very hard before, during and after the meeting.

ANDERSON AND MATTHEWS

Contents

Sponsors of the Symposium

Financial support towards travel and other expenses:

- The Wellcome Trust
- Unilever Research Division and Elida Gibbs
- A. D. International
- Johnson & Johnson
- Cadbury Schweppes
- Coe Laboratories Inc. Chicago
- Reckitt & Colman Products
- Sony (U.K.)
- Carl Zeiss (Oberkochen)
- Data Laboratories
- Digitimer
- Medelec
- Polaroid (U.K.)
- The British Dental Trade Association
- Takiron (U.K.)

Financial support towards the cost of publishing the proceedings:

- Medical Sickness Annuity & Life Assurance Society
- Kodak
- S. S. White
- Parke Davis & Co.

Support in the provision of hospitality:

- The University of Bristol
- Avon Area Health Authority (Teaching)
- John Harvey & Sons
- Avery & Co.
- Berni Inns
- Stephen Carwardine & Co.
- Taunton Cider Co.
- C. C. Harris (Calne)
- Unigate Foods
- Devon Savouries
- Poupart (Bristol)

Sponsors of the Symposium

Speakers

PROFESSOR J. AHLGREN,
University of Lund, Department of Orthodontics, School of Dentistry, S–214 21, Malmö S, Sweden

PROFESSOR D. J. ANDERSON,
Department of Physiology (Oral Biology), University of Bristol Medical School, University Walk, Bristol BS8 1TD, U.K.

DR M. BRATZLAVSKY,
Polikliniek Voor Neurologie, Ruksuniversiteit, Ghent, Belgium

PROFESSOR A. J. BULLER,
Department of Physiology, University of Bristol Medical School, University Walk, Bristol BS8 1TD, U.K.

DR G. H. DIBDIN,
M.R.C. Dental Unit, Lower Maudlin Street, Bristol BS1 2LY, U.K.

DR C. EARL,
The National Hospital, Queen Square, London W.C.1, U.K.

DR L. GOLDBERG,
University of California, School of Dentistry, The Center for the Health Sciences, Los Angeles, California 90024, U.S.A.

DR G. GOLDSPINK,
Muscle Research Laboratory, Department of Zoology, University of Hull, Cottingham Road, Hull, U.K.

DR A. G. HANNAM,
Department of Oral Biology, Faculty of Dentistry, University of British Columbia, Vancouver 8, British Columbia, Canada

MR I. H. HESLOP,
Queen Mary's Hospital, Roehampton, London SW15 5PN, U.K.

DR KAREN M. HIIEMAE,
Anatomy Department, Guy's Hospital Medical School, London Bridge, London SE1 9RT, U.K.

DR C. HINRICHSEN,
The University of Tasmania, Box 252C, G.P.O., Hobart, Tasmania 7001, Australia

DR U. L. KARLSSON,
Dental Science Building, University of Iowa, Iowa City, Iowa 52242, U.S.A.

DR G. LINDBLOM,
Annero, 570 15 Holsbybrunn, Sweden

DR J. P. LUND,
Département de physiologie, Faculté de médecine, Case postale 6208,
Succursale A, Montréal, Québec, H3C 3T8, Canada

DR B. MATTHEWS,
Department of Physiology (Oral Biology), University of Bristol Medical School,
University Walk, Bristol BS8 1TD, U.K.

DR E. MØLLER,
The Royal Dental College, Copenhagen, Denmark

MR R. I. NAIRN,
Royal Dental Hospital, Leicester Square, London WC2H 7LJ, U.K.

PROFESSOR D. C. A. PICTON,
University College Hospital Medical School, Dental Department, Mortimer Market,
London WC1E 6JD, U.K.

DR D. F. G. POOLE,
M.R.C. Dental Unit, Lower Maudlin Street, Bristol BS1 2LY, U.K.

DR B. J. SESSLE,
University of Toronto, Faculty of Dentistry, 124 Edward Street, Toronto,
M5G 1G6, Canada

DR A. STOREY,
University of Toronto, Faculty of Dentistry, 124 Edward Street, Toronto,
M5G 1G6, Canada

DR R. SUMINO,
Section of Physiology, Institute of Stomatognathic Science, School of Dentistry,
Tokyo Medical and Dental University, 1-chome, Ushima, Bunkyo-ku, Tokyo, Japan

PROFESSOR A. TAYLOR,
Sherrington School of Physiology, St. Thomas's Hospital Medical School,
Lambeth Palace Road, London SE1 7EH, U.K.

DR A. J. THEXTON,
Department of Physiology, The Medical College, St. Bartholomew's Hospital,
London EC1M 6BQ, U.K.

DR H. THOMSON,
57 Portland Place, London W1N 3AH, U.K.

PROFESSOR D. M. WATT,
Department of Restorative Dentistry, 23 George IV Bridge, Edinburgh EH1 1EN,
U.K.

DR R. YEMM,
Department of Dental Surgery, University of Bristol Dental School, Lower Maudlin
Street, Bristol BS1 2LY, U.K.

Other Participants

A. B. Acton, Manchester, U.K.
S. Alldritt, Belfast, U.K.
M. Ash, Michigan, U.S.A.
R. M. Basker, Birmingham, U.K.
J. F. Bates, Cardiff, U.K.
J. Besford, London, U.K.
G. A. S. Blair, Belfast, U.K.
J. de Boever, Ghent, Belgium
F. Bosman, Utrecht, Holland
N. Brill, Copenhagen, Denmark
Y. Buyle-Bodin, Lyon, France
R. K. F. Clark, Chislehurst, U.K.
P. Cleaton-Jones, Johannesburg, South Africa
B. L. Dahl, Oslo, Norway
A. I. Darling, Bristol, U.K.
J. C. Davenport, Birmingham, U.K.
B. Derfler, Los Angeles, U.S.A.
G. C. Downer, Bristol, U.K.
E. Dransfield, Bristol, U.K.
S. C. Ericsson, Stockholm, Sweden
H. Fujii, Copenhagen, Denmark
M. Funakoshi, Gifu, Japan
H. I. Gill, Melbourne, Australia
M. Gillon, Kingston-on-Hull, U.K.
P. O. Glantz, Göteborg, Sweden
M. S. E. Gould, London, U.K.
H. Graf, Bern, Switzerland
B. E. Greenfield, London, U.K.
L. F. Greenwood, Toronto, Canada
M. J. Griffiths, Bristol, U.K.
T. Hamada, Hiroshima, Japan
R. Heath, London, U.K.
G. R. Holland, Bristol, U.K.
G. L. J. M. Honée, Amsterdam, Holland
C. R. Hansen, Stockholm, Sweden
K. Holmgren, Lidingö, Sweden
A. M. Horsnell, Adelaide, Australia
B. Ingervall, Göteborg, Sweden
G. Jacquart, Dombasle, France
H. J. de Jongh, Groningen, Holland
H. Kato, Copenhagen, Denmark
Y. Kawamura, Osaka, Japan
Y. Kawazoe, Hiroshima, Japan
J. G. Kennedy, Belfast, U.K.
A. Kettenis-Hyle, Heemstede, Holland
S. Landgren, Umeå, Sweden
C. S. C. Lear, Vancouver, Canada
B. Lennartsson, Göteborg, Sweden
P. H. D. Lewars, Exeter, U.K.
A. Lewin, Johannesburg, South Africa
R. W. A. Linden, Bristol, U.K.
S. J. W. Lisney, Bristol, U.K.
A. A. Lowe, Toronto, Canada
D. A. Luke, Newcastle-upon-Tyne, U.K.
A. R. MacGregor, Glasgow, U.K.
B. R. MacKenna, Glasgow, U.K.
I. Maddick, Winchester, U.K.
F. Magni, Genova, Italy
B. R. R. N. Mendis, Sri Lanka
M. W. Meyer, Minneapolis, U.S.A.
A. E. W. Miles, London, U.K.
C. Molin, Stockholm, Sweden
B. J. Moxham, Bristol, U.K.
C. M. Munson, London, U.K.
D. J. Neill, London, U.K.
A. J. Newton, Liverpool, U.K.
K. A. Olsson, Umeå, Sweden
H. J. Orams, Parkville, Australia
R. Orchardson, Bristol, U.K.
B. Öwall, Malmö, Sweden
P. D. A. Owens, Belfast, U.K.
B. J. Parkins, London, U.K.
O. C. Rasmussen, Sørborg, Denmark
J. Rayne, Oxford, U.K.
D. Rens, Randwick, U.K.
J. M. Richards, London, U.K.
C. Riise, Stockholm, Sweden
K. Rusiniak, Newcastle-upon-Tyne, U.K.
A. Scheikholeslam, Stockholm, Sweden
E. A. Scher, Hove, U.K.
W. D. Schwarz, Bromley, U.K.
S. Simard-Savoie, Montréal, Canada
B. H. Smith, Newcastle-upon-Tyne, U.K.
G. Solberg, U.S.A.
G. H. Sperber, Edmonton, Canada
G. D. Stafford, Cardiff, U.K.
J. P. Standlee, Santa Barbara, U.S.A.
C. D. Stephens, Bristol, U.K.
H. Sussman, London, U.K.
B. Thilander, Göteborg, Sweden
T. Troest, Århus, Denmark
G. Tryde, Copenhagen, Denmark
K. S. Turker, Glasgow, U.K.
T. O. Tweedie, Liverpool, U.K.

M. Vivier, Nancy, France
E. G. Walsh, Edinburgh, U.K.
J. G. Waterson, Nedlands, Australia
R. Watson, Sevenoaks, U.K.
W. A. Weijs, Groningen, Holland

S. E. Widmalm, Tranås, Sweden
J. van Willigen, Amsterdam, Holland
J. D. van Willigen, Groningen, Holland
G. Wreakes, Farnham, U.K.
P. S. Wright, Petts Wood, U.K.

Paper No. 1

Evolution of Mastication

D. F. G. Poole

Some three or four hundred million years ago, chordate animals evolved in the sea and eventually gave rise to the group we know as vertebrates. The basic chordate stock was characterized by the possession of a notochord, a hollow dorsal nerve cord and a pharynx with gill slits which formed a filter-feeding apparatus for collecting plankton. In general, plankton gathering is neither the quickest nor easiest way of obtaining food and, therefore, it is not surprising that in the succeeding vertebrates a number of evolutionary experiments occurred, aimed at changing from a microphagous to a macrophagous habit. Such experiments included, for example, the development of a horny-toothed oral sucker and tongue for rasping away the flesh of other animals which still survives today in the lampreys.

However, the one really successful experiment was the modification of the skeleton supporting the two anterior gill arches to form opposable jaws. Jaws themselves were derived from the first (mandibular) arch and the attachment support was derived from the second (hyoid) arch. This change produced the true fishes which can fairly be claimed to be one of the most successful groups of vertebrates. As far as we can tell, teeth arose simultaneously with jaws by the simple modification of the dermal denticles surrounding the mouth and, in most fishes, teeth occur on all jaw bones as well as on several palatal bones and on the rod-like tongue. In cartilaginous and bony fish, both upper and lower jaws are movable, being suspended posteriorly from the hyoid apparatus. Jaw action consists of the approximation of upper and lower jaw elements with specialization of the teeth into cutting dentitions, such as in sharks, or crushing dentitions, with flattened teeth, as in the rays and skates. In bony fishes there are many variations on this basic theme and in all fishes the teeth are replaced continuously throughout life.

In crossopterygian fishes, which accomplished the evolutionary transition from water to land, the upper jaw, as in all succeeding groups of land vertebrates, was firmly attached to the skull, with the lever-like mandible pivoted posteriorly. This freed the hyoid skeleton to assume new functions such as acting as the first ear bone, the stapes, and in providing support to structures in the throat and pharynx. Of course, on land the support once provided to both prey and predator by water is lacking and this may be related to the firmer anchoring of the upper jaw and also to the development of a flexible neck. Even so, the range of mandibular movements remains remarkably limited throughout the amphibians which succeeded the fishes and the reptiles which succeeded the amphibians.

*Scrutiny of the jaw activities of amphibians and reptiles shows that they are all relatively simple up and down movements with little lateral displacement. Jaws are essentially grasping, restraining devices and the simple, spiky teeth are able to puncture the skin or cuticle of living prey and thereby facilitate the penetration by gastric juices. Tearing of food may occur but never mastication and the living prey are usually seized head first and swallowed whole, facilitated by a tongue stiffened by a rod-like hyoid and by a large gape and huge pharynx. Sometimes accessory devices are present such as an anteriorly-hinged, sticky tongue in frogs and toads to pick up insects and their larvae, and movable (kinetic) skull bones in lizards and snakes. The mandible is compound, consisting of six separate bones. Thus, the mouth and jaws gather food which begins to be digested in the stomach.

In mammals, the one great advantage of mastication is the breaking down of food at the beginning of the gut, that is the mouth. This results in an enormous gain in digestive efficiency, necessary for the rapid metabolism associated with homoiothermy which also evolved with the mammals. Furthermore, true salivary glands appeared for the first time in mammals, their secretions moistening and lubricating food and adding chemical degradation by enzymes to mechanical degradation. Supersaturation with essential ions enabled the buffering of natural and bacterial acidity to occur, as well as the restoring of the integrity of tooth surfaces by means of remineralization.

Mammalian teeth changed from being grasping and piercing devices to elements of a cutting and milling machine. Wholesale modifications of the skull and mandible included the reduction of the mandible to a single bone, the dentary, and this occurred simultaneously with the rebuilding of the middle ear where all the ear ossicles have evolved from jaw bones. Also, profound restructuring of the craniomandibular musculature took place together with the emergence of new structures such as the lips and cheeks which, together with the tongue now powered entirely by intrinsic muscles, assist in positioning food between the teeth and in swallowing. The development of a secondary palate and an elaborate mechanism which safeguards the entrance to the pharynx, make it possible for breathing to continue during prolonged mastication.

Thus, in summary, one sees that the cranial revolution which accompanied the evolution of mammals from reptiles left its mark on almost every structure in the head and has resulted in the ability to satisfy the mechanical requirements of the mastication of food (56, 137). And from the point of view of this symposium, there are several points which are worth emphasizing.

The first mammals were small, heterodont creatures with high pointed cusps for piercing, as may still be seen in modern hedgehogs and shrews, survivors of insectivorous mammals. Rapid diversification from this stock occurred with corresponding diversification of teeth and temporomandibular articulation. There was a reduction in tooth numbers and an adaption of crown form towards either sectorial teeth for cutting or broad-crowned teeth for grinding. Clearly, basic crown form is genetically determined but functional form depends on use. Thus, the carnassials of cats are basically elongated, molariform teeth but the production of the functional cutting edge is achieved by working lower against upper, cross-biting being avoided by means of 'locked' canines. The transverse condyles prevent the jaws from slipping forwards. Contrarily, the flattened surfaces of the grinding molar battery of ungulates are produced by sliding the lower teeth across

*This part of the paper was illustrated by a cine film.

the upper, effected by means of a universal temporomandibular joint; some or all of the canines are absent.

It may be asked, if they are going to be ground flat, what are the advantages to the ungulate of developing molars with elaborate crown forms including high, complex cusps? The answer is that a simple crown form would result in the flattened surface consisting of dentine surrounded by a simple ring of enamel which, with grinding, would become deeply saucerized with a raised edge of enamel subject to mechanical risk. By retaining, and even elaborating, the developmental form of the cusps, the functional surface consists of dentine broken up by infolded enamel, often with cementum filling the folds, resulting in a markedly advantageous grinding surface. Such a complex grinding surface is a direct consequence of the initial crown form and we can therefore conclude that the presence of developmental cusps does not necessarily indicate a function, in biting or chewing, of those cusps as cusps.

These examples demonstrate how the pattern of occlusion in any animal is a function of basic crown form plus masticatory wear. Furthermore, exposure of dentine in mammals is accompanied by peritubular infilling and secondary dentine formation (features not present in reptiles) which maintain the efficacy of the grinding surface. This function is supported by the remineralizing role of saliva.

We can now understand the evolutionary basis for continuous tooth replacement in reptiles being succeeded by just two generations, and sometimes only one, in mammals. A few teeth missing from a well-spaced row in a reptile would make no real difference to the functioning of that dentition. On the other hand, the loss of a carnassial in a cat, or of a molar from the grinding battery of an ungulate, would result in considerable handicap, requiring a long eruption time plus wearing-in time before full function could be restored. In other words, continuous tooth replacement is incompatible with a dentition in which correct function is only achieved by attrition and, in some mammals, this wearing-in process is so important that it occurs in utero so that the dentition is functional at birth (9, 56).

If this, then, is the biology of mastication, how do we define the natural pattern of occlusion in man? We are all aware of what is currently accepted as the ideal occlusion in Western man, but how natural is it? An examination of contemporary Eskimo or Aborigine dentitions, reveals that, by early teenage, some molar wear has occured and that, by the age of about 20, all the teeth are flattened with the grinding surfaces of molars resembling those of ungulates. Similar observations have been made on Anglo-Saxon dentitions (396). Likewise, a contemporary Javanese dentition has all the appearance of a milling machine (56) and, even though the incisor crowns are obliquely flattened, they still retain a good functional edge for biting.

These degrees of wear are undoubtedly considerable but there are compensatory devices such as peritubular infilling and secondary dentine formation which, bolstered by the remineralizing role of saliva, provide new grinding surfaces. Furthermore, tooth eruption and alveolar bone growth continue, and maintain facial height (420). It would be of interest to know if there are any regulating (feedback ?) mechanisms associated with these compensatory devices. From recent investigations in man (134), it seems clear that peritubular infilling is governed more by the rate of attrition than by age. On the other hand, slow eruption and alveolar bone growth continues with age so that, in the absence of attrition and periodontal disease, facial height increases (10).

Consideration of a great deal of evidence of the kind outlined above forces me to the conclusion that the true, natural occlusion in man is one in which teeth are ground quite flat. Indeed, the persistence of cusps into early middle age has only occurred in Western man for the last two or three hundred years (99, 100, 398). On these grounds, the currently accepted ideal occlusion must be regarded as unnatural. My own feeling is that, in the control of clinical disease and oral rehabilitation we should aim at the true, natural occlusion.

Discussion CHAIRMAN: D. J. ANDERSON

ORAMS: What do you think is the purpose of cusps on teeth ?
POOLE: The only advantage I can think of in man is the establishment of the teeth in the correct position in the first place. I think this is probably true of all mammals, when one looks at the careful way the dentition develops and the way the teeth come through in a certain sequence. I see the cusps as guiding the teeth into the right position to start with. Having got them there, I personally believe that the cusps should go.

Other points which have been put to me by Mr Ian Maddick are, first, that the presence of pointed cusps will reduce the load required to puncture and shear food during eruption when the periodontal support mechanism is developing and, secondly, that the production of flattened, planar surfaces on all teeth might be achieved more easily by starting with conically shaped cusps which initially have relatively small surface areas.
GREENWOOD: When you showed the film of the toad eating you commented that it used its eyes to assist in swallowing. Would you like to elaborate on that ?
POOLE: There is no secondary palate, or bony floor to the orbit, so the eye muscles form the roof of the mouth. The eyes are very large compared with those in mammals and, in swallowing, all the muscles associated with the mouth, including the eye muscles, are brought into operation. Thus, every time the animal swallows, its eyes are brought downwards into the mouth.
SUSSMAN: In fact, it is not using the eyes to push the food. The muscles which push the food also happen to move the eyes.
PICTON: I just wonder whether there is an additional advantage of cusps during development. It could be advantageous that they are very small when tooth development starts. The cusps begin to mineralize first and at this stage are bunched up together and therefore occupy little space in a small developing creature. As tooth formation progresses, so the cusps are spread out.
POOLE: Yes, with eruption the cusps spread out as the mandible enlarges. Although some of the pig teeth are very large and they don't have quite the dramatic slope we saw on the hippopotamus teeth.
PICTON: During development the cusps are very close together in pigs, about one-half the width of the eventual tooth.
LEWIN: I wonder whether the purpose of cusps perhaps is purely to increase the surface area of the tooth.
POOLE: No, I think that, in grinding animals, the cusps ensure a greater area of enamel on the grinding surface.
GOLDBERG: What do you regard as the consequences of not flattening cusps ?
POOLE: Dental caries, because of the persistence of fissures and interstitial stagnation sites. Also some forms of temporomandibular joint dysfunction and perhaps periodontal disease.

Paper No. 2

The Diet and Dentition of New Kingdom Pharaohs

A. Storey

For the major portion of my presentation I will attempt to relate diet to dentition in the period of the New Kingdom in ancient Egypt encompassing dynasties XVII to XX and extending from about 1600 B.C. to 1100 B.C. This period encompasses the time of Joseph's sojourn in Egypt, the zenith of the Cretan Empire, the Israelite exodus and the sacking of Troy. I will be limiting my discussion to the pharaohs of the New Kingdom since we have studied them most thoroughly (*Table* 2.1). With the exception of Tutankhamun, all the mummies identified by + are in the Cairo museum. Dr James E. Harris of the University of Michigan has directed these investigations which were designed to study the craniofacial characteristics of the pharaohs but not caries or periodontal disease. These studies have been extended to include a full body radiographic survey of the pharaohs and their queens. The results of these studies will appear in an atlas to be published by the University of Chicago Press. A popular account of the study has been published (247).

In the X-rays I will be showing you the radiation source was a 90 KVP dental X-ray unit mounted on a specially designed adjustable tripod. The mummies were

Table 2.1 A list of the pharaohs of New Kingdom Egypt. With the exception of Tutankhamun, all the pharaohs marked with a + are in the Cairo Museum and were X-rayed by the Michigan expedition

	Seventeenth Dynasty (–1546BC)		*Nineteenth Dynasty (1319–1199)*
+	Sekenenre		Rameses I
	Kamose	+	Sethi I
+	Ahmose	+	Rameses II
		+	Meneptah
	Eighteenth Dynasty (1546–1318BC)	+	Siptah
+	Ahmenhotep I	+	Sethi II
+	Thutmose I		
+	Thutmose II		*Twentieth Dynasty (1200–1085)*
	Hatshepsut		Sethnakht
+	Thutmose III	+	Rameses III
+	Amenhotep II	+	Rameses IV
+	Thutmose IV	+	Rameses V
+	Amenhotep III	+	Rameses VI
	Amenhotep IV - Akhnaton		Rameses VII
+	Smenkhare		Rameses VIII
+	Tutankhamun	+	Rameses IX
	Eye		Rameses X
	Harmhab		Rameses XI

radiographed within their wooden caskets so that the X-rays always passed through one side or the bottom of the casket. The lateral head X-rays were taken with the central beam passing through both external auditory meati at a source–target distance of 60 inches. The conventional cephalometer with its rigidly linked X-ray tube and cephalostat could not be used so that correct orientation of the central beam to the head was obtained using a specially designed gun sight on the X-ray head and an aligning device temporarily placed over the mummy's head. A right-angled gun sight oriented along the exit path of the central beam was mounted on a threaded collar which could be readily attached to and removed from the X-ray head. A metal channel carrying four arms, tipped with small silvered balls in a straight line, was mounted near the mummy's head so that two balls were positioned inside the casket opposite the meati and the other two were outside the casket (*Fig.* 2.1). The X-ray tube was positioned so that the central beam was aligned with the two balls outside the casket. The sighting and aligning devices were removed after proper orientation and the film cassette placed inside the casket to the left of the head for exposure. For anteroposterior views the X-ray head was positioned 60 inches above the head and the cassette placed underneath the wooden casket. With the exception of two of the pharaohs, this was their first radiographic examination.

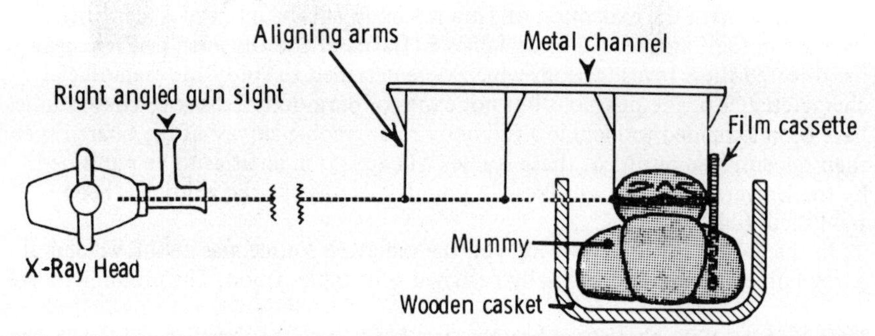

Fig. 2.1. The device used to align the central X-ray beam with both external auditory meati in the taking of lateral cephalograms. The tips of the aligning arms adjacent to right and left porion are in a straight line with those outside the casket. The X-ray tube was aligned with the tips of the two outside aligning arms by means of a removable right-angled gun sight oriented along the exit path of the central beam. The source–target distance was fixed at 60 inches. The aligning devices were removed prior to taking the cephalogram.

Some pharaohs had excellent dentitions: others, badly deteriorated dentitions. Sethi II (*Fig.* 2.2*a*), the last pharaoh of the XIXth dynasty, who probably died in his mid twenties, had a full complement of 32 teeth in good health. Rameses V (*Fig.* 2.2*b*), a pharaoh of the XXth dynasty who likewise died rather young, possibly of smallpox, also had a fine dentition. In an anteroposterior view it is possible to see normal healthy alveolar bone around the mandibular incisor teeth. In contrast to these pharaohs the oldest mummies displayed badly broken-down dentitions. Rameses II (*Fig.* 2.2*c*), the oldest and the most famous of the New Kingdom pharaohs, was over 50 when he died. The lateral head X-ray shows considerable destruction of tooth substance both by wear and other processes and there is clear evidence of periodontal breakdown in the region of the mandibular first molars. Meneptah (*Fig.* 2.2*d*), son of Rameses II and pharaoh of the Exodus, had a very derelict dentition at the time of his death, estimated at fifty years.

Many of his teeth are missing and bone loss is excessive around the remaining maxillary first bicuspid and the mandibular second molars. In the pharaohs and queens of the XVIIth and XVIIIth dynasties we discovered Class II, division 1 malocclusions. The mummified queens of the last pharaoh of the XVIIth dynasty both had Class II malocclusions, suggesting that pharaoh Ahmose liked a Leslie Caron face. Amenhotep I, the son of one of these queens, also demonstrates a Class II, division 1 malocclusion.

The dentitions of Rameses II and Meneptah indicate that the services of a dentist could have alleviated the severe dental problems of these two men in their

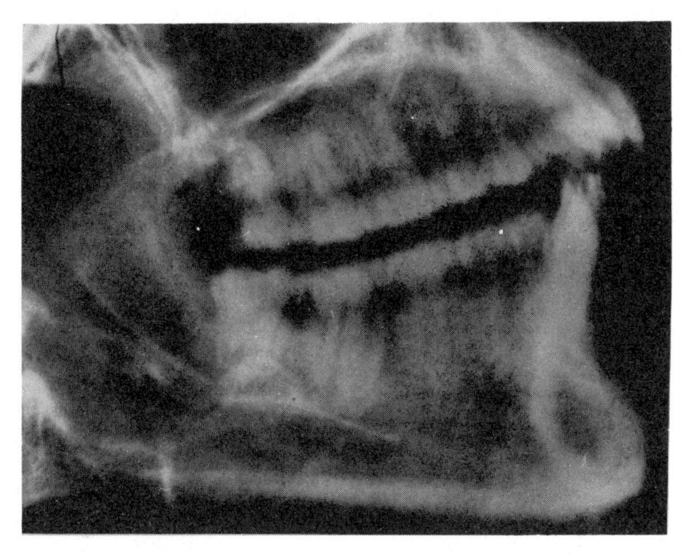

a

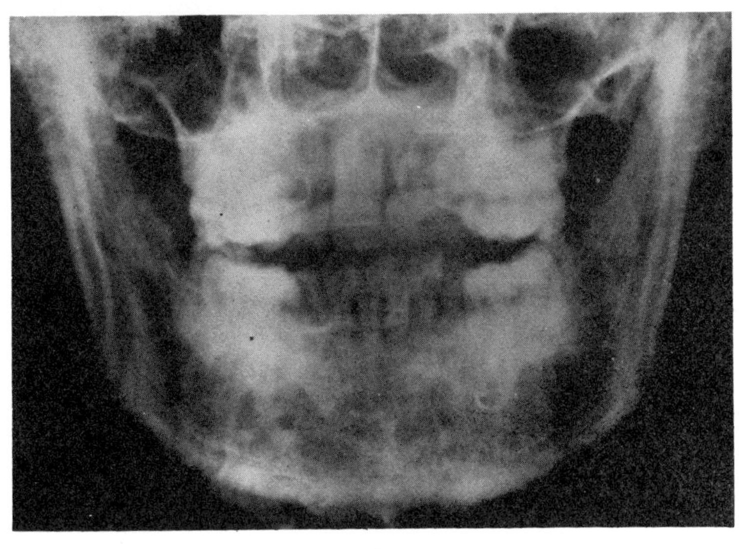

b

(*see* p. 8 for legend)

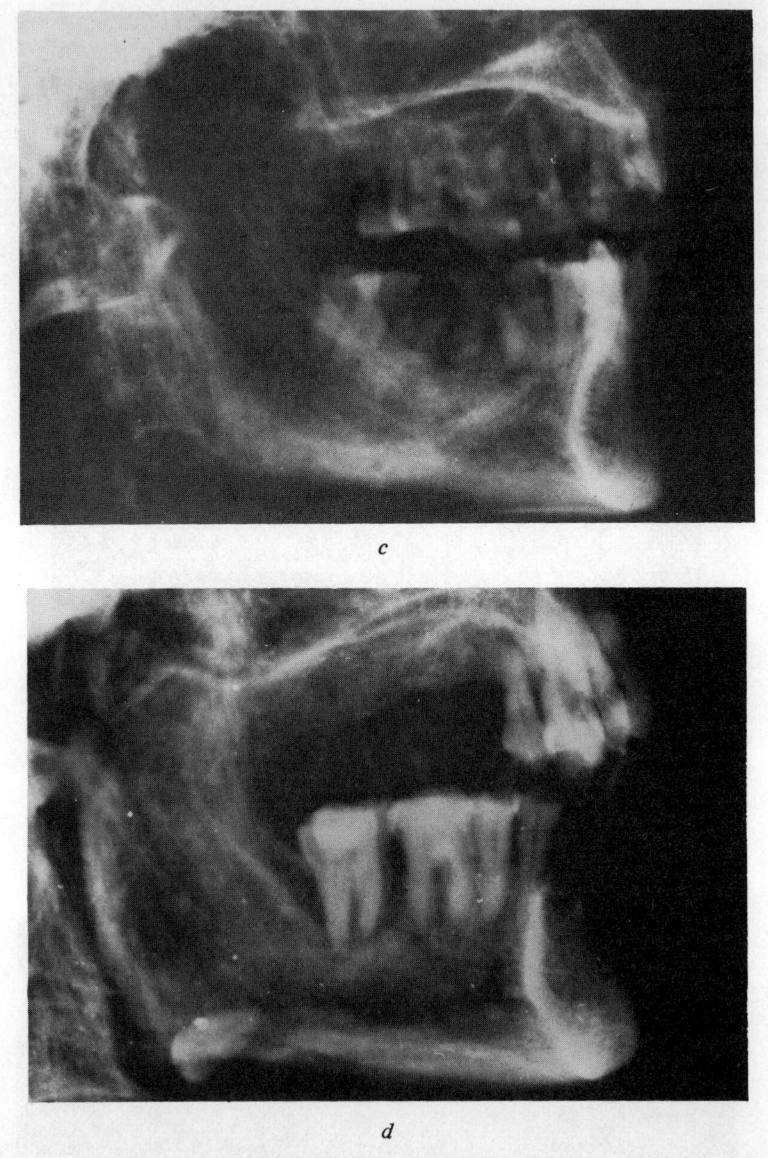

c

d

Fig. 2.2. Status of the dentitions of four New Kingdom pharaohs. Views of the dentitions were taken from lateral and anteroposterior cephalograms. *a,* Sethi II, *b*, Rameses V, *c*, Rameses II, *d,* Meneptah.

later years. There is ample textual evidence that dental practitioners existed in ancient Egypt—five references in the Old Kingdom and one in the XXVIth dynasty (91). The most famous evidence for dentists in ancient Egypt comes from the beautiful Hesi Re stele of the 3rd dynasty (*Fig.* 2.3). This is one of five beautifully carved and coloured wooden panels from Hese Re's tomb at Saqqara. The titles of interest in this stele are in the upper right corner of the panel. The hieroglyphic symbols of the bird, tusk and arrow have been interpreted as

'Chief of the Dentists and Physicians' or 'Chief Physician of the Teeth'. The tusk, a symbol for the tooth, also appears in the third register to the left above the symbol of the eye. This designation has been interpreted as 'Chief Treater of the Teeth' or 'Tooth Worker'. To some, this has suggested there may have been two types of dentists in Old Kingdom Egypt. Perhaps the former type may have specialized in oral medicine and the second in dental prosthetics. There is abundant evidence in the medical papyri of a variety of prescriptions for the treatment of dental and oral ills. As Leek (346) has pointed out, until recently there has been no extant evidence of prosthetic intervention. This view must now be revised with the recent discovery of two skulls of the Old Kingdom showing wires connecting a dummy incisor to two adjacent teeth. However, we found no evidence of dental treatment in any of the pharaohs or queens which we examined. Since there is no textual evidence of dentists in the New Kingdom and since there is no evidence of dental treatment perhaps we may conclude that the prosthetic art had fallen into decline.

Having described in general the status of the dentitions of the pharaohs of the New Kingdom I would like to turn to a consideration of their diet. Information about foods eaten in Ancient Egypt comes from five sources. Temple inventories of offerings of food, such as those on the outer walls of the temple at Medinet

Fig. 2.3. One of the wooden panels from the tomb of Hesi Re at Saqqara. Hesi Re served many important functions including court dentist during the reign of a pharaoh who reigned early in the Old Kingdom (about 3000 B.C.). The three hieroglyphs of bird, tusk and arrow in the upper right corner of this stele (indicated by ←) identify Hesi Re as 'Chief of the Dentists and Physicians' or 'Chief Physician of the Teeth'.

Habu and in the Harris Papyrus, give us an indication of the foods which were thought suitable for the gods and which were presumably eaten by the priests and retainers of the temple. Routine daily offerings recorded at Medinet Habu list 3220 loaves of bread, 24 cakes, 144 jugs of beer, 32 geese and several jars of wine. On festival days the offerings were very much larger. Tomb paintings of the same period give an indication of the food offered to the protecting god and also required by the deceased in the life after death. Paintings from the tomb of Nakht (a scribe of the XVIIIth dynasty) depict Nakht and his wife offering grapes, wine, geese, lettuce and beef to his god; while grapes, figs, melons, eggs, geese, ducklings and lotus flowers are being offered to Nakht and his wife for sustenance in the life hereafter. Ration lists of food distributed to the army are helpful in comparing the diets of the privileged with the less privileged. Orders for the preparation of a banquet for a Rameside pharaoh (101) gave detailed information about specific meals, as do the remains of funerary repasts such as those described by Winlock (630) and Emery (161). The Winlock paper describes the remains of a funerary banquet for eight persons at the time of the separate burial of the materials used at the embalming of King Tutankhamun. The identity of the eight banqueters is a mystery. The meal consisted of meat, fowl, bread and cakes washed down with copious amounts of wine or beer. From the bones which remain, it is possible to identify the shoulder blade and connecting bones of a cow, the ribs of a sheep or goat, the skeletons of nine ducks and four geese. Most of the serving dishes were deliberately broken at the end of the meal. Flower collars worn at the banquet were stuffed into large jars along with the broken dishes and two brooms used to sweep away the footprints of those who attended these final ceremonies for King Tutankhamun. In contrast to this New Kingdom feast, the repast described by Emery was left intact, presumably for the use of the deceased in the life hereafter. The meal was discovered undisturbed since the time of the burial almost 5000 years earlier. The meal consisted of:

Triangular loaf of bread made from emmer wheat on a pottery dish.
A form of porridge made from ground barley on an alabaster dish.
An unidentified liquid containing some sort of fatty substance.
A cooked fish cleaned and dressed with the head removed on a pottery dish.
Pigeon stew on a pottery dish.
A cooked quail cleaned and dressed with the head tucked under one wing on a pottery dish.
Two cooked kidneys on a pottery dish.
Ribs and legs of beef on a pottery dish.
Unidentified material containing cut ribs of beef.
Stewed fruit, probably figs, in a pottery bowl.
Fresh nabk berries (from the sidder; rather like cherries in appearance) in a diorite bowl.
Small circular cakes sweetened with honey on a pottery dish.
Three small jars containing some form of cheese.
Grape wine in a large jar.

As in the previous burial feast, many stone vessels had been deliberately broken, the most plausible reason being that the vessels had to be 'killed' so that they could be carried with their deceased owner into the after life. From such varied sources it is possible to compile the list of foodstuffs found in *Table* 2.2 (163, 487, 665).

Table 2.2 A list of foodstuffs from New Kingdom Egypt. The list has been compiled from temple inventories, tomb paintings and inclusions, ration lists, menus and funerary banquets

Foods of New Kingdom Egyptians	
Beef	Watermelon
Fowl	Grapes
Fish	Figs
Milk	Dates
Eggs	
	Mandrake
Beans	Carob
Peas	Dom-Nuts
Onions	
Leeks	Honey
Lettuce	Beer
Cucumber	Wine
Bread	

The two items in this list of most relevance to the discussion of diet and dentition are honey (because of its relationship to dental caries) and bread (because of its relationship to wear of the teeth). As Lucas (364) has shown, honey would be the only significant source of concentrated sugar in New Kingdom Egypt. Honey is listed as one of the offerings at the Temple of Medinet Habu. It forms one of the rations of a king's messenger and standard bearer of the XIXth dynasty. In the tomb of Rekhmire, governor of Thebes in the early XVIIIth dynasty, combs and jars of honey are depicted. Other tomb paintings record the harvesting of honey and its processing. Although there is ample evidence for the demand for honey in ancient Egypt, it is difficult to estimate the magnitude of the supply. In dynastic Egypt, honey would have been collected from wild bees; it is not until the Ptolemaic period that bees were domesticated and kept in pottery jars. Lucas (364) reported examining two small pottery jars from the tomb of Tutankhamun each marked in hieratic with 'honey of good quality'. In contrast to the large number of jars containing wine, oil and precious unguents the small number of vessels containing honey suggest a limited supply (249). The incidence of caries would tend to bear this out, in spite of the misleading statements of Elliot-Smith (159) and Ruffer (488) that caries was rampant in Dynastic Egypt. Elliot-Smith's (160) fantastic attribution of the appearance of caries in the Old Kingdom Egypt to an Alpine intrusion was effectively refuted by Derry (144) at the time it appeared. Our X-ray studies have repeatedly proven Elliot-Smith's dental observations to be incorrect. These misleading statements about caries emphasize the care with which a diagnosis must be made in ancient skulls to differentiate the lesion from attrition, abrasion and erosion. There is very little evidence of caries in any of the pharaohs, with the possible exception of Rameses II and Meneptah. These observations agree with those of Leek (346) which showed a low caries incidence in skulls of the pre-dynastic period, as well as of the Old and Middle Kingdom. I am inclined to attribute this low incidence to a limited supply of honey in Ancient Egypt. Another explanation might be that the sugars in honey are less cariogenic, but Bodden et al. (64) have recently shown that honey is 1·5 times more cariogenic than sucrose.

Although caries is less common in the New Kingdom compared with our society, wear of the teeth is much more extensive. In the very young pharaohs wear is minimal. Between 20 and 40 years of age, wear becomes moderate; and over 50 years of age, extreme. The dentition of Rameses V (*Fig.* 2.4*a*) has minimal wear, that of Thutmose IV (*Fig.* 2.4*b*), moderate wear and that of Sekenenre (*Fig.* 2.4*c*), extreme wear. As others have noted, the wear may be of a flat or of a cupping nature, but the mechanism accounting for one type rather than the other seems to be unknown. Ruffer (488) and Leek (346) attribute the major cause of this wear to abrasives in the flour used to make bread. Bread was very popular in ancient Egypt. The Harris Papyrus (162) speaks of 30 types of bread being offered in the temples. The Egyptian soldier's ration was four pounds per day. Leek (347) has demonstrated, by radiological examination, the presence

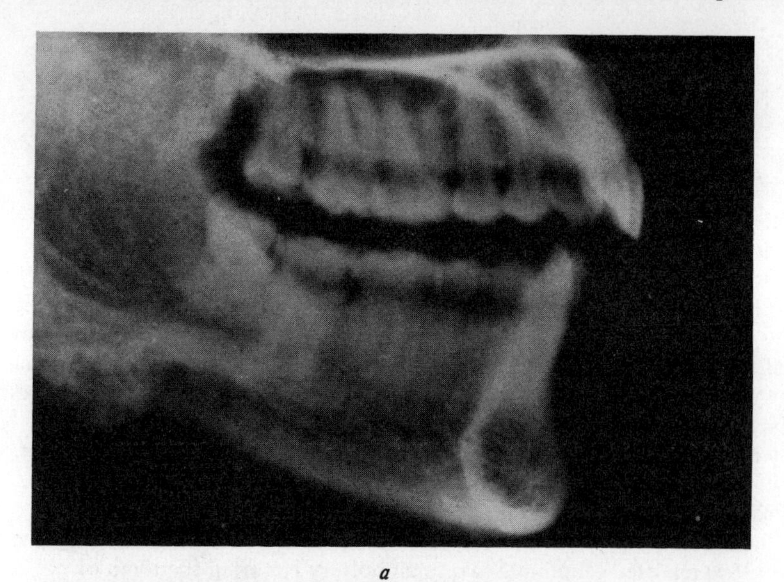

a

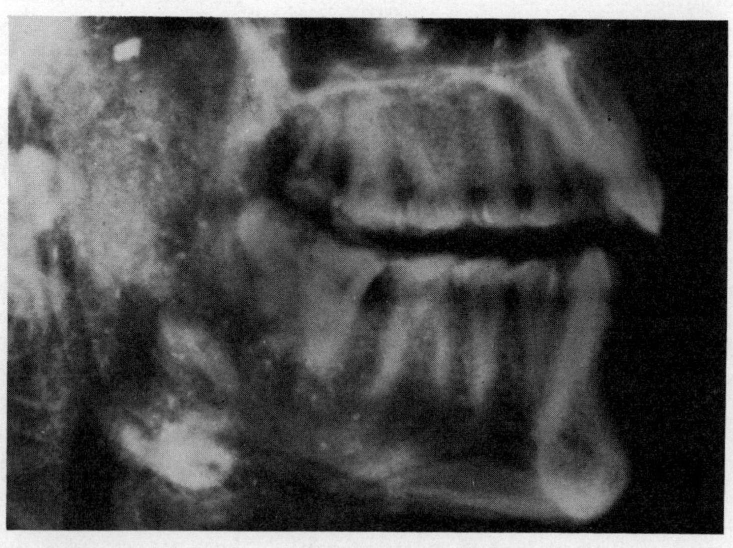

b

(*see* p. 13 for legend)

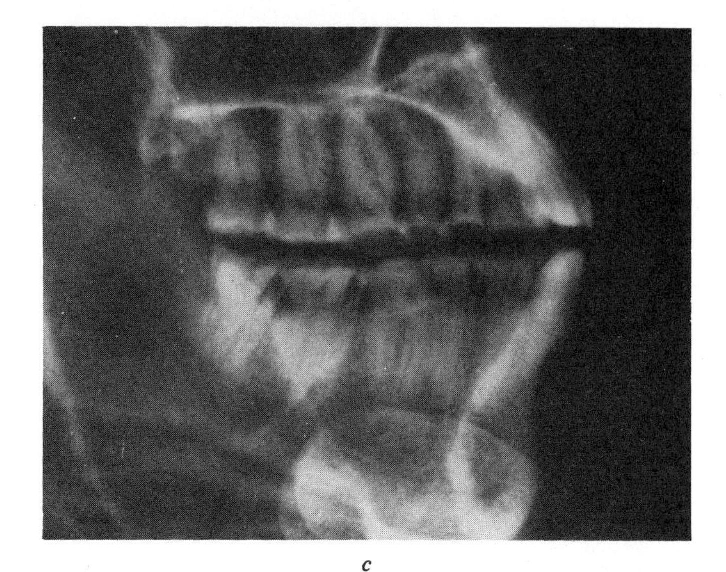

c

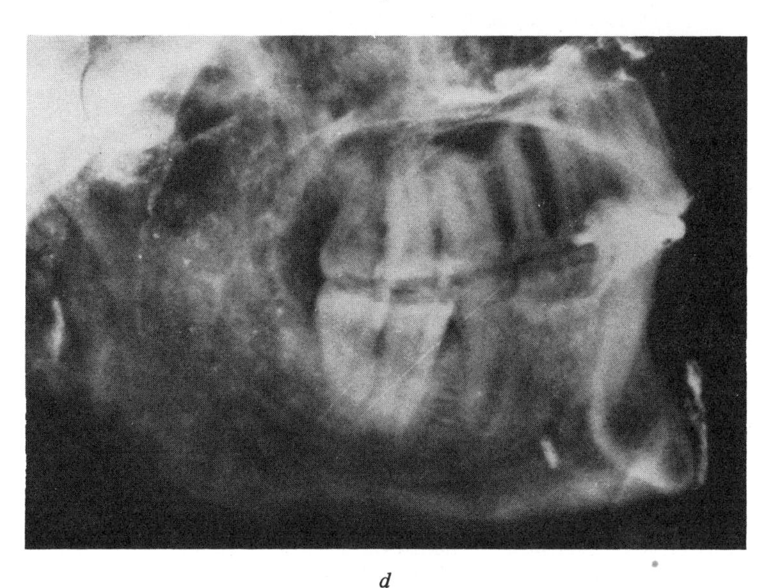

d

Fig. 2.4. Extent of tooth wear in the dentitions of four New Kingdom pharaohs. Views of the dentitions were taken from lateral cephalograms. *a*, Rameses V, *b*, Thutmose IV, *c*, Sekenenre, *d*, Ahmose.

of inorganic material in a number of samples of bread from the New Kingdom. Further testing demonstrated that the inorganic material in Egyptian bread was primarily wind-blown desert sand. Since the grain was ground into flour using either large stone mortars and pestles or saddle-stones and querns it seems likely that some of the inorganic material could have come from these stones.

Another possible mechanism of wear could be non-functional grinding of the teeth (bruxism), but this is difficult to determine. Dr W. M. Krogman has estimated the age of each pharaoh from the full body radiographs and I have plotted the degree of wear against his age estimates in order to see how well the degree of wear related to age. As can be seen in *Table* 2.3 the degree of wear correlates reasonably well with age except for the young pharaoh Ahmose (*Fig.* 2.4*d*). Was Ahmose a bruxer ? Since Ahmose is credited with re-unifying Upper and Lower Egypt after the Second Intermediate Period, perhaps his reign was more stressful than that of some of his successors. Or is it possible that he consumed excessive amounts of abrasive food ? Although it is by no means certain, one of the most probable periods for Joseph's sojourn in Egypt is the end of the Second Intermediate Period or the beginning of the New Kingdom (408). You may recall that, at the time of Joseph's incarceration, the pharaoh's baker had been jailed with him for making excessively gritty bread. Is it possible that Ahmose was the pharaoh whose dreams Joseph interpreted correctly ?

Table 2.3 A comparison of degree of wear with age in New Kingdom pharaohs. Wear has been classified into three categories, and age into five categories. The estimates of age were made by Dr W. M. Krogman from full body X-rays. As would be expected, the extent of wear increases with age. Ahmose, who was estimated to be 30 years of age at death, has more wear than would be expected at this age

Relation of Dental Wear to Age in New Kingdom Pharaohs

*Age**	*Wear*		
	Minimal	*Moderate*	*Extreme*
<20	Smenkhare	Thutmose I	
21–30	Thutmose II Amenhotep III Siptah Rameses V	Amenhotep I Sethii II	Ahmose
31–40		Thutmose IV Rameses III	Rameses IV Rameses VI
41–50		Amenhotep II	Sekenenre Thutmose III Sethi I Meneptah
>50			Rameses II

* Estimated from full body radiographs by W. M. Krogman.

Excessive wear on the teeth, particularly when it is rapid, may have dire consequences for the health of the dentition. Wear which exceeds the deposition of secondary dentine may lead to pulp exposure, pulp infection and thence to pulp death and periapical pathology. Such destructive wear is very common in the skulls of the ancient Nubian and Old Kingdom nobles buried at Gizah. *Fig.* 2.5 demonstrates how wear may destroy most of the crown of the tooth and, on exposing the pulp, lead to pathology of the supporting bone. Wear of this severity was seen only in the oldest pharaohs such as Rameses II and Meneptah. This observation suggests that pharaohs tended to die young, as Krogman's ageing suggests, or that their diets contained less abrasive material.

In summary, it would appear that the rulers of New Kingdom Egypt enjoyed a highly nutritious and varied diet, with bread which was more abrasive than ours. The pharaohs demonstrated increasing tooth wear with age, which is more marked than in our society, but less than that seen in less privileged contemporaries.

Although honey was certainly available, the quantities consumed were so small as to have little effect on caries incidence. Periodontal disease ravaged the dentitions of the older pharaohs. Class II, division 1 malocclusions are found in several of the pharaohs and their queens of the pre-Rameseid period.

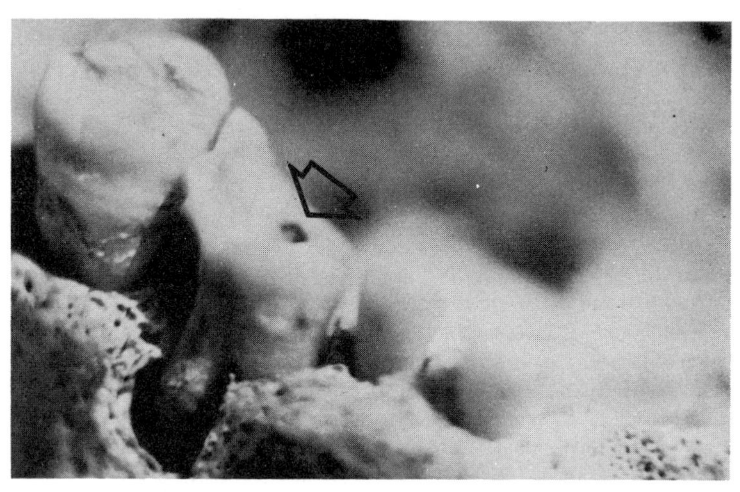

Fig. 2.5. An example of excessive wear of the teeth leading to loss of the clinical crown of two mandibular bicuspids and exposure of the pulp in the first molar. Pulpal exposure (indicated by ←) resulted in pulp death and a periapical abscess around the distal root. Photo is of the dentition of a Nubian of the early dynastic period. (*By courtesy of Dr J. E. Harris.*)

Discussion CHAIRMAN: D. J. ANDERSON

THOMSON: From this symposium so far, it would seem that a little wear is a good thing for the teeth. I should like to state the point of view that our cusps are with us for the trituration of food without their touching, and to touch only on swallowing to guide the path of the mandible and prevent us having to grind ourselves into a position we cannot find.

CHAIRMAN: Thank you for saying that: I was wondering whether I would have anything to argue with anybody about on Wednesday.

SPERBER: Could I ask if you have encountered any evidence of dental work having been done on the teeth of mummies.

STOREY: We didn't, and we were looking for it, but you have to recognize that a cephalogram would not be the ideal way of finding it. One would want to look in the mouth and this is impossible. Perhaps, at some time, it may be possible to do a survey using say the pantograph, which might provide better evidence than we have obtained from the cephalogram. We didn't see any evidence of intervention, and wondered why, but have assumed that there wasn't any.

SPERBER: I am particularly referring to the partially edentulous jaw which you displayed. I wondered if this was due to extraction.

STOREY: That is possible, but the teeth may have been lost because of periodontal destruction. Did someone actually pull the tooth out for the pharaoh in the end, or did he pull it out himself ? There is no way of knowing.

Paper No. 3

Fibre Types in the Muscles of Mastication

A. Taylor

It is now known from the study of limb muscles that muscle fibres are homogeneous within each motor unit (157), but show structural and histochemical specializations between units (e.g. 112). Research into these specializations has flourished in the past ten years because there is, on the one hand, a conviction that it should throw light on the basic biophysical mechanisms of muscle, and on the other a desire to see to what extent muscle is adapted to efficient use in a wide variety of situations (252). The latter is particularly relevant to this symposium, but in presenting an account of the fibre types in the jaw muscles I shall also pay some attention to the former, in so far as we can exploit certain unusual features to supplement the main area of knowledge which relates to other muscles.

PRINCIPLES OF MOTOR UNIT CLASSIFICATION

After the work of Ranvier (471, 472), redness of muscle was generally associated with slowness of contraction and this view was repeated as late as 1929 by Denny-Brown (143) and 1930 by Cooper and Eccles (119), despite the production of clear evidence of important exceptions by Meyer (395) in 1875 and by Paukul (448) in 1904. Interestingly, these exceptions were noted in jaw muscles. Thus, rabbit masseter which is red, histologically resembled the majority of white muscles, and its twitch speed was similar to that of gastrocnemius. Though the terms 'red', 'white' and 'intermediate' are still used to some extent, the density of myoglobin and lipid droplets on which the obvious coloration depends are seldom measured and it seems better now to rely on specific histochemical reactions.

Difficulties in classification of units have arisen because workers have generally concentrated either on structural or on physiological features, but in their desire to emphasize the interrelationships have used nomenclature which is not strictly justified. However, it now seems that the most reliable classification makes use of a combination of histochemical tests for myosin ATPase (MATPase), oxidative enzymes (such as succinic dehydrogenase, SDH or NADH tetrazolium reductase) and glycolytic enzymes and of physiological measures of contraction dynamics and fatiguability. Thus, the MATPase reaction at pH 9·4 (442) divides fibres into two groups: weakly reacting and strongly reacting—the types I and II respectively of Dubowitz and Pearse (148, 149), and Fenichel and Engel (169). It has been shown that large scale MATPase activity correlates well with contraction speed (36) and single unit studies have confirmed that type II units are fast in normal cat muscles (97).

While the type I fibres appear to belong to homogeneous populations of small, slow units (the S units of Burke) (96), the type II group may be further subdivided by reactions for SDH (423) and NADH tetrazolium reductase (499) into oxidative poor and rich (148). The latter have a particularly characteristic peripheral distribution of the reaction product. The justification for this second histochemically-based division is that it correlates well with a second physiological characteristic, namely fatiguability (97, 437), and it is now accepted that the dense uniform distribution of mitochondria rich in oxidative enzymes in type I fibres, and the dense peripheral distribution in a proportion of type II fibres are essential structural correlates of resistance to fatigue. Just how completely the fast (type II) units may be divided into fatiguable and non-fatiguable types is at present unsettled. In some cases (97) the division into large fatiguable and small non-fatiguable fast units (FF and FR respectively) has seemed to be sharp and reliable, in others (543) fast units have shown a continuous range of fatiguability and size with no clear-cut division. The explanation of the differences probably lies in the realization that the oxidative capabilities of the fibres and their fatiguability can be modified by usage so that 'FF and FR' units are essentially interconvertible (37, 38, 156). By contrast, the type I/type II classification seems to be more fundamental and probably depends on the properties and connections of their respective motor neurones (94, 171).

Certain other histochemical reactions often employed give supplementary information, but do not disturb the generalization stated above. Glycolytic activity is strongest in the type II fibres with low oxidative enzymes (164) and glycogen content is highest in type II fibres with high oxidative capacity (186), and low in type I. The MATPase reaction has been modified to reveal differential lability of the enzyme in different fibre types at pH 4·5 and pH 10·4 (637), but it seems that its general application may lead to confusion because of species differences.

No entirely satisfactory system of nomenclature has been devised to be comprehensive yet not to incorporate unjustified implications regarding histochemical and physiological correlations. We shall accept the type I/type II division coupled with division of the latter into non-oxidative and oxidative. These appear to correspond exactly to the types A and C originally defined by Stein and Padykula (539), and the terms will be used interchangeably, as may also be said for type I or B fibres. In this scheme, A is equivalent to fast glycolytic (FG), B to slow oxidative (SO) and C to fast oxidative glycolytic (FOG) of Ariano et al. (22)

HISTOCHEMISTRY OF THE JAW MUSCLES

The recent data relevant to jaw muscles are limited (260, 481, 482, 483, 484, 485, 567, 601). All accounts agree that the jaw-closing muscles are heterogeneous. With respect to MATPase, Ringqvist (482) recognized three levels of staining in human biopsies from masseter and temporalis. Weak and strong staining fibres were identified with types I and II of biceps brachii, but moderately staining ones were believed to be a special group which was called 'intermediate'. It appears that Ringqvist's intermediate fibres were probably type C. Identification by this feature may be unreliable since not all the muscles showed it and we have seen that it may or may not appear in cat gastrocnemius. It is also better not to use 'intermediate' in this sense because of possible confusion with the 'red'/'white' scheme in which intermediate identifies with type I. It is most likely that the 'intermediate' staining units in human masseter and temporalis belonged to the subdivision of type II with high oxidative activity (C), especially since they had the relatively high phosphorylase reaction associated with these units in a variety of animal species and called 'FOG' by Ariano et al. (22), and this will be assumed in what follows.

In a study of three jaw closing muscles of normal adult cats, Taylor et al. (567) were only able regularly to distinguish two levels of MATPase staining. Nevertheless, in serial sections stained for SDH, it was clear that the type II fibres could be divided into those rich in oxidative enzymes (C) and those with poor oxidative activity (A). The SDH staining alone would usually have been sufficient to identify the three types, though the MATPase reaction gave greater confidence in separating C from B.

The most striking difference between human and cat fibre type distribution is in the proportions of type I (B) present as shown in *Table* 3.1. Human temporalis contained about 40 per cent in contrast to only 2 per cent in the cat. Human masseter had between 9 and 55 per cent as against 10 per cent in the cat. In the cat, the pterygoid muscle contained the highest proportion of B fibres (29 per cent) but no comparable figure is available for man. There was little difference between species in the proportions of type C in corresponding muscles, and the difference in type B was reflected in a converse difference in type A.

Table 3.1 Proportions of fibre types in human and cat jaw-closing muscles

	A%	*B%*	*C%*	*A+C%*	*A/C*
Human					
Masseter*	58	26·4	15·6	73	3·7
Temporalis†	23	42	25	48	0·92
Cat					
Masseter	82	10	8	90	10·3
Pterygoid	59	29	20	71	2·6
Temporalis	72	2	26	98	2·8

*Ringqvist (482).
†Ringqvist (484).

Type I taken to be equivalent to B, type II to A and 'intermediate' to C. Other data from Taylor et al. (1973). (Note that Ringqvist (personal communication) has reservations regarding this interpretation of her data.)

The quantitative significance of the proportions of the various types is better seen in estimates of their contribution to the cross-sectional area of the muscle, since this will indicate their approximate relative force-generating capabilities. *Table* 3.2 has been compiled from published estimates of average fibre dimensions and numbers. In the cat data fibre cross-sectional areas were calculated as circles of diameter equal to the mean of maximum and minimum fibre diameter, whereas using Ringqvist's data we have calculated the areas from minimum fibre diameters. Clearly this will have underestimated areas for the human fibres but will not have had much effect on the estimated percentages of total areas occupied by the different fibre types. Attention is drawn particularly to the very large proportion of types A plus C in cat muscles, of which most are type A. The situation is most marked in temporalis. There is a much larger proportion of type B fibres in man than in cat. This is illustrated in *Fig.* 3.1 with MATPase sections from temporalis of cat and normal man and sections of cat gastrocnemius and human biceps brachii for comparison. Sections stained for SDH are not shown but the type II oxidative fibres (C) identified by SDH in adjacent sections are marked by stars. In the work of Hiiemae and Houston (260) on the rat, 'light' and 'dark' fibres were separated

Table 3.2 Estimated mean fibre cross-sectional areas and percentage areas for the fibre types in human and cat jaw closing muscles

	Mean areas μm^2		*Estimated % areas*	
Masseter	*Man*	*Cat*	*Man*	*Cat*
A	201	4100	29·0	90·5
B	830	1400	54·6	3·8
C	419	2670	16·4	5·7
(A+C)			45·4	96·2
A/C			1·8	15·9
Pterygoid				
A		3380		67·2
B		1300		14·7
C		2330		18·1
(A+C)				85·3
A/C				3·7
Temporalis				
A	377	3770	32·2	79·7
B	1359	871	65·4	0·5
C	499	2590	2·4	19·8
(A+C)			34·6	99·5
A/C			13·4	4·0

Origin of data Ringqvist (483, 484), Taylor et al. (567).

by Sudan black staining and the light fibres regarded as probably type A and dark as B and C. There were significant differences in distribution of the two types in the different parts of the musculature and it was particularly clear that the dark (presumed 'tonic') type were rare in the posterior part of temporalis and most plentiful in anterior temporalis and deep masseter.

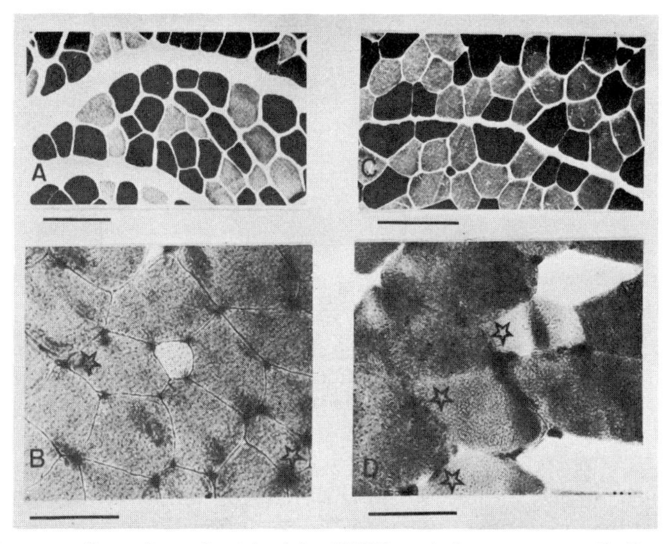

Fig. 3.1. Frozen sections of muscle stained for MATPase. A, human temporalis (from Ringqvist, 1974). B, cat temporalis. C, human biceps brachii (from Ringqvist, 1973). D, cat gastrocnemius, stained as B. In B and D, stars mark SDH rich fibres as identified in adjacent serial sections. Calibration bars 100 μm.

The size distribution of fibres is also notable. Cat jaw-muscle fibres are arranged A, C, B in descending order of size with type B very much the smallest, particularly in temporalis. By contrast, the human fibres are ordered C, B, A with A much the smallest, particularly in masseter.

In the histochemical observations of Baker and Laskin (647) in rabbit, cat, dog and monkey whole muscles were characterized as red, white or mixed on the basis of the ratio of numbers of fibres staining heavily for SDH and phosphorylase. Their results agreed with the above, regarding the association of large fibres with low oxidative activity in the cat, but in other respects there were differences. The conclusion was reached from the high oxidative capacity of cat jaw muscles that they were not fast contracting, and the conclusion rationalized by unsupported proposals regarding feeding behaviour. It is clearly dangerous to assume correlation of high oxidative enzyme content with slowness of contraction. The evidence is that they are largely independent.

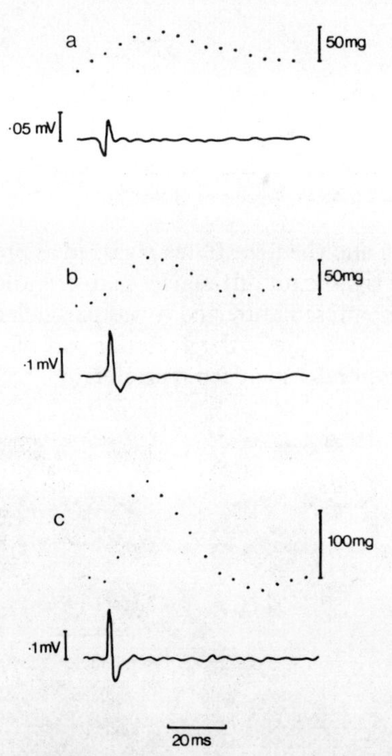

Fig. 3.2. Three examples of averaged single unit twitches obtained from cat temporalis by the technique of Stein et al. (1972). The force calibration refers to closing force at the symphysis menti. Lower traces; unit EMG in each case, delayed 3 ms.

In summary, the histochemistry indicates that the same three types of fibre found in limb muscles are also present in the jaw-closing muscles of both man and cat; but that the proportions differ. MATPase-rich fibres (A and C) make up 48–73 per cent by number and 35–45 per cent by area in man. In the cat the corresponding figures are 90–98 per cent by number and 96–99·5 per cent by area.

FUNCTIONAL ASPECTS

Following the arguments introduced in the first section, we would hope to be able to see correlations between the histochemistry and the contraction characteristics of the jaw muscles, as elsewhere. Our attention was first drawn to this subject by the observation (568) that the jaw closers in the cat had bulk twitch speeds approaching those of the extra-ocular muscles. Subsequently, more systematic measurements have been made on strips partially separated from temporalis and masseter (567) so as to achieve a more satisfactory mechanical situation than is possible when recording force on the lower jaw. The mean times to peak twitch tension (Tp) for masseter and temporalis were 13·1 (SD 2·27; n = 18) msec and 11·4 (SD 2·11; n = 21) msec respectively. The corresponding times from peak to half relaxation were 12·8 (SD 2·49) msec and 9·81 (SD 1·84) msec. Tetanus/twitch ratios were high as in other fast muscles and tetanic fusion appeared complete at 100 per sec though the rate of rise of tension in a tetanus continued to increase up to 400 per sec. These findings are fully in accord with the classification of the majority of the fibres as type II (A and C). However, note that though whole segments of temporalis and masseter have values of Tp as short as the shortest single units in cat gastrocnemius, the MATPase of the jaw-muscle fibres is distinctly weaker than that of gastrocnemius type II fibres when treated identically (*see Fig.* 3.1 *B* and *D*). Clearly, MATPase staining at pH 9·4 is not a quantitative correlate of speed from one muscle to another even in the same animal.

Whole muscle contraction characteristics represent the majority unit properties unless there is a very clear bimodal population (61), so that it is not surprising to see no sign of the minority type I slow contraction in bulk muscle twitches. To look for evidence of the minority fibres we have recently made single unit recordings from cat temporalis by the averaging method of Stein et al. (541). Under light anaesthesia a modest tonic stretch reflex can be recorded and force on the lower jaw averaged with a computer triggered with single unit EMG spikes. The units which fire tonically under these conditions have small action potentials and small, slow averaged twitches (*Fig.* 3.2*a*). When larger stretches are applied, particularly when the animal is aroused as much as possible, higher threshold, more phasic units are recruited. These yield larger, faster averaged twitches (*Fig.* 3.2*b*). The range of twitch speeds and forces so far observed is shown in *Fig.* 3.3, which also reveals the usual correlation between the two. These results indicate that there are indeed small, slow units present in temporalis and that they are the ones recruited first in the tonic stretch reflex. It is difficult to escape the conclusion that they are composed of the minority type I (B) fibres.

The other physiological characteristic of obvious interest is fatiguability. The only information available is from the muscle strip experiments. Adopting the stimulus routine favoured by Burke et al. (97) of tetani lasting 0·3 sec every second at a frequency related to the twitch speed (i.e. 55 per sec for temporalis), contraction strength fell to 25 per cent in 1 minute and thereafter much more slowly. One might speculate that the rapid phase is due to the loss of function in the type A fibres and that the slower loss thereafter represents the progressive fatigue of the type C. Type B fibres should be able to survive this routine much longer, but are present in too small numbers for their contribution to be expected to be obvious. In relation to fatigue, it is interesting to see that the jaw muscles are unusually well supplied with capillaries compared with gastrocnemius (*Fig.* 3.1*b*). The latter has its capillaries sited predominantly next to oxidative fibre types B and C, but temporalis has a capillary at practically every angle of the polygonal fibres irrespective of their types. Thus, although the blood supply is apparently rich and uniform in this muscle, the small B fibres still have a particularly rich supply per unit volume, because of their small cross-sectional area.

An attempt has been made to relate structural features to function in human jaw muscles by Ringqvist (482, 483, 484) by looking for changes which might accompany defects in dentition and variations in bite force. The size of type II fibres correlated with maximum bite force in masseter, and this was taken to support the view that these fibres are the ones recruited at high contraction strengths. Also type II fibres were significantly smaller in temporalis in subjects wearing defective dentures than in others with normal dentitions. This may mean that discomfort prevents the former subjects from ever exerting large bite force and so leads to relative atrophy of the high threshold units. Such findings are consistent and interesting but serve to emphasize the need for quantitative studies of contraction characterization of human masticatory muscles at the unitary level.

Summarizing this section on the functional aspects, histochemical characteristics appear to have the same significance in masticatory muscles as in limb muscles; speed being associated with MATPase, and fatigue resistance with SDH activity. Cat temporalis is exceptional in being almost entirely fast with a few small, slow units recruited at low threshold. If the presence of type B fibres is to be correlated with tonic activity, then clearly the temporalis and masseter are relatively much more concerned with tonic postural regulation in man than in the cat.

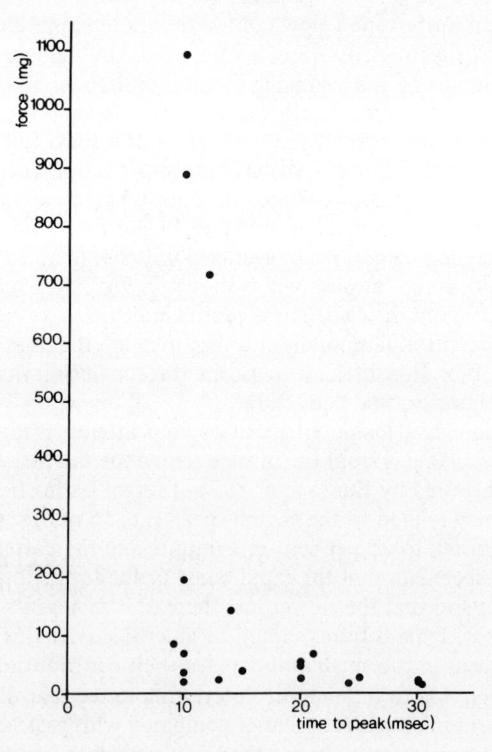

Fig. 3.3. Plot of twitch force against time to peak for 20 single units recorded from cat temporalis.

This work was assisted by grants from the Wellcome Trust and the Medical Research Council, and was carried out in collaboration with Dr F. W. J. Cody and Dr L. M. Harrison.

Table 3.3 The numbers of 'A' (fast contracting, fast fatiguing) and 'B' and 'C' (slow and fast contracting fatigue resistant) fibres in samples (cat) and whole jaw muscles (rat) expressed as a percentage of the total fibre count

		A (FF) Fibres		*B(S) Fibres + C (FR) Fibres*	
Temporalis					
Cat*		72%		28%	
Cat†		50%		50%	
Rat‡		72%		28%	
Masseter					
Cat*		82%		18%	
Cat†		41%		59%	
Rat‡	Deep	70%	75% overall	30%	28% overall
	Superficial	81%		19%	
Pterygoids					
Cat*		51%		49%	
Cat†		44%		56%	
Rat‡		67%		33%	

*Taylor et al. (567).
†Baker and Laskin (647).
‡Hiiemae (656).

Discussion CHAIRMAN: D. J. ANDERSON

LUND: Could it also be that the type B fibres are associated with grinding ? Maybe if you looked at ruminants you would find a larger proportion of them than in cat or man.
TAYLOR: Yes, the proposition being that the type B fibres are engaged in more sustained contractions. However, if the sustained contractions were sufficiently powerful to require a high proportion of the available force from a muscle, the larger, faster units would be recruited also. What we need here of course is more unit data from human muscles showing the way in which the different types are recruited. I hope we shall be hearing something of that from Dr Yemm.
BULLER: In comparing the human and cat data it seems that you have also got to make some statement about the relative muscle mass and the weight of the jaw. The postural necessity for the slow type of fibre will not be the same in both species.
TAYLOR: I quite agree. It would be very interesting to do a dimensional analysis. As the linear dimensions increase, the weight would increase as the cube and the cross-sectional area of the muscle (other things being equal) would increase as the square. Thus, with increasing size, an animal would be progressively at a disadvantage in providing tonic force.
BULLER: Have you done any quick sums of that type ?
TAYLOR: No, we haven't as yet.
HIIEMAE: Earlier this year I came across a paper by Baker and Laskin (647) which purports to give a comparison of the proportions of 'red' and 'white' fibres in the

major jaw muscles of the macaque, cat, dog and rabbit, based on the levels of succinic dehydrogenase and phosphorylase. Their results for the cat were so much at variance with yours, and with mine for the rat, that I recalculated my figures for comparison. It turns out that Baker and Laskin's figures are not based on true histochemical profiles of individual fibres and must be discounted, but to my complete surprise I found that allowing for the anatomical differences in the muscles and equating levels of Sudan black B with SDH, and I appreciate that they are not in fact the same, my figures are, for temporalis and masseter, within 1 percentage point of yours (*Table* 3.3). This does, to some extent, answer two of the other points that have been raised. There is a large difference in jaw proportions and weight between the cat and rat but Professor Taylor's figures for SDH in the cat masseter and temporalis are to all intents and purposes the same as mine for the rat. And, in relation to Dr Lund's comment, rats definitely grind!
TAYLOR: Very intriguing.

Paper No. 4

The Properties of their Motor Units, and Length-tension Relationships of the Muscles

R. Yemm

As the prime movers of the mandible, the jaw-opening and -closing muscles are of fundamental importance in the process of mastication. The objective of the present paper is to outline some important aspects of muscle function of a general nature and to examine the present knowledge of the properties of the jaw muscles.

SOME GENERAL ASPECTS OF MUSCLE CONTRACTION

It is well known that the tension developed by a muscle varies with its length. The contractile unit of a skeletal muscle fibre is a sarcomere, and above and below the optimum sarcomere length, the tension developed during a contraction declines. Gordon et al. (213) and others have shown that the optimum length for tension depends upon a critical overlap of the contractile proteins. The optimum length of a muscle is thought to coincide with the length at which the muscle normally functions in the body (510).

The fibres of which a muscle is composed are organized into functional units, all innervated by the same motoneuron. Under physiological conditions a steady force can be developed isometrically by asynchronous firing of a number of units, and increase in force is produced by an increase both in the number of active units and in the rate of firing of already active units. Evidence at present available suggests that at low forces, recruitment of additional units is the main factor, while at high force levels, changes in firing rate become more important (401).Recruitment may be more significant when movement is occurring, especially with rapid movements during which there is limited scope for changes in firing rate. During a graded contraction of a hand muscle under isometric conditions, recruitment has been shown to be an orderly process; small units developing small tensions being recruited first (400).

Motor unit size, which depends upon the number and diameter of its constituent fibres, varies within a single muscle. In general, muscles involved in finely controlled processes tend to have smaller units, whereas muscles not usually involved in skilled activity tend to be composed of larger units (510).

LENGTH–TENSION RELATIONSHIP OF JAW-CLOSING MUSCLES

For some time it has been accepted that maximum forces are developed by the jaw-closing muscles when they are at the length determined by the occlusal contact of the natural dentition. This was proposed by Boos (68) as a result of experiments on the maximum voluntary biting power of human subjects at varying jaw positions. More recent experiments have supported this (591), but O'Rourke (439) and others

(70) have criticized these experiments on the grounds that the results were influenced by many uncontrolled variables. The criticism seems justified as the results could have been influenced by sensory feedback from numerous receptors as well as by the motivation of the subject. Furthermore, the data do not apply to a single muscle, since all the elevator muscles were involved.

The only experiments designed to examine the length–tension relationships of jaw-closing muscles have been those of Nordstrom and co-workers, in which the principal muscle investigated was the masseter muscle of the rat. Using direct stimulation of the main belly of the masseter, it was found that the maximum twitch tension did not occur with the teeth in occlusion. The position at which the peak tension was recorded with an opening of between 7 and 12·2 mm between the incisors (435). Associated histological studies (433, 434) demonstrated, as expected, that the sarcomere length of the masseter and temporal muscles increased as the mouth was opened, but that the magnitude of the increase differed between the two muscles. Assuming that the optimum length of the sarcomeres of the two muscles does not differ, it follows that at most jaw positions the two muscles will operate at a different point relative to their optima.

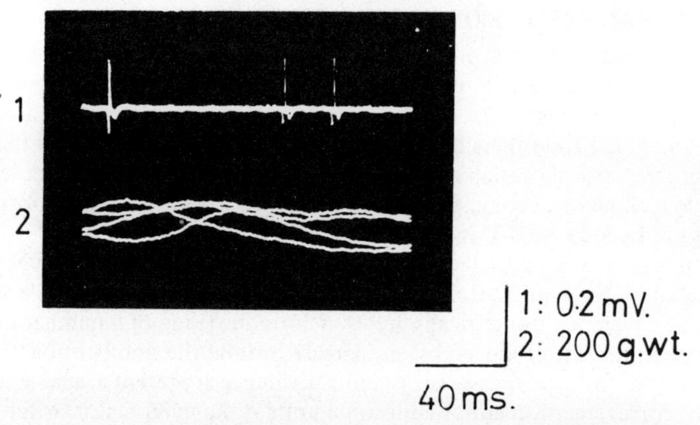

Fig. 4.1. Four single oscilloscope sweeps of motor unit activity and voluntary force fluctuation. The motor unit action potentials are superimposed towards the left of the record on the upper trace. On two occasions the unit fired a second time during the sweep. The random nature of the fluctuations in the force level are seen on the lower part of the record. A mean biting force of 3 kg was exerted by the subject in order to activate the unit.

STUDIES ON MOTOR UNITS OF HUMAN JAW MUSCLES

The methods developed to examine motor units of a human hand muscle (399, 400, 401) are currently being used to investigate the motor units of the masseter and temporal muscles. The experiments are incomplete, but so far over 100 units have been studied in 4 subjects.

The technique depends upon simultaneous recording of the force of a voluntary contraction (in this case the subject clenching his teeth on a force transducer), the electrical activity of single motor units by means of a bipolar needle electrode, and the interference electromyogram using surface electrodes. The force exerted in a voluntary contraction exhibits small rapid fluctuations (*Fig.* 4.1), some of which are due to the unfused twitches of the motor units participating in the contraction.

By averaging the force transducer output immediately following each action potential recorded from a single motor unit, the force contributed by that unit was singled out from the contributions of other units which occurred randomly with respect to the firing of the unit under examination. Data are recorded on magnetic tape for periods of three minutes during which the subject is asked to maintain the minimum force at which a unit fires with a regular, slow discharge. More than one unit may be sampled at a single electrode position if the waveforms permit subsequent discrimination; alternatively, the electrode is moved to a new site. In a single session between 5 and 10 units are sampled.

In analysing the results, 512 events are usually averaged to produce a record representing the time course of the twitch of a single unit (*Fig.* 4.2). Two criteria must be satisfied before the record can be assumed to represent the twitch of one unit. The first is that there should be no tendency for the discharge of other units

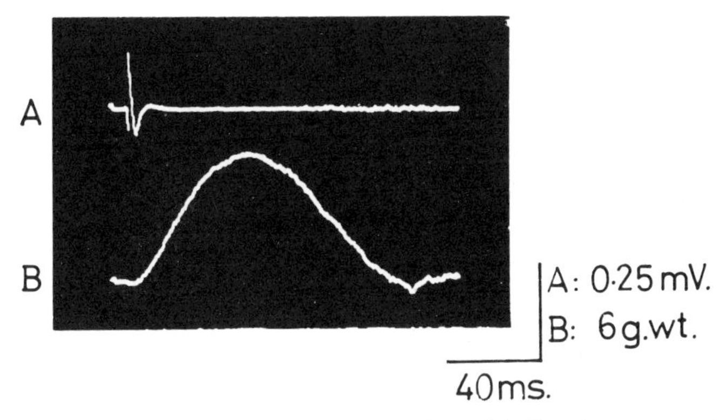

Fig. 4.2. The average of 512 sweeps similar to those of *Fig.* 4.1 (data from same unit). The unit fired a second time during many sweeps; the minimum interval between the first and second action potentials was 80 msec and the most common interval was 95 msec. Therefore during the apparent relaxation phase, the record of the twitch is distorted by the next twitch.

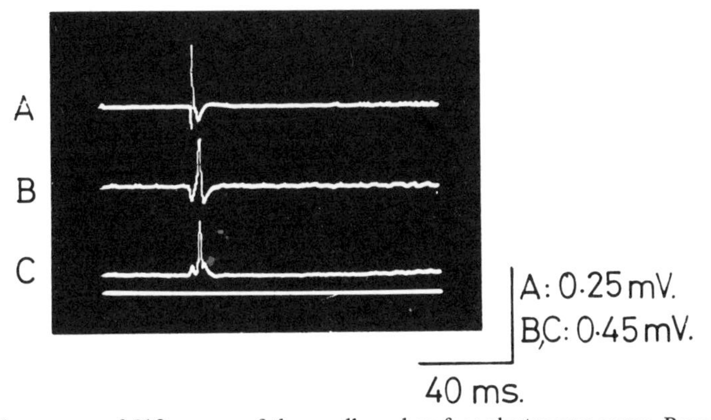

Fig: 4.3. The average of 512 sweeps of the needle and surface electromyograms. Record A is from the needle electrode and Record B is from the surface electrodes. Record C, with zero, is the surface electromyogram full-wave rectified before averaging. Apart from the component attributable to the single unit, the record is parallel with the zero, and indicates that other units are tending not to synchronize with the unit being studied (data from same unit as *Figs.* 4.1 and 4.2).

to be synchronized to the unit under study. As in the earlier experiments on the hand muscle, this was checked by averaging the surface electromyogram, and no evidence of synchronization of motor units was found in any of the subjects (*Fig.* 4.3). The second is that, during the voluntary contraction, the rate of discharge is such that the individual unit is sufficiently slow so that fusion of the contractions of the single units does not occur. This possibility has not been examined directly. However, the discharge rates and apparent time courses of the twitches of masseter and temporal motor units were similar to those recorded from hand muscle units which showed no evidence of fusion of contractions.

In the experiments so far the properties of the motor units have been examined at a single jaw position. It has been found that units recruited to produce low forces exhibit smaller twitch tensions than those recruited at higher force levels. The recruitment thus appears to be orderly and the relationship between the twitch tension of a unit and the force at which it is recruited (force threshold)

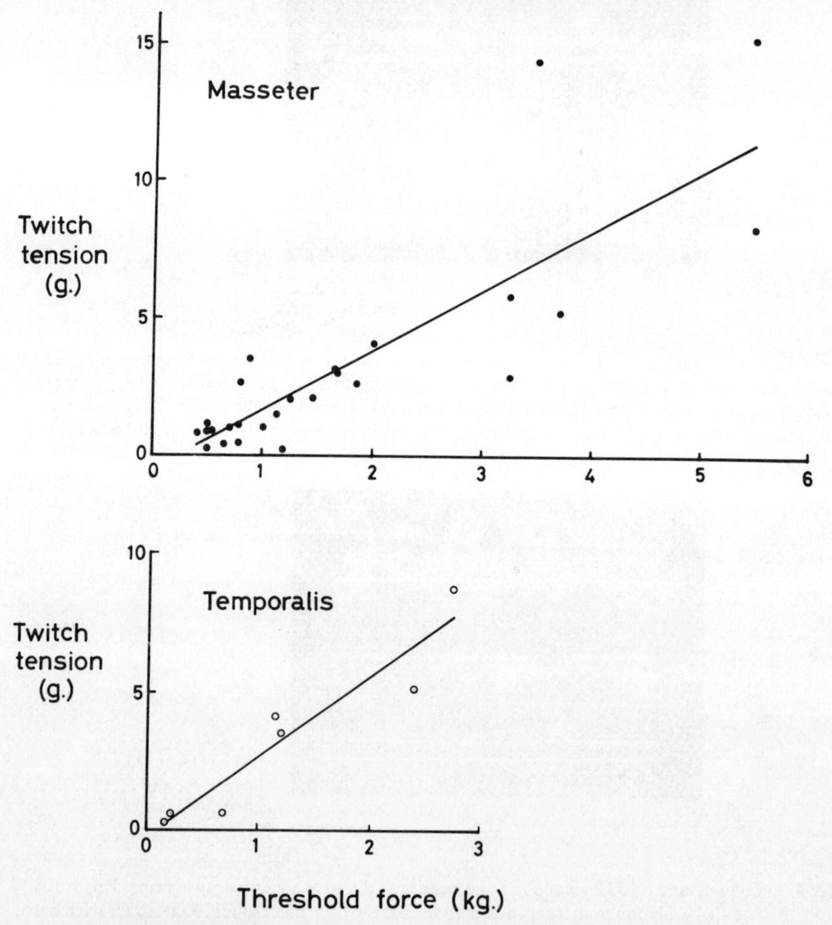

Fig. 4.4. The relationship between the force threshold of units and their twitch tension, for the masseter and temporal muscles of one subject. The lines have been fitted statistically, assuming a linear relationship. The correlation coefficients are 0·84 (masseter) and 0·95 (temporal muscle), which are very significantly different from zero.

appears to be linear (*Fig.* 4.4). The smallest unit detected so far had a twitch tension of 0·2 g; the largest, 33 g. The relationship between twitch tension and contraction time (the interval between the onset of the action potential and the peak tension) of the units shows more variability (*Fig.* 4.5), although the evidence suggests that larger units tend to be slower in reaching peak tension than smaller units. The range of times-to-peak is 24–91 msec.

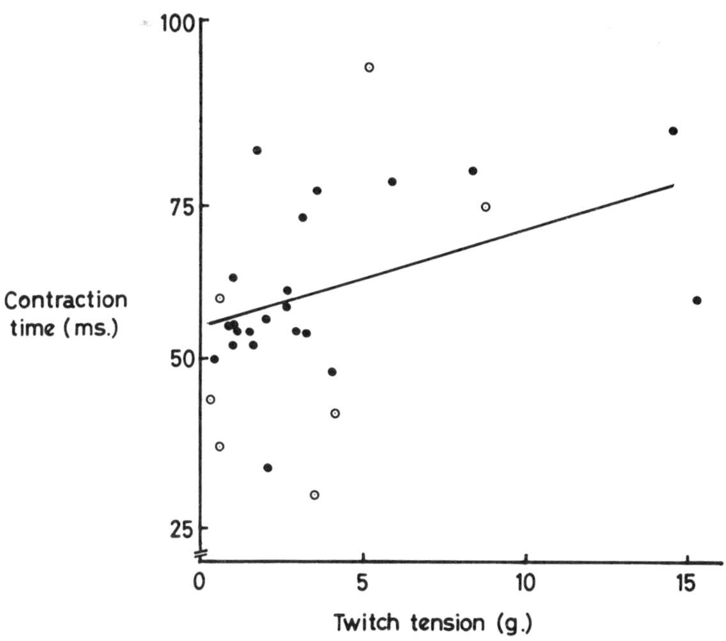

Fig. 4.5. Measured times to peak tension for the masseter (filled circles) and temporal muscle units for the same subject as in *Fig.* 4.4. The line has been fitted for the masseter units only, and the correlation coefficient of 0·46 is significantly different from zero at the 5 per cent level.

The action potential associated with the firing of every unit that has been isolated with the needle electrode could also be detected in the surface electromyogram although usually this required averaging. While there was some evidence for a direct relationship between the measured twitch tension of a unit and its amplitude in the surface electromyogram (*Fig.* 4.6) other factors must be involved, such as the position and impedence of the electrodes, and the depth of the unit. Thus the surface potentials from some units, especially the smaller ones, are well below the normal amplifier noise level.

Further experiments are planned, once these techniques have been established, to study the length–tension characteristics of units in the jaw muscles and look for possible regional differences within a muscle. In experiments performed to date, the needle electrodes have sampled units from the greater proportion of the masseter muscle, but only from the anterior two-thirds of the temporal muscle.

As will be reported in a subsequent paper no evidence has been found of tonically active units which could be responsible for active control of mandibular rest position. So far, all the units firing at low forces have exhibited small twitch tensions. It is possible that further experiments may detect the large, low threshold units postulated by Wyke (636) to be present in the temporal muscle.

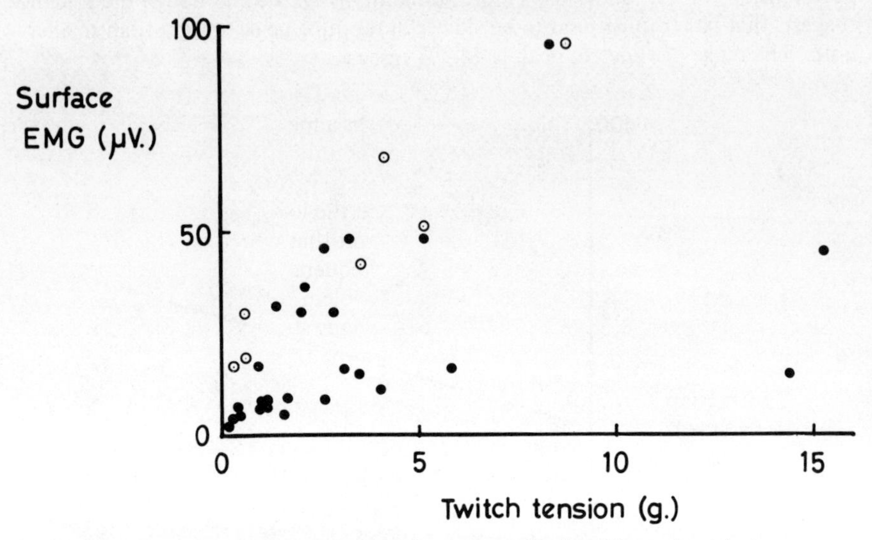

Fig. 4.6. The relationship between twitch tension and amplitude of the surface electromyogram (same subject as in *Figs.* 4.4, 4.5; filled circles masseter, open circles temporalis).

Conclusions

The available techniques for the investigation of the contractile properties of skeletal muscles have not been applied extensively to the jaw muscles. The relationship between optimum length and the length of the muscles with the teeth in occlusion is uncertain. There is conflicting evidence from studies of human voluntary biting and a single study on rat masseter. The motor units of masseter and temporal muscles in man are recruited in an orderly fashion. The twitch tensions of the units ranged from 0·2 g to 33 g and their times to peak ranged from 24 to 91 msec.

Acknowledgement

Thanks are due to Mr K. Robbins for assistance with the preparation of the illustrations.

Discussion CHAIRMAN: D. J. ANDERSON

LUND: In the experiment in which you showed the length–tension relationship in rat masseter (435), did you measure tension in the long axis of the muscle, or at the end of the mandible ?

YEMM: The muscle was left untouched, apart from reflection of a small superficial part of it. Wire electrodes were inserted into the muscle and the muscle was stimulated with the mandible fixed in different positions. The tension transducer was attached to the lower incisor teeth.

LUND: Did you make any corrections for the fact that in some cases the muscles were pulling in different directions ?
YEMM: We aligned the measuring apparatus so that its long axis was parallel with the mean direction of the masseter fibres.
LUND: When you talk about size of the motor units you are talking about the size of the spike, are you not ?
YEMM: No, I am talking about the twitch tension. The amplitude of the spike recorded by the needle electrode is meaningless as a measure of unit size. However, as I have shown, there is a tendency for the larger units to contribute most to the amplitude of the surface EMG.
GOLDBERG: You mentioned that you had used the lowest biting force which produced a continuous firing of the particular unit that you were interested in studying. Do you have any data relating firing frequency to recruitment position in terms of force, or did you not look at that problem ?
YEMM: No, I haven't looked at this. In our previous study on the hand we asked the subject to track varying forces and measured the firing rates. We have not repeated this in the present study yet.
WEIJS: Are you aware of the fact that the force of the masseter will be divided between the teeth and the temporomandibular joint and so if you measure the force on the teeth, with a unidirectional transducer, there could be some bias in the results you get.
YEMM: This is referring to the rat study in particular ? Yes I accept that.
WEIJS: The relationship between length and tension shown by you in the rat could have been affected by the different distribution of the reaction forces on the teeth and temporomandibular joint in different jaw positions.
THEXTON: We have measured the twitch tension in the opossum jaw musculature (579, 657), measuring the force as torque, and found that the temporalis produced its peak force with the mouth pretty well wide open, masseter somewhere a little bit further closed and it was the medial pterygoid that produced the force when the teeth were nearly in occlusion. We also found evidence that there is considerable hysteresis, that is, the response depended on whether you measured the twitch during a sequence of closing or opening movements. Have you any similar results ?
YEMM: No, we used a random order of jaw positions, to avoid, for instance, progressive effects of fatigue.
MØLLER: Did you record the action potentials with bipolar needle electrodes simultaneously with the force in the human experiments ?
YEMM: Yes.
MØLLER: It has been shown that differences in contraction time may depend on the excitability of the surface membrane of the muscle fibres, in that action potentials of fast contracting fibres have shorter rise times than those of slow fibres (650).Since you have recorded the action potentials with bipolar needle electrodes you would be able to go back and look at that.
YEMM: Frankly I regard the signal picked up by the bipolar needle electrode simply as a useful means of fixing a time for triggering the averaging process. I think that the signal one detects with the needle electrode depends very much on where the electrode tip is positioned in relation to the unit, at least with the very small electrodes that I have been using.
MØLLER: But what you should be concerned about is the rise time. Would it be possible to go back over your recorded data ?
YEMM: It might indeed.

MØLLER: With relation to the earlier questions, did you find that the length–tension relationship was consistent, for both the temporal and masseter muscle ?
YEMM: No, we only examined the masseter muscle.
MØLLER: As I recall from some of your previous work, there was a difference between the length–tension relationships.
YEMM: The results I showed you here from one animal were consistent for other animals. The optimum jaw position for maximum tension in the method we used ranged from about 8–12 mm of opening. We were stimulating only the main part of the masseter muscle. We didn't stimulate the temporal muscle at all. I think what you are referring to is the fact that we did some concurrent histological studies in which we looked at the way in which the sarcomere length changes in the masseter and temporal muscles with changes in jaw position. The results are very difficult to evaluate but there is a suggestion from these results that the masseter and temporal muscles may have different optima for operation.
MØLLER: Your measurements were from the superficial part of the masseter muscle in the rat ? Is that correct ?
YEMM: No, the most superficial part of the masseter muscle was removed; that is, the very small, oblique part of the muscle. We stimulated the rest of the muscle. We began by applying stimuli of gradually increasing intensity until we got a plateau response, so that we could be reasonably certain we were stimulating the whole muscle. I should like to stress that I don't regard that experiment as being the last word in studies of length–tension relationships in the masticatory muscles. We need to learn more.
TAYLOR: Did I understand you to say that the large units were the slow ones ? If so isn't that rather surprising in relation to other findings ?
YEMM: Yes. I haven't had time to sit and evaluate that observation. These data have only been available for a short time. It seems consistent in all the muscles of all the subjects we have studied.

ACTIVITY OF SINGLE MOTOR UNITS IN THE MASSETER MUSCLE OF HUMAN SUBJECTS*

L. J. Goldberg

In studies in our laboratory on the activity of single motor units (SMU) in the masseter muscle of human subjects we have observed tonic activity of motor units when the lower jaws of the subjects were in the rest position.

The experiments were performed on 12 subjects, 20–30 years of age with normal occlusions and no history of myofacial pain dysfunction syndrome. Bipolar recording of SMU activity was obtained with two silver wires (75 μm diameter) coated with teflon. The wires were passed through the shaft of a standard 22-gauge injection needle. The ends of the wire, which protruded through the bevelled end of the needle, were bent to form two prongs of approximately 1–2 mm length. The separation between the ends of the wire was approximately 1 mm. Recordings were obtained from the bare tips of the cut ends of the wire. The electrode wire and needle were autoclaved prior to use. The subject sat in a dental chair for insertion of the recording electrode. The needle was held parallel

* In the Symposium, Dr Goldberg presented this material immediately before his paper on 'Changes in the excitability of elevator and depressor motoneurons produced by stimulation of intra-oral nerves'.

to the occlusal plane and parallel to the buccal surfaces of the molar teeth and was inserted through the mucous membrane into the muscle belly. The needle was inserted approximately 5–7 mm into the muscle. The needle was then withdrawn from the oral cavity while the prongs held the electrode wire stationary in the muscle. At the conclusion of the experiment the wires were simply pulled out of the muscle.

The subjects were seated in a straight-back chair without head support and instructed to sit upright and look straight ahead at an oscilloscope screen which was positioned at eye level. All subjects reported no discomfort from the electrodes and appeared to be at ease and not apprehensive. The position of the lower jaw at this time could therefore be considered to approximate the subject's normal rest position when sitting upright. Under these conditions 12 single motor units were identified and recorded from 7 of the 12 subjects. The spikes had low amplitudes ranging from 30 to 160 μV (70 ± 37 μV; mean and SD) and fired at fairly constant rates. An example of one of these units is shown in *Fig.* 4.7B.

The subjects were then instructed to relax their jaw muscles and let their jaws go 'slack'. Normally, in rest position, there was approximately 2–4 mm between the occlusal surfaces of the upper and lower teeth. When subjects were instructed to allow their jaws to go 'slack' the distance between the occlusal surfaces increased by approximately 5 mm. Units which were recorded in the rest position immediately ceased firing when the subjects went into the 'slack' position.

After recording in the rest and 'slack' positions a force transducer was placed in the subject's mouth between the upper and lower premolar teeth on the side ipsilateral to the electrodes and isometric interocclusal force production was recorded along with SMU activity in the masseter muscle. The subjects produced

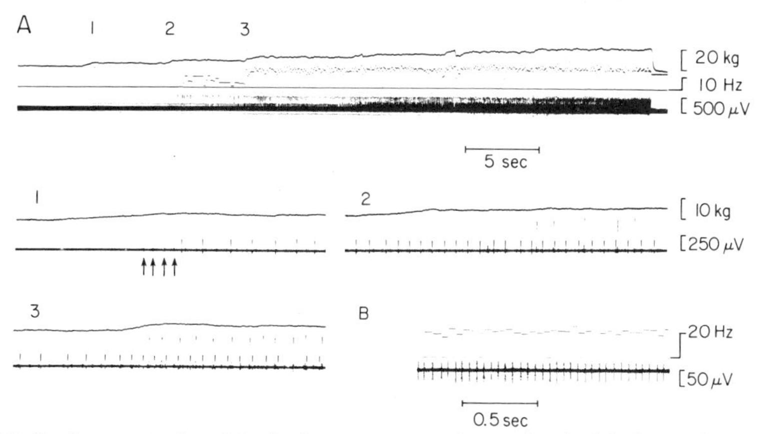

Fig. 4.7. Single motor unit activity in the masseter muscle correlated with changes in interocclusal force production. A, the upper trace shows the output of the force transducer placed between the upper and lower premolar teeth; the lower trace is the recording of single motor unit activity in the muscle; in the middle trace each line or dot represents the instantaneous frequency of the large single motor unit illustrated in the lower trace which first becomes active at level 2. 1, 2 and 3 are taken from the record in A, but with a faster time base and increased gains in the force and EMG records to show the recruitment of two units in 1 (arrows indicate onset of firing of small unit), sporadic firing of large unit in 2, and full recruitment of large unit in 3. All units once recruited continue to fire throughout the trial. B, single motor unit and instantaneous frequency record from another subject while lower jaw is in the rest position. Time base in B is the same for 1, 2 and 3. EMG and frequency calibrations in 2 are the same for 1 and 3.

increasing levels of force by moving the beam of the oscilloscope which they were viewing. The beam was connected to the output of the force transducer and by increasing or decreasing biting force the subjects could raise or lower the oscilloscope beam. *Fig.* 4.7A is an example of single motor unit activity and bite-force production in one subject. In this case the force increased in 2·7-kg steps from 2·7 to 19 kg. Three SMU were recruited in this series; the first two at 5·4 kg (*Fig.* 4.7A1) and the third at 8·4 kg (*Fig.* 4.7A2). It is obvious that the recruitment of increasingly larger amplitude units was correlated with increasing interocclusal force production.

In conclusion, the results of this study indicates that there are motor units in the masseter muscle which are tonically active in rest position. These units presumably contribute to the maintenance of the position of the lower jaw against the force of gravity. Motor units active in rest position invariably have low amplitudes. When isometric interocclusal forces are produced increasingly larger amplitude motor units are recruited. This evidence supports the view that the 'size principle' of motoneuron and motor unit recruitment first described in spinal cord systems (400, 655, 667) also applies to the trigeminal motor pool.

Utilizing the concept of the 'size principle' one could hypothesize that in rest position, where no interocclusal forces are being produced and the mandible is simply in a relatively fixed postural position, the excitatory drive to the masseter muscle motor pool is at a minimum. Under these conditions only the smallest motoneurons are active. To produce interocclusal force, the excitatory input to the motor pool must be increased and the larger, higher threshold motoneurons are then recruited.

This investigation was supported by USPHS Grants NS 08604 and DE 04166.

Paper No. 5

The Structure and Distribution of Muscle Spindles and Tendon Organs in the Muscles

U. L. Karlsson

INTRODUCTION

Histological studies and studies on reflexes indicate the presence of nervous receptor organs in the masticatory muscles of man and animals similar to those in limb muscles although their embryological origin (branchial rather than somatic) and central nervous organization (cranial rather than spinal) differ. The physiology and pathology of mastication are integral parts of diagnostic and therapeutic considerations in clinical dentistry. This presentation will attempt to outline the morphological characteristics of proprioceptive muscle receptors in general, but with special reference to the main masticatory muscles. This paper will not include a full review of the literature, but I refer you to the recent reviews by Matthews (388) on mastication and to the proceedings of two recent Symposia on muscle receptors (20, 33). My object is to familiarize you with the receptors which are generally assumed to partake in the control of masticatory activities.

A HISTORICAL PERSPECTIVE

A typical skeletal muscle contains muscle spindles, various different types of musculotendinous organs, notably Golgi tendon organs, and different types of Pacinian corpuscles. Muscle also contains receptors associated with pain. Since only muscle spindles and Golgi tendon organs have been systematically studied in terms of function I will restrict my discussion to those two receptors. They were discovered in the nineteenth century, and their sensory role was postulated at an early stage but was not proven until this century (218).

Masticatory muscle receptors were first described by Baum (42). Muscle receptors have so far not been described in all muscles involved in mastication and particularly the question of their presence in some of the depressor muscles still seems unsettled. We are unable, therefore, to arrive at a general conclusion as to the functional role of these receptors in mastication. This also applies to many other muscle groups in the craniofacial area. The literature, however, constantly provides new evidence for muscle receptors in the head muscles and it therefore seems wise to bear in mind that the absence of receptors in a given muscle may be apparent, due to technical imperfections, rather than real. The problem is accentuated by the fact that physiological evidence is also lacking for some of the muscles, for example, the depressors, where muscle receptors are thought to be few or absent (388).

I will proceed to discuss the distribution of muscles spindles and Golgi organs in elevator and depressor muscles and then give an account of their structure.

PRESENCE AND DISTRIBUTION OF MUSCLE SPINDLES

All jaw elevator muscles in man have been found to contain muscle spindles (176, 185, 274). For a long time there was doubt regarding the presence of spindles in the human lateral pterygoid muscle but the studies of Honée (274) and Gill (185) must be taken as indications for their existence. This controversy, however, raises questions about the validity of earlier studies and invites a cautious attitude towards the techniques used (*see below*). Similar controversies appear in the literature on other species (107, 173, 299, 300, 336, 530). From the literature it is difficult to avoid the impression that the lateral pterygoid actually differs from the other muscles in terms of its muscle spindle organization. The elevator muscles contain many muscle spindles and should be classed between larger limb muscles, containing relatively few spindles, and muscles responsible for fine finger movements. These muscles also occupy an intermediate position in terms of their speed of contraction (388). By contrast, the jaw depressor muscles are not found consistently to contain spindles (69, 599); and recent studies (Lennartson, personal communication) support these findings. It is remarkable that there is even inconsistency in the results for these small muscles.

There are several possible explanations for the inconsistencies mentioned above. One is that some muscles truly lack spindles. In that case the reflex control of the muscles must either be very different from current concepts, or the activity of the spindles in one muscle may influence neighbouring muscles. A second possibility is that the muscles contain receptors which function as spindles in some muscles but which are not always recognizable as such. A third is that some of the tissue-processing and light-microscope techniques employed would not have permitted a critical identification of receptors.

Positive identification of contacting nerve endings in spindles requires a resolution of at least 10 nm which can only be achieved with the electron microscope. The latter technique has now been developed to the point at which large volumes of tissue can be analysed with adequate resolution. This requires proficient sectioning techniques and is helped by a scanning transmission attachment to the electron microscope. This attachment provides resolution down to 10 nm with 1 μm thick sections (477) and most of the objections raised against this three-dimensional technique can be overcome by careful attention to procedure. Careful analysis of a few selected muscles in this way would either provide a positive identification of atypical muscle spindles or strengthen the argument that there are no receptors there.

There is good evidence in the literature that muscle spindles are not as randomly distributed in jaw muscles as they appear to be in many larger limb muscles. The masseter muscle in man and animals appears to have the highest concentration of spindles in its deep portion (299, 336, 598). All the spindles in the temporalis muscle of the Japanese shrew-mole are present in the vertical and horizontal portions (336). The medial pterygoid in animals is reported to have most of its spindles concentrated to its medial portion (299, 336). In the lateral pterygoid in man the spindles appear to be present in the mid-portion (274).

Muscle spindles in different parts of a muscle give different responses to a given stimulus (251). The apparent concentration of spindles in certain locations of the elevator muscles implies that the central nervous system uses information from those parts preferentially. Several spindles grouped together in a muscle (presumably giving similar responses) may therefore have some relation to the mesencephalic neurons each having a relatively limited surface area for synaptic connections. In order to obtain adequate reflex connections, a degree of redundancy (several spindles in the same location) could therefore be necessary. We do not know, however, if all the spindles in a group are identical. If they are not, functional specificity is implied.

STRUCTURE OF THE SPINDLES

Our present knowledge about spindle structure is derived from both light and electron microscopy, but very few structural studies have been published on masticatory spindles (185, 299) and those involved only light microscopy. One ultrastructural study is forthcoming (Hinrichsen, personal communication). In the absence of specific information on the spindles in jaw muscles I shall discuss spindles in other muscles. There is at present no evidence that the spindles in jaw muscles differ in their structural or functional properties from spindles elsewhere (114, 388).

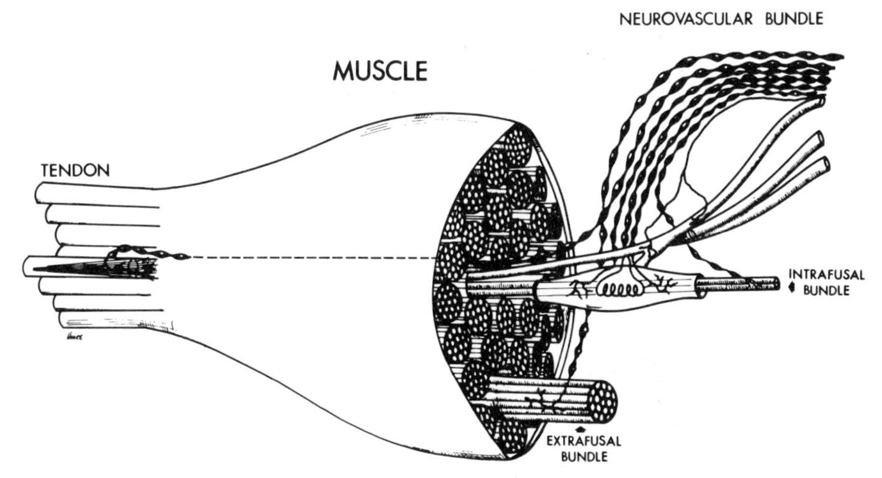

Fig. 5.1. Diagram of the relationship of a muscle spindle and a classic tendon organ to the extrafusal muscle bundles and neurovascular bundle of a muscle. Several myelinated nerve fibres (varicose chains) enter the spindle capsule while others innervate the intrafusal bundle, the extrafusal bundles and the tendon organ. Some unmyelinated nerve fibres (solid line) approach some vessels entering the muscle. They represent the sympathetic nerve supply to the muscle.

Diagrams summarizing spindle and tendon organ anatomy are shown in *Figs.* 5.1–5.3. Each spindle contains several thin intrafusal fibres some of which (the nuclear bag fibres) have more nuclei in the equatorial region of the spindle than others (the nuclear chain fibres). These intrafusal fibres may be attached to the inside of the spindle capsule, to extrafusal fibres, or to the muscle tendon. All of them display marked reduction of myofilamentous material in the equatorial region (*Fig.* 5.2), where the nuclear chain fibre diameter is smaller than in the polar regions (304). This indicates that the stretchability is higher in this central region where the primary sensory endings attach (305) and it implies that the dynamic response of the spindle endings may emanate from here (279). The bundle of intrafusal fibres in this region is surrounded by a substantial internal capsule (in which capillaries may be present) and the extracellular space contains elastic fibres (190) which may be highly significant in spindle function (305).

It should be added that the whole bundle of intrafusal fibres in this central region is connected to the same primary afferent neuron (304) and this indicates that they function as a unit. The form of the nerve endings in the central region is probably not purely spiral (497), contrary to the implication of the original name 'annulospiral ending' (*see Fig.* 5.2).

The two peripheral or polar regions of the intrafusal fibres are slightly larger in diameter, contain fewer nuclei, but more myofilaments than the central regions. There are secondary afferent nerve endings on one, both or neither of the polar

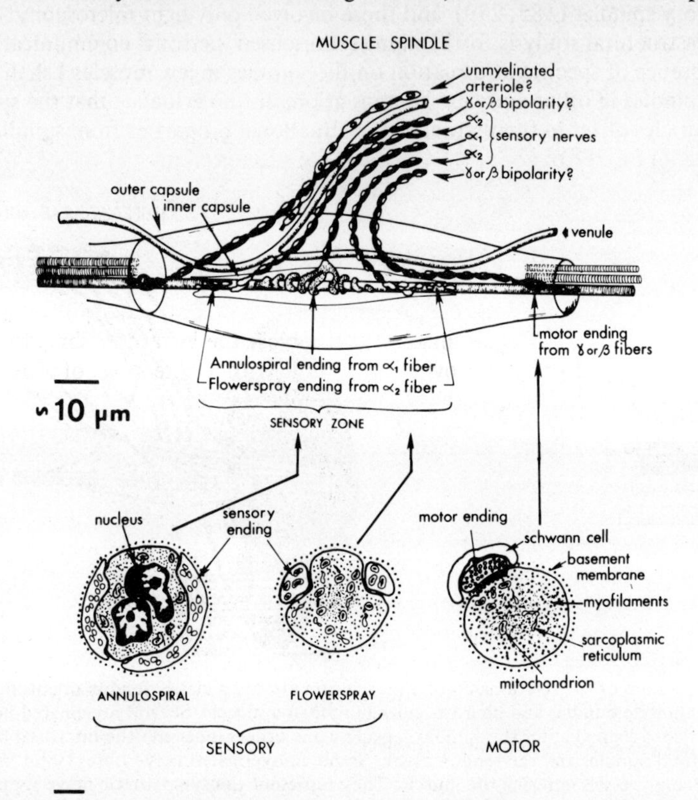

Fig. 5.2. Diagram of spindle structure, including one intrafusal muscle fibre. Note that in the upper drawing, the inner capsule surrounds the intrafusal fibre at its sensory zone. The outer capsule may enclose at least part of the motor zone. Vessels are commonly present within the outer capsule as are unmyelinated nerve fibres. The central part of the intrafusal muscle fibre is less conspicuously cross-striated than the polar zones. Here there is at least one nerve terminal, termed annulospiral, in the central part and flowerspray endings on one or both of the juxtacentral portions. The nerve fibres supplying the flowerspray endings are usually somewhat thinner than that for the annulospiral ending. The juxtacentral portions of the intrafusal fibre contain more myofilaments than the central portions. The motor zones may be supplied by thin myelinated fibres. It is not known if bipolar innervation exists. Different types of motor nerve endings are known to exist. In the lower part of the drawing are shown three cross-sections of an intrafusal fibre corresponding to central, juxtacentral and motor zones. Because of an accumulation of nuclei there are relatively fewer myofilaments here than elsewhere in the fibre. Note that the basement membrane is interposed between the muscle and the motor ending but not between the muscle and the sensory endings. This arrangement may have functional significance.

regions (*Fig.* 5.2) and these regions are not as conspicuously wrapped in internal capsule layers. At the border between central and peripheral regions there is also a high concentration of satellite cells (305), the function of which is poorly understood (302). There is less extracellular material in this region than in the equatorial region. Motor nerve endings, commonly called fusimotor (35), contact the intrafusal fibres both inside (303), and outside (35) the external capsule (*Fig.* 5.2). This may be of some significance since pharmacological effects at the two locations should differ because of the barrier afforded by the capsule. It is still not established whether motor nerve endings occur on both sides of the sensory region of the same intrafusal fibre. Such an arrangement could be of importance to the time course of the transduction at the sensory endings since the contraction wave would arrive from both or only one direction. Barker et al. (35) indicate that some motor nerves to extrafusal fibres may be branches from those innervating the intrafusal fibres. This is of significance since we have found that, in frog spindles, the zone corresponding to that contacted by the primary afferent can stretch when the two adjoining zones contract. This results in deformation and, presumably, transduction in these endings. Positive afferent responses to motor stimulation in amphibian muscle spindles have previously been demonstrated (165). Several types of motor endings can be differentiated on a structural basis in muscle spindles (35, 301). Presumably these subject the intrafusal fibres to different stimuli and thereby provide the spindle with a wide range of sensitivities, in a similar manner to that envisaged by Leksell (348).

PRESENCE OF TENDON ORGANS

This is presently a controversial area. Although Szentagothai (556) claimed to have observed Golgi tendon organs in the cat masseter muscle there has since been only one study (Lund, personal communication) with a similar conclusion. The consensus appears now to be that neither tendon organs, nor activity attributable to such receptors, have been unequivocally demonstrated in the masticatory muscles (388). There are several problems associated with the structural identification of Golgi tendon organs. Two different types of tendon organ have been described, one as discovered by Golgi and another at the end of muscle spindles located close to the tendon (34). The former should be easier to identify than the latter since, at the resolution of the light microscope, there may be confusion with intrafusal motor endings or other spindle nerves. A technique similar to that proposed above for the detection of muscle spindles would settle this problem.

STRUCTURE OF TENDON ORGANS

The tendon organs named after Golgi surround the collagen bundles which are coupled in series with the extrafusal muscle fibres at the end of a muscle (*Figs.* 5.1 and 5.3). They are invested by a capsule similar to that of the spindle, about 0·1 mm in diameter and sometimes over 1 mm in length. An ultrastructural description has been published (393) and a three-dimensional light microscopical one (86) (*Fig.* 5.3). The nerve endings are invested in a basement membrane and the afferent nerve is of the 1b type. The receptors respond classically to increased tension in the tendon during muscle contraction.

Another type of tendon organ is found at the end of muscle spindles (34) and there are also accounts of Pacinian receptors in or on a muscle (218).

Transduction in the Golgi organs is assumed to result from compression of the nerve endings when the collagen bundles straighten during muscle contraction.

The receptors also fire during physiological muscle stretch. It appears that these receptors fire during most phases of muscular activity and that the central connections are more widespread than was previously believed (548). Their role in mastication is unknown (388). There appears to be no electrophysiological evidence for their presence in masticatory muscles but, if they respond to stimuli other than muscle contraction, the physiological criteria for their identification, based upon functional characteristics, may be too narrow.

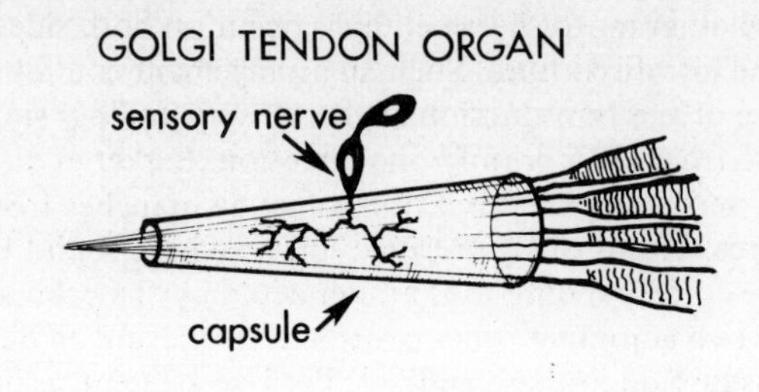

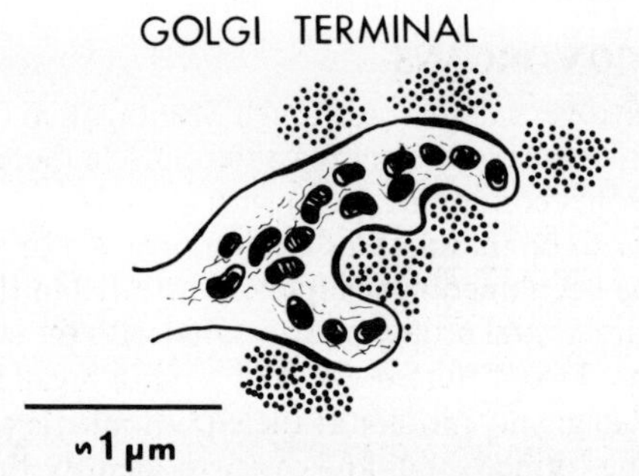

Fig. 5.3. Diagrams of Golgi tendon organ structure derived from the work by Merrillees (1962) and Bridgman (1968).

SUMMARY

The presence, structure and distribution of muscle spindles and tendon organs in masticatory muscles has been reviewed. Little information is available. The masticatory system to display certain structural characteristics (localized distribution of muscle spindles, mesencephalic representation of afferent neuron somata and relative absence of tendon organs generally, and spindles in certain muscles) which attract attention as a field for future research. Identification of receptors is a central problem for which an ultrastructural approach is proposed.

Discussion CHAIRMAN: D. J. ANDERSON

MØLLER: I should like to ask how you expect the presence of many spindles in these muscles could compensate for the absence of Golgi tendon organs ?

KARLSSON: I can suggest two mechanisms. One is that the equatorial zone in the muscle spindles can, and has been shown to, react positively to extrafusal muscle contraction, as the Golgi tendon organs have been thought classically to act. A remote possibility is that a spindle with its central reflex connections and its localization in a specific muscle may act in a Golgi tendon organ fashion for antagonistic muscles.
MØLLER: I could not quite follow that—the spindles and the Golgi tendon organs act differently.
KARLSSON: I do not think that I actually suggested that many spindles would compensate for the absence of Golgi tendon organs. What I did imply, however, was that there are mechanisms in muscle spindles which would respond positively to those forces which were classically thought of as stimulating only the Golgi tendon organs. The second point is that there is so much complexity in the reflex connections of the masticatory muscle spindles that the reflex effects usually associated with Golgi tendon organs could still take place. I do not profess any knowledge in this area.
LUND: We have recently found Golgi tendon organs in masseter and temporalis muscles of kittens (673) and are presently continuing the investigation in adult animals. Reconstruction of the neonatal muscles from serial sections showed that, in the temporalis, the tendon organs were found at the insertion into the coronoid process, while in the masseter they existed close to the origin. In both muscles, tendon organs were always in association with muscle spindles and within the deeper parts of the muscles.
MATTHEWS: May I ask Dr Karlsson if there is any physiological evidence that mammalian spindles are excited when alpha-motoneurons are stimulated and extrafusal fibres contract ?
KARLSSON: Barker believes that there are motor nerves to extrafusals which give branches to intrafusal fibres, but I have no personal evidence for this.
TAYLOR: If I may just anticipate something I shall be saying tomorrow in this connection; in recordings from spindles in normally behaving cats there is never any evidence of the spindles speeding during actual contraction of the muscles, and so in the normal situation there is no evidence for this.
KARLSSON: I suppose I should resort to structure.

Paper No. 6

Periodontal Mechanoreceptors

A. G. Hannam

The morphological and physiological characteristics of periodontal mechanoreceptors have received considerable attention over the past forty years. In some respects, the variety of experimental approaches which have been adopted has been to our disadvantage, because it has made correlation of the various studies difficult. For example, although it is possible to demonstrate the presence of different types of receptor in the periodontal tissues by histological means, it is not always possible to assign a functional role to each of them. Furthermore, when nerve impulse discharges are recorded, one cannot always be certain that the receptors involved are actually located in the periodontal ligament. Finally, the interpretation of human sensations when forces are applied to the teeth must take into account the central organization of the variety of neurons concerned with relaying tactile information to the cortical region, and this makes any comparison between sensory experience and single, peripheral nerve fibre responses a complex affair. Despite these difficulties, it is possible to make some observations and to speculate about periodontal mechanoreceptors.

Morphological studies of the periodontal innervation have been carried out by many workers, and in such studies there can be little doubt about the locations of the nerve endings. Terminals have been observed in both the gingival tissues (231) and the periodontal ligament (16, 226). In addition, the histology of mechanoreceptors in the periosteum of the frontal, nasal, maxillary bones and the mandible has been described (491). Distilling the available descriptions of probable mechanosensitive nerve endings in these tissues into a meaningful form, is no mean task. It seems reasonable to assume that mechanosensitive nerve endings with various morphologies exist in gingival tissues, periodontium and periosteum. Complex unencapsulated, complex capsulated and free endings have been described, and while it can be assumed that the complex endings are mechanoreceptors, it is also likely that some of the free-fibre endings are mechanosensitive, at least in the periosteum (494).

In the gingival tissues, both capsulated Krause-like and Meissner-like endings are found in the papillary and subpapillary lamina propria. Compact and loose unencapsulated whorls are also found in these regions (231). Variations are considerable in the papillary lamina propria, and fibrillar extensions may leave the whorls and pass into, and even through the epithelium to reach the outer layer of cells. Free gingival fibres often seem to form loops and may give off intra-epithelial fibrils which end in knob-like swellings.

Nerve terminals in the periodontal ligament which are probably mechanosensitive, have essentially the same structure as those described in gingival tissue. There is, however, a strong possibility that some species differences exist (349). It seems clear that the organized endings, whether they are capsulated or unencapsulated, are served by the wider diameter fibres, probably between 4 and 14 μm in diameter (16, 225). Griffin and Malor (226) believe that there are two types of encapsulated mechanoreceptors in the human periodontal ligament. One of these is a discrete unit consisting of an encapsulated myelinated fibre which pursues a tightly coiled or spiral course, loses its myelin sheath, and divides into many smaller unmyelinated terminals. This end organ is only loosely encapsulated, and is surrounded by loose connective tissue (*Fig.* 6.1). The other type is more complex, consisting of three or more nerve fibres which lose their myelin sheaths and encircle one another with their terminal branches. A diagrammatic representation of one of these receptors is shown in *Fig.* 6.2.

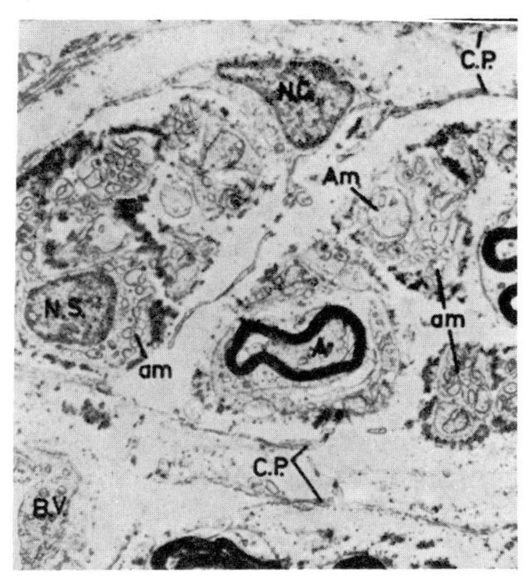

Fig. 6.1. Encapsulated periodontal nerve ending in man, C.P., capsular cell processes; N.C., nucleus of capsular cell; N.S., nucleus of Schwann cells; E, endoneurium; Am, large unmyelinated nerve fibres; am, small unmyelinated nerve fibres; A, myelinated nerve fibres; B.V., blood vessel. (Uranyl acetate × 9000.) *(Griffin, 1972.)*

Although the alveolar periosteum contains some complex unencapsulated and free nerve endings which may be mechanosensitive, the majority of mechanoreceptors appear to be of the Golgi–Mazzoni type (491). Most of these are composed of one corpuscle and one terminal and are very similar to receptors found in joint capsules. They are found in either the inner or outer layers of the periosteum, either singly or in clusters, and are supplied by fibres from 3 to 10 μm in diameter.

Analysis of observations on periodontal mechanoreceptors in function, whether from recordings of the activity of neurons, or from studies on human sensation, is generally hampered by the problem of locating the receptors involved with any degree of precision. It is, however, possible to gain some idea of receptor location during peripheral nerve recording to the extent that most of the gingival and mucous membrane receptors may be excluded from the sample. This is particularly true when the nerve to the canine tooth in cat and dog is selected for recording (139, 237, 492). To a certain degree, distinction between the nerves

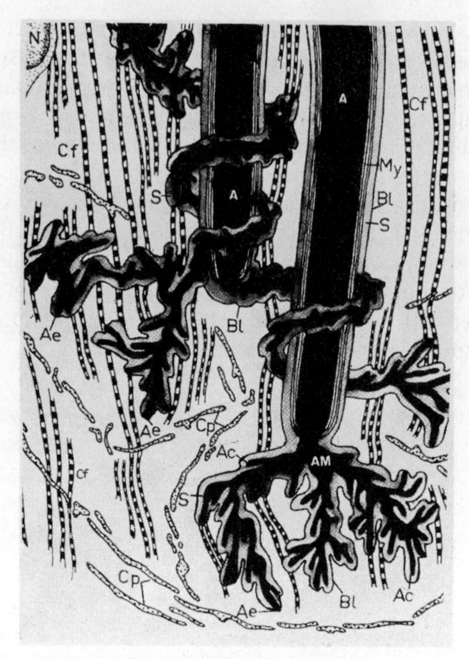

Fig. 6.2. Reconstruction of compound periodontal mechanoreceptor. A, axons; My, myelin sheath; S, Schwann cell cytoplasm; Bl, basal lamina of Schwann cell; M, unmyelinated axons; Ac, enclosed, or lacuna nerve ending; Ae, exposed nerve endings; Cf, collagen fibrils; Cp, cytoplasmic processes of capsular cells. (*Griffin and Malor, 1974.*)

supplying true periosteal receptors and those supplying true periodontal receptors is also possible on the basis of their response characteristics (*vide infra*).

The results of single neuron recording from peripheral dental nerves in cat, rabbit, and dog, have demonstrated the presence of two basic receptor types in the tooth-supporting structures, both of which respond to movement of the teeth following the application of an external force. One of these is rapidly adapting, that is, it generates a transient discharge to a sustained stimulus, and the other, which is more commonly described, is slowly adapting, in that it continues to generate nerve impulses for longer periods, often throughout steady-state tooth displacement. Variants have been described in terms such as 'very slowly adapting', and 'spontaneously discharging' (237), and the examples of all four types are illustrated in *Fig.* 6.3. While it is probable that the slowly adapting receptors (*see* Anderson et al. (16) for review) are located within the ligament itself, it is by no means certain that all the rapidly adapting responses which have been reported can be attributed to periodontal mechanoreceptors. Sakada (491) has shown that the majority of periosteal mechanoreceptors are of the Golgi–Mazzoni type, and are rapidly adapting. It is quite possible that many of the rapidly-adapting responses previously attributed to periodontal receptors may in fact have originated from the alveolar periosteum. With hindsight, it is a pity that the periosteum was not removed in some of our previous studies. From a functional point of view of course, it hardly matters where the receptors lie, so long as they respond to tooth movement.

On the assumption that both rapidly- and slowly-adapting mechanoreceptors may be found in the periodontal ligament, it has been proposed (225, 226) that the discrete encapsulated ending with its loose capsule and surrounding tissue may be associated with rapidly adapting neural responses, and that the more complex ending, responding more tonically to changes in periodontal tension, may be responsible for the slowly-adapting, and perhaps spontaneous, neural discharges. However, spontaneous resting discharges are a characteristic of many afferent nerves elsewhere in the body, and it is still unclear whether spontaneous discharges from periodontal mechanoreceptive neurons are generated as a result of some feature of the environment of the receptor apparatus, or whether they are an inherent property of the receptors.

The apparent tendency for rapidly-adapting mechanoreceptors in the tooth-supporting structures to have higher thresholds than slowly adapting endings (236, 453) not only suggests that there is a morphological, and perhaps functional distinction between the two kinds of ending as has been suggested (226), but also indicates the possibility of a difference in the regions of origin of the responses. More intense mechanical stimulation of a tooth could be required to excite the rapidly-adapting receptors in the alveolar periosteal tissues than is needed to stimulate the majority of slowly-adapting receptors actually within the periodontal ligament itself.

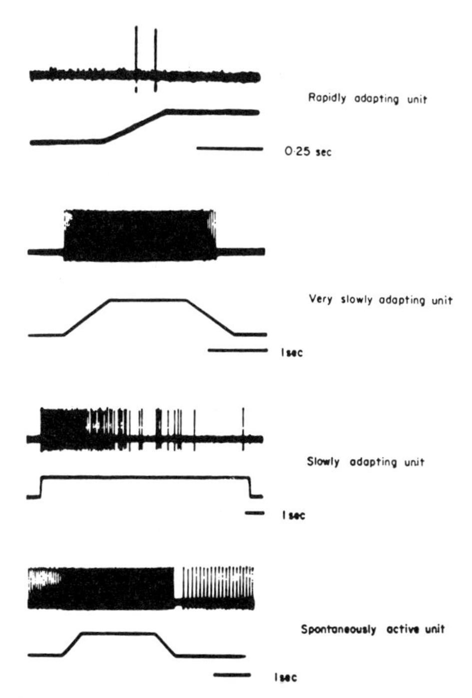

Fig. 6.3. Examples of the responses of single periodontal mechanoreceptor units. The upper trace in each case represents the response, and the lower trace represents the force applied to the teeth. The force used in the top record was 120 g. All other forces were 100 g. (*Hannam, 1970.*)

The thresholds of mechanoreceptors within the tooth-supporting structures to stimulation of the teeth have been reviewed elsewhere (16). Displacements of the order of 2–3 μm would seem to be necessary to evoke a response in single neurons. The forces involved therefore depend upon the mobility of the particular teeth involved, and are of the order of a few grams (1–5 g). A comparison of the literature on single neuron studies in animals, and studies in human sensation, reveals a certain disparity with regard to threshold phenomena. It is clear that the threshold for human sensation measured for the incisor teeth is of the order of 0·5 g (183, 383, 623) and this is considerably lower than that reported for single neurone responses. Garton's (183) observations that the presence or absence of ear plugs affected measured thresholds offers one explanation for this curious difference, in that contributions from receptor sites other than those around the teeth may assist in the perception of a tooth tap in man (16).

The directional characteristics of periodontal mechanoreceptors have received considerable attention. These have already begun to cause a certain amount of confusion in the literature, mostly because there has been an unfortunate tendency to equate the results of single neuron responses with those involving human sensation. It has been found (431, 453) that single, peripheral periodontal neurons showed a maximal response to movement of a tooth in one direction which decreased as the direction of stimulation was moved to either side of this direction. However, it is also known that receptors, or the receptor systems of single neurons, frequently respond vigorously to stimulation of a given tooth in more than one direction (237, 493). The percentage distribution of the response of such neurons to stimulation in five directions is shown in *Fig.* 6.4. It is also clear that the receptive fields of single first-order periodontal neurons commonly extend beyond individual teeth (16, 237, 493), and may involve several teeth and adjacent soft tissue. This phenomenon can be explained by hypotheses involving the spatial location of the receptor in question, and perhaps by invoking the possibility of branched nerve terminals shared by the same neuron, but such speculations require confirmation.

We have little information concerning the perception of the direction of stimuli to individual teeth in man, but we have some information regarding the ability to discriminate between stimuli of different intensity applied to single teeth (67, 71, 535). Here, it is pertinent to consider the marked sensitivities of single periodontal mechanoreceptors to the magnitude of applied stimuli (16). It is known that the slowly-adapting receptors in particular are capable of signalling small changes in the force applied to the teeth. Recent experiments designed to quantitate the discriminatory ability of such single periodontal neurons have shown that these mechanosensitive units are capable of transmitting 3·6 bits of information, or 12 discrete levels, when 128 different levels are applied to teeth over a 200 g range (239). Most of the useful information about force amplitude is transmitted over the first 100 g, that is from threshold to the force which produces a near maximum response although there are specific ranges for each receptor.
On the basis of experiments such as these, one would predict that human sensory discrimination would be best at low force levels, but this does not appear to be the case. Bonaguro et al. (67) have shown that the optimal functioning discriminatory range for human central incisors is between 50 and 500 g, despite the fact that we know human incisor thresholds are of the order of 1 g or less. This discrepancy implies that a more elaborate mechanism for force discrimination must be at work in sensory perception than the simple relay of peripheral single neuron information. It is conceivable that the complex organization of receptor fields and the variable directions of sensitivity of a population of periodontal mechanoreceptors may result in conflicting inputs to the central nervous system at low intensities of stimulation, thereby lessening the overall information

transmitting capacity of the system. Perhaps only at levels in excess of 50 g does the afferent pattern allow central synapses to produce better stimulus differentiation.

Bonaguro et al. (67) have also demonstrated that human discrimination of applied forces is about the same whether they are applied in the long axis of the tooth or 90° to the long axis, and that the ability to discriminate differences in the intensity of force is essentially the same in mandibular incisors, canines and first premolars. There is no information on the ability to discriminate between stimuli applied in different directions.

It is known that the cell bodies of first-order, peripheral periodontal neurons are located in two sites, namely, the Gasserian ganglion and the mesencephalic nucleus of the fifth cranial nerve. An idea of the functional role of periodontal mechanoreceptors may be gained by considering their further projections from these sites. The projections have been reviewed (16, 292). The neurons which have their cell bodies in the Gasserian ganglion project to the main sensory and spinal nucleus of V. From there, the information may be relayed at segmental level to influence trigeminal motoneurons reflexly. As a result, both excitatory and inhibitory reflex effects can be seen in the jaw-closing musculature. These and allied phenomena are dealt with more properly elsewhere in the Symposium. Neural activity originating from periodontal mechanoreceptors may also be relayed via the main sensory and spinal nucleus of V to the ventrobasal nuclei of the thalamus, and from there to the cerebral cortex.

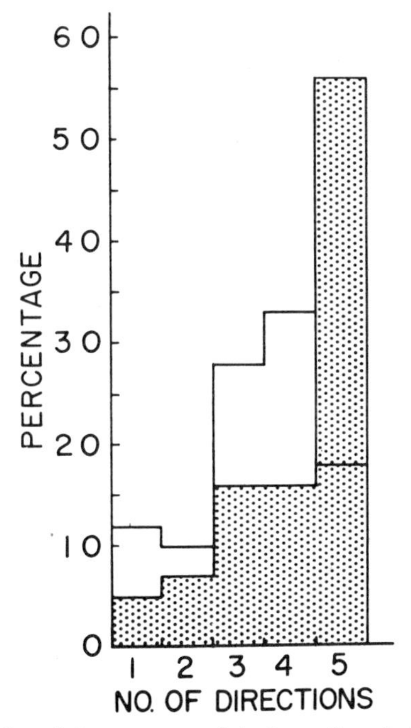

Fig. 6.4. Percentage distribution of the responses of single-tooth periodontal mechanoreceptor units following stimulation of the tooth in five directions. The stippled area represents the distribution for 140 rapidly adapting receptors and the unfilled area that of 120 slowly adapting units. (*Sakada and Kamio, 1971.*)

Periodontal neurons which have their cell bodies located in the mesencephalic nucleus of V are also involved with reflex activity in the muscles of mastication. There are strong projections from the mesencephalic to the motor nucleus of V, many of which are also concerned with relaying afferent information from muscle spindles. Other potential projections of periodontal neurons from this site, for example, to the cerebellum, or cerebral cortex, are at present ill-defined. The known projection pathways of the neurons serving periodontal mechanoreceptors are summarized diagrammatically in *Fig.* 6.5.

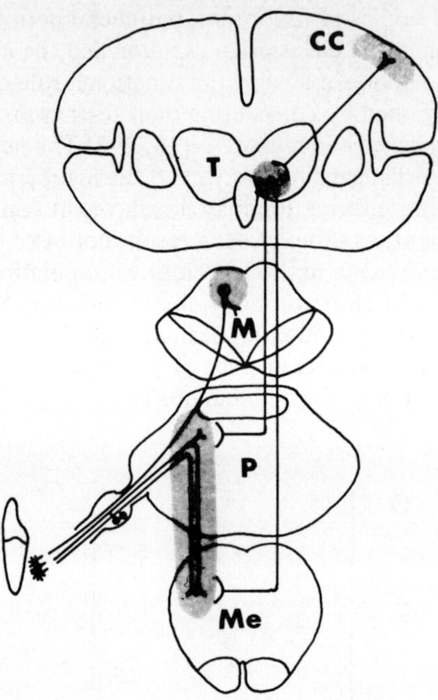

Fig. 6.5. Diagrammatic representation of the central projections of periodontal mechanosensitive neurons. Primary cell bodies lie either in the Gasserian ganglion or the mesencephalic nucleus of V. CC, cerebral cortex; T, thalamus; M, midbrain; P, pons; Me, medulla. Shaded areas in the midbrain, pons and medulla have reflex connections at segmental level with the muscles of mastication.

The ascending network of neurons which is associated with relaying tactile information from the teeth, has prompted the suggestion that functionally, the periodontal ligament has a dual innervation, one set of receptors being involved in the mediation of brain-stem reflexes, and the other with the conscious perception of forces on the teeth (292). It is clear that this assumption is unwarranted. While it is true that brain-stem reflexes can be evoked by the stimulation of periodontal mechanoreceptors and that these receptors are involved with sensory experience, there is as yet no proof that two functionally different sets of neurons exist. Moreover, recent evidence demonstrates that the periodontal innervation may have a positive feedback effect upon jaw-closing motoneurons at the cortical level, in addition to its influence in the brain-stem (369, 371). When other integrative actions of periodontal input are considered, for example the interaction at brain-stem level with laryngeal inputs (514), and at the cortical level with muscle afferents (371), it is obvious that the periodontal innervation has a more complex role to play than that of a simple innovator of uncomplicated brain-stem reflexes in the jaw muscles and a transducer for conscious sensory experience.

Discussion CHAIRMAN: D. J. ANDERSON

TAYLOR: In your consideration of the relationship between the unit characteristics and perception of forces as a whole, aren't you ignoring the advantage the animal would get of convergence of input from a number of units ?
HANNAM: Yes, I accept that. Convergence would refine the input a great deal. However, perceptual discrimination in man does not reach optimum values until forces of over 50 g are applied to the teeth, and this finding does not appear to be consistent with the large number of responsive, low-threshold periodontal neurons found in peripheral studies. Although convergence and other synaptic events must influence perception, discriminability seems to be less than optimum at low levels of applied forces, and I find this surprising.

Paper No. 7

Temporomandibular Joint Receptors

A. Storey

Most mammalian temporomandibular joints (human, monkey, cat, rabbit, guinea-pig and rat) contain three types of receptors common to other joints of the body (527), namely (1), unencapsulated spray type endings, which have been called Ruffini-like receptors when located within the joint capsule and Golgi-tendon organs, when located in the ligaments, (2), encapsulated Pacinian-like corpuscles, and (3), free nerve endings (54, 174, 326, 581). These reports, depicted in *Fig.* 7.1, have been assigned a variety of names and numbers but in spite of the various designations, their identity is usually obvious. As in other joints of the body, the free nerve endings are most abundant, with Ruffini, Golgi and Paciniform following in that order. Most authors (with the exception of Kawamura et al. (318), agree that the Pacinian-like corpuscles are very few in number. The Golgi tendon organs are innervated by the largest articular fibres, the Ruffini and the Paciniform endings by medium-sized myelinated fibres and the free endings by the smallest myelinated and unmyelinated fibres. The receptor densities are greatest in the lateral and

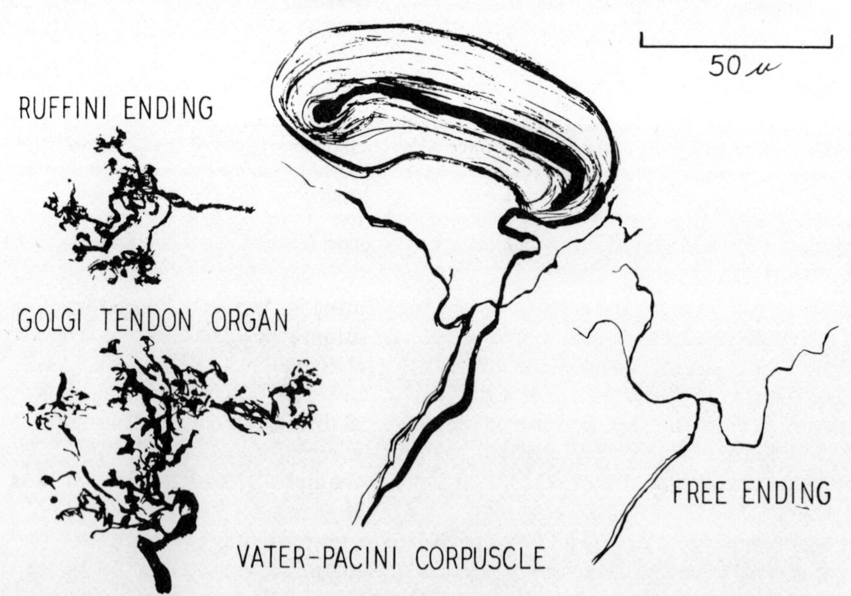

Fig. 7.1. The three types of receptors found in the adult temporomandibular joint. The two varieties of spray ending are represented on the left, the encapsulated Pacinian-like corpuscle in the middle and the free nerve ending to the right. (*After Thilander, 1961.*)

posterior parts of the capsule of the joint which is innervated by the auriculo-temporal nerve. No nerve endings are found in the synovial tissue nor the meniscus.

What functions do these receptors serve ? According to one theory, each type of receptor would be responsible for transmitting one kind of information. The way in which a receptor discharges gives an indication of the kind of information it may be providing. In *Fig.* 7.2 are shown the first successful recordings from single fibres innervating the temporomandibular joint (314).The responses of three types of units from the cat temporomandibular joint are illustrated. The upper trace is of action potentials recorded from a single sensory unit in the auriculotemporal nerve. The lower trace indicates movement of the isolated condyle head of the joint containing the receptor. The figures under the lower trace indicate the relative position of the condyle to the fossa. Trace A illustrates a rapidly-adapting unit which fires only when the condyle head is rotated through a specific position.

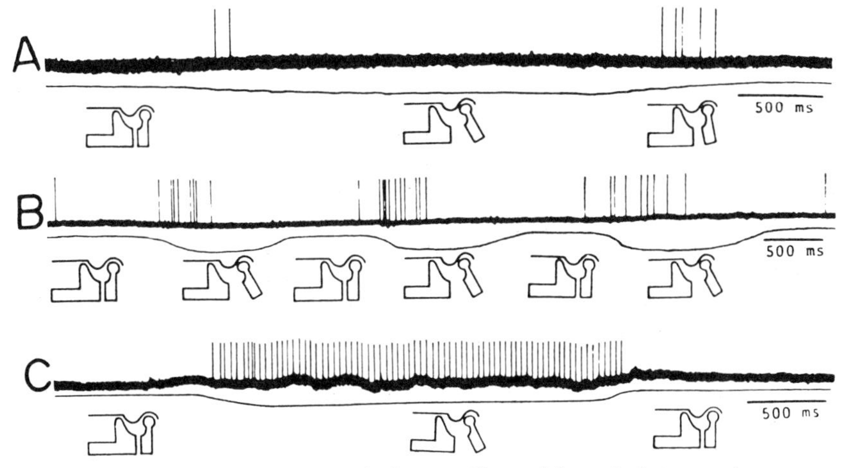

Fig. 7.2. Action potential discharges in single nerve fibres of the auriculotemporal nerve innervating rapidly adapting (A) and slowly adapting (B and C) receptors in the isolated temporomandibular joint of a cat. Upper trace is the oscilloscope record of the receptor discharge at varying degrees of joint rotation, indicated in the lower trace and in the sketches. Downward movement of the lower trace would represent jaw opening. The condyle and adjacent mandibular ramus have been severed from the rest of the mandible to eliminate excitation of muscle receptors. Time scale 500 milliseconds. (*From Kawamura and Abe, 1974.*)

Traces B and C record the activity of slowly-adapting receptors which continue to fire while the condyle head is in a specific position. It seems likely that units of the type A would convey some information relative to movement of the mandible whereas units such as B and C would convey information relative to the position of the mandible. Because of the technical difficulties of recording such units and at the same time identifying the specific receptor from which the discharges are arising, it is impossible to assign unequivocally any of these patterns to a specific temporomandibular joint receptor. However, on the basis of recordings from joints elsewhere in the body, the temporomandibular joint receptors have been assigned the functional designations given in *Table* 7.1. Since Ruffini endings and Golgi tendon organs are slowly adapting receptors elsewhere in the body, they are designated static mechanoreceptors. Since the free nerve ending has been associated with pain, it has been designated a pain receptor. It is tempting to continue this line of thought one step further and attribute the discharge in trace A in *Fig.* 7.2 to a Vater-Pacini corpuscle and the discharges in traces B and C to Ruffini endings or Golgi tendon organs. One further step is made in assigning reflex roles to these receptors in *Table* 7.1. The Ruffini ending becomes the receptor for posture, the Vater-Pacini corpuscle for regulation of movement and the Golgi tendon organ and free ending receptors for protection of the muscle and joint from damage.

Table 7.1 Anatomical and functional designations for temporomandibular joint receptors and their inferred reflex roles (*From Storey, 1973, by permission of C. V. Mosby Co.*)

Anatomical designation	*Functional designation*	*Reflex role*
Ruffini ending	Static mechanoreceptor	Posture
Vater-Pacini corpuscle	Dynamic mechanoreceptor	Movement accelerator
Golgi tendon organ	Static mechanoreceptor	Protection (ligament)
Free ending	Pain receptor	Protection (joint)

There are two hazards in attributing the single nerve discharges in *Fig.* 7.2 to specific endings. The first is the sampling problem. All anatomical investigations have indicated that free endings are much more numerous than the organized endings. Sampling would tend to favour recording from the most numerous receptors, i.e. the free endings. The fact that the type A unit was only one of its kind found in a sample of 29, suggests that this was a Vater-Pacini corpuscle which was also encountered infrequently in histological studies. Measurements of conduction velocity would provide insight into the size of the nerve fibres innervating the receptors and provide some evidence as to their identity. Since free endings elsewhere in the body are now known to be sensitive to thermal and mechanical stimuli as well as noxious stimuli, the discharges in B and C could have arisen from free nerve endings. Free endings in the cornea, skin and the mucosa of the pharynx and larynx are exquisitely sensitive to mechanical stimulation and show both rapidly and slowly adapting characteristics. Both the small myelinated and unmyelinated fibres innervating joints could serve a mechanoreceptive role but this has yet to be explored.

Even though the specific roles of individual receptors in the temporomandibular joint may not be assigned with certainty, we may ask what roles they play collectively in mastication. Several studies (109, 223, 318) have demonstrated that

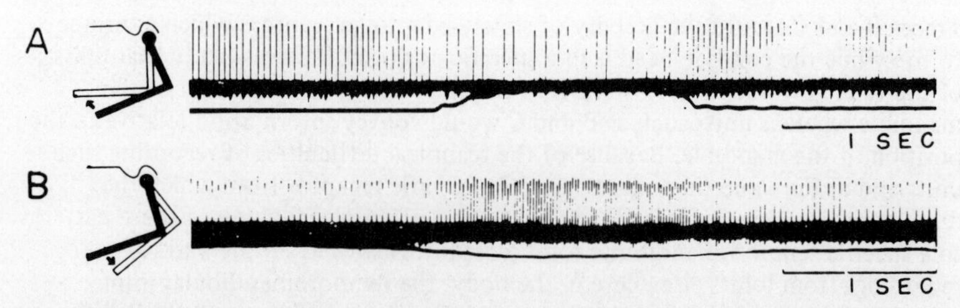

Fig. 7.3. The alterations in firing of a masseter motoneuron in the cat on passive rotation of the ipsilateral condyle. The condyle and adjacent mandibular ramus had been severed from the rest of the mandible. A closed position (A) resulted in a sustained decrease in firing while an open position (B) produced a sustained increase in firing. (*From Kawamura et al. 1967.*)

isolated inputs from the temporomandibular joints of the cat can alter the ongoing activity of mandibular muscles or the motoneurons innervating them. The possibility of a muscle input was excluded by isolating a segment of the mandible from its muscle insertions. Further confirmation of the joint origin of these effects is provided by observations that the effects are reversibly abolished by infiltration of the capsule of the joint with local anaesthetic. The single unit recording in the top trace of *Fig.* 7.3 A and B is from the trigeminal motor nucleus in the region of the masseter motoneurons. Mandibular rotation in a closing direction, indicated in each lower trace by a movement upward, inhibited the masseteric discharge as long as the mandible remained in that position. The effect must have been mediated by the temporomandibular joint receptors since all other sensory inputs were removed. Maintaining the mandible in an open position resulted in an increased discharge of the masseter motoneuron. Since the altered firing of the tonically discharging masseter motoneuron is sustained, the evidence suggests that the temporomandibular joint receptors could be playing a role in the regulation of posture. Some recent studies performed in our laboratory (363) suggest that temporomandibular joint receptors also play a role in tongue posture.

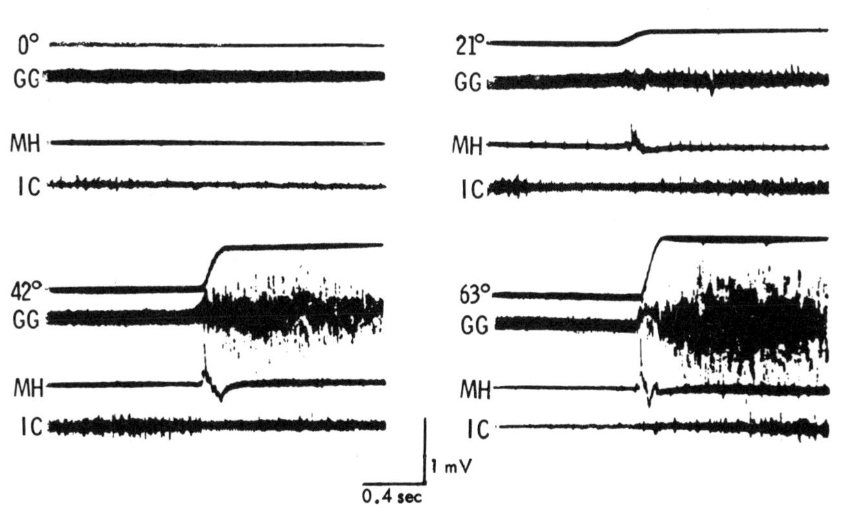

Fig. 7.4. The effects of various degrees of jaw opening on muscle activity in the cat. Top trace indicates the angular degree of jaw opening. The lower three traces are electromyograms from genioglossus (GG), mylohyoid (MH) and inferior constrictor (IC) muscles. Genioglossus activity appears at an opening of 21° and is marked at openings of 42° and 63°. The effect is abolished reversibly by infiltration of both temporomandibular joints with local anaesthetic. Time marker 0·4 seconds and amplitude marker 1 millivolt. (*Kindly provided by Dr Lowe.*)

Fig. 7.4 shows the effects of various degrees of jaw opening on the activity of the genioglossus muscle. With the teeth in occlusion, the genioglossus muscle in the cat is silent; at 21° of opening the genioglossus demonstrates activity in a few motor units; at 42° and 63° of opening the amount of genioglossus activity is quite marked. Infiltration with local anaesthetic into both temporomandibular joint capsules abolished the response completely. Of more relevance to mastication, is the degree to which receptors in the temporomandibular joint might modify ongoing mandibular movement. Greenfield and Wyke (223) found that motor unit responses recorded in electromyograms of masticatory muscles adapted rapidly at the onset of movement. Wyke (635) has speculated that these reflexes serve to accelerate movement at its onset in order to overcome the inertia of the mandibular complex. Recent work on the knee joint of the cat suggests that joint receptors sense resistance to movement as well as angle of rotation (230) and that rapidly-adapting receptors are conveying more information relative to acceleration than velocity (167). Future studies on the temporomandibular joint should incorporate more precise control of movement and should monitor torque as well as angle of rotation. In animals with lateral function some measurement of translation will have to be incorporated also.

Receptors within the temporomandibular joint can also give rise to reflexes protecting the joint from damage. Such reflexes might be expected to come into play to restrict extreme opening movements such as yawning and to prevent subluxation of the condyle. In cases of pathology of the joint, reflexes would be expected to spare the joint both in posture and during movement. Klineberg (326) has reported that both jaw-opening and jaw-closing reflexes are produced by electrical stimulation of the nerve supplying the joint. Shwaluk (525) and Kawamura (313) on the other hand, have only been able to evoke jaw-opening reflexes. In *Fig.* 7.5 is illustrated the response of digastric motor fibres to

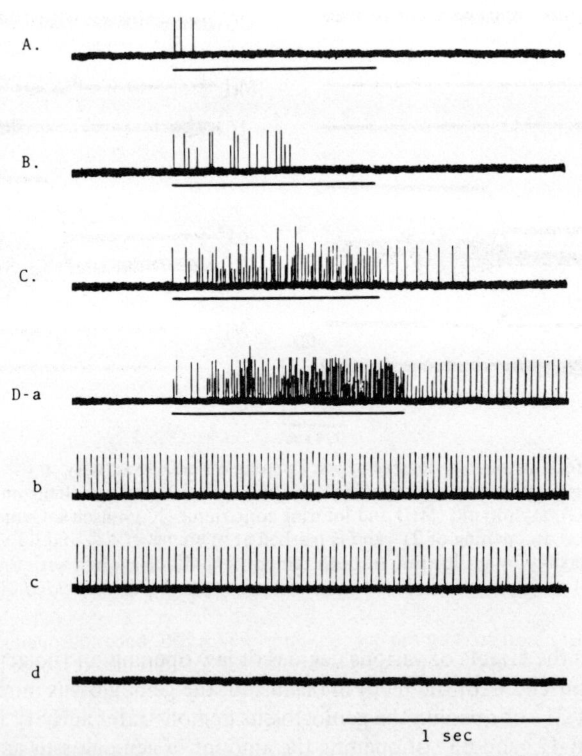

Fig. 7.5. Discharges of digastric motor nerve fibres to mechanical stimulation of increasing intensity of the ipsilateral temporomandibular joint of the cat. Strength of stimulation is increased through A to D: stimulation in D is assumed to be noxious since the discharge of one digastric fibre continues after the stimulation (indicated by the horizontal bar) has ceased. Time marker 1 second. (*From Kawamura, 1974.*)

increasing pressure applied to the joint capsule with a glass rod. The duration of stimulation is indicated by the horizontal bar: strength of stimulation increases from A to D. Note that strong stimulation induces an after-discharge that may last as long as 2 minutes. If the temporomandibular joint, as Klineberg (326) claims, is capable of initiating both jaw-closing and jaw-opening reflexes one has a mechanism for the peripheral initiation of cyclic jaw movements similar to that advanced by Jerge (291). According to Klineberg, the nature of the response is dependent on the joint nerve stimulated, the strength of stimulation and the initial jaw position. He speculated that receptors in the lateral joint capsule initiate jaw-opening reflexes while receptors in the anterior capsule regulate closing forces. Although the receptors in the joints are thought not to monitor stresses carried by the bony elements it now seems likely that joint afferents are sensitive to muscle tension (527). Receptors in muscles and joints may come to be regarded as complementary components of a single afferent system rather than two separate systems. This raises questions regarding the role of joint sensory systems in patients who use heavy masticatory forces or continually clench and grind their teeth. Since most of the physiological studies on temporomandibular joint function have been performed on cats, little insight has been gained on reflexes which might protect the joint in horizontal movements. Since the mandible frequently deviates to one side during protrusion in temporomandibular joint disorders, lateral reflexes of joint origin could be expected. Lund et al. (370) have shown that the application of firm labial pressure to the upper incisor teeth of the anaesthetized decerebrate rabbit will result in a contralateral displacement of the mandible, but the possibility of joint receptors being involved has not been excluded.

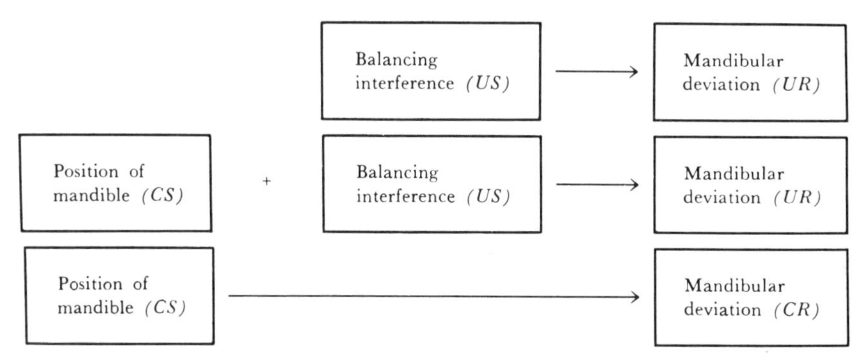

Fig. 7.6. The postulated learning sequence for avoidance of a balancing-side occlusal interference. A balancing-side interference (US) gives rise to mandibular deviation (UR). This is an unlearned reflex. Mandibular position near tooth contact, probably detected by mandibular joint receptors (CS), coupled with balancing-side contact (US) result in continued mandibular deviation. Eventually mandibular position near tooth contact (CS) gives rise to mandibular deviation (CR) without striking the occlusal interference. US, unconditioned stimulus; UR, unconditioned response; CS, conditioned stimulus; CR, conditioned response. (*From Storey, 1973.*)

Another possible role of temporomandibular joint receptors is in providing a conditioned stimulus for learned reflexes initiated from other receptor sites. For example, some kinds of occlusal contacts may, under certain conditions, give rise to learned reflexes avoiding further contact with the interfering tooth. Some receptor systems must cue the centres controlling mandibular closure to alter the normal closure path. Temporomandibular joint receptors could serve this function very well, although receptors in muscle and possibly the skin overlying the joints may also participate. The sequence of events I have in mind is illustrated in *Fig.* 7.6. Since Schaerer et al. (501) have shown that balancing side interferences will lead to reflexes resulting in mandibular deviation one can identify the balancing interference as the unconditioned stimulus for mandibular deviation, the unconditioned response. Some previously neutral input, monitoring position of the mandible occurring in close proximity with the balancing interference, leads to that previously neutral input, designated the conditioned stimulus, initiating the mandibular deviation now called the conditioned response.The closure is effected without contact on the balancing interference. I will not dwell further on this point since tooth contacts and their effects is the topic of the programme on Wednesday morning.

Much of the work I have cited above on the role of temporomandibular joint receptors in mandibular posture and movement has come from studies on experimental animals, particularly the cat, in which the masticatory patterns are very different from those in man. To what degree can one extrapolate to the human subject ? What evidence is there to indicate the temporomandibular joint receptors in man play a role in mastication ? With respect to posture, I am told by Sigurd Ramfjord that, on a visit to the Royal Dental School in Malmö some years ago, he was told of a student project conducted under Posselt which demonstrated that infiltration of the temporomandibular joint with local anaesthetic resulted in a change in postural position. Apparently the study was never published, which is unfortunate, since the reported results are most interesting. Schaerer et al. (500) published a brief paper on 2 patients in which the effect of anaesthesia of the gingiva, periodontium and temporomandibular joints was investigated on chewing movements. The anaesthesia reduced the patient's ability to control the food bolus but the performance of chewing movements was not affected. This led the authors to conclude that chewing movements in man seemed to be influenced 'only slightly by the sensory information from the oral cavity'. The results would indeed seem to suggest that input from the temporomandibular joint as well as from the teeth are of little significance in mastication. I am inclined to agree in the case of subjects with normal temporomandibular joints and occlusions free of threatening interferences or teeth with highly elevated sensibilities. On the other hand if stresses are brought to bear on any part of the stomatognathic system which threatens its well being, alterations in masticatory movements will occur. Unconditioned reflexes originating in the temporomandibular joint certainly do occur in pathological conditions involving the joint. Clinical observations of patients with inflammatory disorders of the temporomandibular joint support this view. Receptors in the temporomandibular joint may serve to protect other sites in the system by sensing mandibular position and warning of imminent contact with the offending structure. Recent work suggesting that the receptors of the joints and muscles function together as a unit is worth keeping in mind. An understanding of bird migration was not possible until it was realized that birds navigate using both the sun and a magnetic compass.

Discussion CHAIRMAN: D. J. ANDERSON

KAWAMURA: We have obtained various pieces of neurological evidence about the

sensory information from the temporomandibular joint in the cat and I have always been trying to emphasize the functional importance of this information. However, as you mentioned, there are obvious species differences. Recently we have seen 3 young patients who have had fractures of the condylar head on both sides as a result of traffic accidents. Later both condylar heads were ankylosed in the fossae and the patients have pseudo-joints on both sides of the mouth. However, they can chew very well without the sensory information from the joints. This morning Dr Hannam discussed the sensory functions of the periodontal ligament. However, the full denture wearer can chew very well without periodontal ligament sensory function. The sensory feedback control in the stomatognathic area relies on information from both these sources, but once we have lost the sensory functions of the periodontal ligament, the muscle or temporomandibular joint sensory functions may cover the loss. We must always consider sensory functions of muscles, tendon and joint and periodontal ligament all together. This morning nobody mentioned the mechanical sensory functions in muscle other than muscle spindles and tendon organs. In the capsule of the muscle there are sensory receptors. These may be stimulated during chewing. So if anybody has ideas or suggestions about these receptors in the muscle capsule I should be glad to hear them.
STOREY: You have raised two points: first of all that there may be interactions. There has been a fair bit of work on joints elsewhere in the body which suggests that the input from the joint is to some degree determined by what the muscles are doing. Millar (663) raised this recently, but other people have identified it elsewhere in the body. The other point is the adaptation which occurs when you lose the input from one system. I think that those are two important points we ought to keep in mind.

Paper No. 8

The Concept of Occlusal Vertical Dimension and its Importance in Clinical Practice

R. I. Nairn

When the preliminary programmes for this meeting appeared I was listed to speak on the 'concept of a rest position and its importance in clinical practice'. I requested a change to the present title, not out of sheer contrariness but because I am inclined to the view that the rest position (or resting posture) is something of a red herring set to lead us away from the essential importance of the jaw separation established when the teeth are together, and which is known in the current terminology as the occlusal vertical dimension. I will try to explain why.

Provided that there are contacting natural teeth in both jaws, a partial denture, either tooth or mucosa borne, should normally be made at the occlusal vertical dimension provided by the natural teeth. But if there is a complete denture in either or both jaws then the occlusal vertical dimension must be otherwise determined.

It is the case that the commonest fault in the construction of complete (or complete against partial) dentures is that of making the occlusal vertical dimension too great—so much so that I have thought it ought to be called 'the height of folly'. I hope in the course of this paper to examine the causes of this difficulty, and to start by looking at the occlusal vertical dimension of the natural teeth.

It has long been said that in development and growth there develops a rest position for the mandible determined by the resting length of the mandibular muscles, or by some basic motor activity which is innate and unalterable throughout life (583).It follows that the space between the jaws is also fixed, and into this space erupt the teeth. The teeth cease to erupt before they have entirely obliterated the space, so that when the jaw is in the rest position a gap remains between the occlusal surfaces of the teeth to the extent of 2–3 mm, the so-called free-way space. So indeed the distance between the jaws when the teeth are together (the occlusal vertical dimension) is a little less than the space between the jaws when the mandible is in the rest position (rest vertical dimension). According to this theory, the removal of the teeth does not affect the rest position of the mandible. Therefore, when the patient has lost all natural occlusal contacts, it is only necessary to put the teeth in at such a level that a free-way space is again produced, and the result will be a restoration of the original occlusal vertical dimension. And this is held to be a desirable end result.

Unfortunately this amiable view is at variance with reality in three ways at least.

First, as shown by Tallgren (560), dentures can be constructed at a variety of occlusal vertical dimensions and all demonstrate a free-way space. And therefore the demonstration of a free-way space is no guide to the occlusal vertical dimension of dentures.

Secondly, it has been the experience with immediate denture patients who have had dentures constructed reproducing the occlusal vertical dimension provided by the natural teeth, that some return in time with all the symptoms associated with wearing dentures made with the occlusal vertical dimension too great. Therefore it seems that the vertical dimension provided by the natural teeth is not necessarily that required in dentures.

Thirdly, it has recently (in the last ten years) been noticed that removal of the natural teeth (or even the occlusal stops) is followed in many people by the adoption of a new rest position, usually with the jaws closer together. Therefore an occlusal vertical dimension made with reference to the edentulous rest position may be different from that which was obtained with the natural teeth.

It is possible to offer a variety of conjectural explanations for these events which hinge about the view that the rest position of the mandible is not stable, but is a position taken up in response to the circumstances, e.g. the amount the teeth have erupted, the size of the lips, the length of the muscles, the size and posture of the tongue, and particularly the tendency of the soft-tissue masses to return to a state of physical rest—by no means a harmonious collection to which each person must adapt. Change the circumstances, namely take out the teeth, and you change the rest position, and in replacing the teeth, the vertical dimension provided must be in harmony with the needs of the other structures.

How large, therefore, ought the occlusal vertical dimension be for a patient who has lost all contacting teeth ? The answer must be that we do not at present know for certain, although I suspect that the view I have just expressed on the nature of the edentulous rest position can offer a justification, and I shall return to this in due course. Clinical necessity and experience suggest that a reliable guide is to be found in the edentulous rest position with neither (or no more than one) denture in the mouth. The dentures are made with a vertical dimension in occlusion which is rather less than the edentulous rest vertical dimension. The occlusal vertical dimension so determined may not be the ideal one for the patient but there are no more certain criteria in our present state of knowledge. As I have explained, this is unlikely to produce the same occlusal vertical dimension as existed with the natural teeth, and it is important to understand that it is not an attempt to produce a free-way space.

The free-way space is a gap which appears between the teeth when the patient, with natural teeth or dentures in place, adopts what purports to be the 'rest position'. I have said that this is a response to the presence of the natural dentition and so I believe it to be in the case of dentures. It has very little clinical relevance.

Furthermore, I think it apparent that not only is the free-way space irrelevant, but that the rest position of the mandible is very much a red herring, both in dentures and in consideration of the natural dentition. The essential functions are concerned with the jaw separation provided by the dento-alveolar structures; in between functions, the mandible returns to whatever position of rest it can find. This position may be influenced by the tendency of the structures to come to physical rest because of their visco-elastic properties, by other physical factors such as the weight of the mandible and the development of a sub-atmospheric pressure in the mouth, or by neuromotor influences such as segmental reflexes (the monosynaptic stretch reflex or responses to afferent information from the oral mucosa, mechanoreceptors in the periodontal tissues or the temporomandibular joint) and influences from higher centres.

But, functionally, it is the jaw separation in occlusion that is the essential factor. In people with dentures it is important that they can speak, chew and swallow comfortably, and that, for example, they can with ease elevate the tongue to the palate to start the bolus on its further path, and that they can easily and comfortably produce an anterior oral seal.

In many people with natural teeth, the jaw separation provided by the dento-alveolar structures allows this degree of mobility. But in some it does not. I would tentatively suggest that, in some people, the discrepancy increases with age and with the development of periodontal disease. I cannot say if this situation produces any functional disability because there is no way of knowing if such people would be better off if their natural occlusal vertical dimensions were reduced.

I am saying that in some people the occlusal vertical dimension, i.e. the tooth size, the extent of tooth eruption and alveolar growth, is not in harmony with the rest of their features. I am not, however, suggesting it should be altered unless there are some very clear indications for doing so.

I think that the importance of occlusal vertical dimension in denture construction has been recognized by those clinicians who seek to record it directly, rather than indirectly with reference to the rest position. I have in mind the judgement of appearance, the assessment of the mandibular position in swallowing, the variation of jaw separation until the patient declares himself most comfortable, or the determination of the jaw separation which allows maximum biting force to be developed. Unfortunately, each of these methods has its own hazards and is subject to more variability than might be acceptable.

The edentulous rest position is one in which the oral structures come to rest without the interference of the teeth, and it provides a measure of the space available into which the artificial replacement can be fitted. It is characterized by the position of the tongue filling the oral cavity and resting against the palate and by the contact of the lips. It seems a reasonable starting point, not because it is a resting posture, but because of its functional characteristics. I must admit I would rather have some positive and direct criteria for determining the occlusal vertical dimension in any case, but while the quantity of scientific information is great, and growing fast, such useful knowledge is hard to come by.

Discussion CHAIRMAN: B. MATTHEWS

BLAIR: I have long been interested as a prosthetist in the problem of finding the correct occlusal vertical face height and I have reported that some manifestations of an over-open bite may even be diagnosed as classic trigeminal neuralgia (649). I succeeded in clearing up a number of cases by reducing the over-open bite to what I thought was the correct vertical face height, but, unfortunately, I can think of two cases where the condition recurred. In one case the patient was in a neurosurgical ward to have a post-Gasserian rhizotomy the next day. We fitted a splint which propped the jaws, as we considered, too far apart. The condition resolved within half an hour ! I find it very difficult indeed to find any satisfactory way of determining correct vertical face height, but we have heard about the change which can occur in muscles and I wonder if Mr Nairn, or anyone else, could

offer me an explanation for what I have found clinically a very distressing condition.
NAIRN: I wouldn't care to offer an explanation. The patient has a problem in deciding whether to eat or put up with the pain !

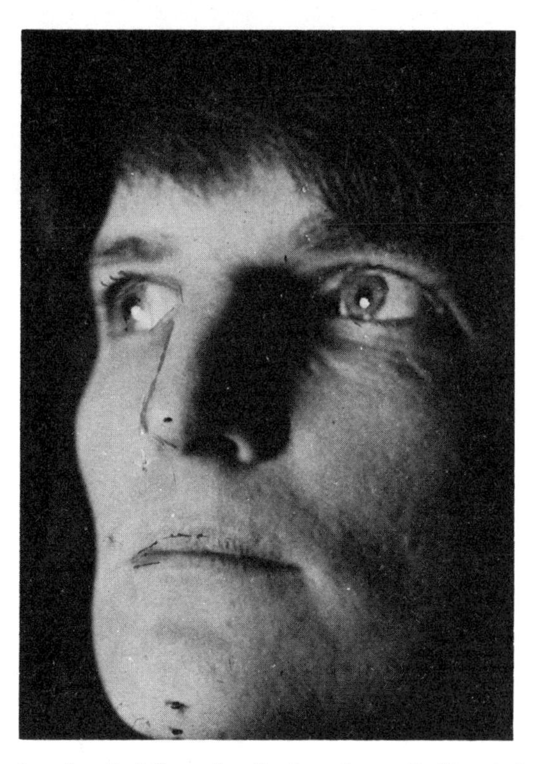

Fig. 8.1. Gross opening of vertical dimension. Dentures in mouth. Lips strained.

SCHER: I should like to compliment the speaker on a beautiful presentation, so very rapidly given. I should like to make some further points. *Fig.* 8.1 shows a lady with a surprised expression; she is wearing complete upper and lower dentures. She could neither smile nor speak properly, and she said she had considerable pain on the side of her face. She was referred to me in the hope that I could do something for her. Using dividers to measure, which is rather primitive, we put a spot on her nose and a spot on her chin (*Fig.* 8.2*c*). This, of course, does not displace the soft tissues of the submental region, as a Willis bite gauge might do. We took a measurement and found that when she had her upper dentures in, the vertical dimension was open slightly (*Fig.* 8.2*b*), but when she had both dentures in, it was open even more (*Fig.* 8.2*a*). Now the reason I show you these illustrations is to tell you of a very simple method by which a judgement can be made, and which is illustrated in *Fig.* 8.3. It comes from advice given to me by Professor Chick, who was once a senior lecturer at the University of Bristol. The method is very interesting because it is so simple. He said to me, 'You are a small chap Scher, you will be able to peep very easily, and all you have to do is to get the patients to stay with their jaw at rest. Rehearse them—and say to them "I am going to very gently part your lips and have a look at what is inside,

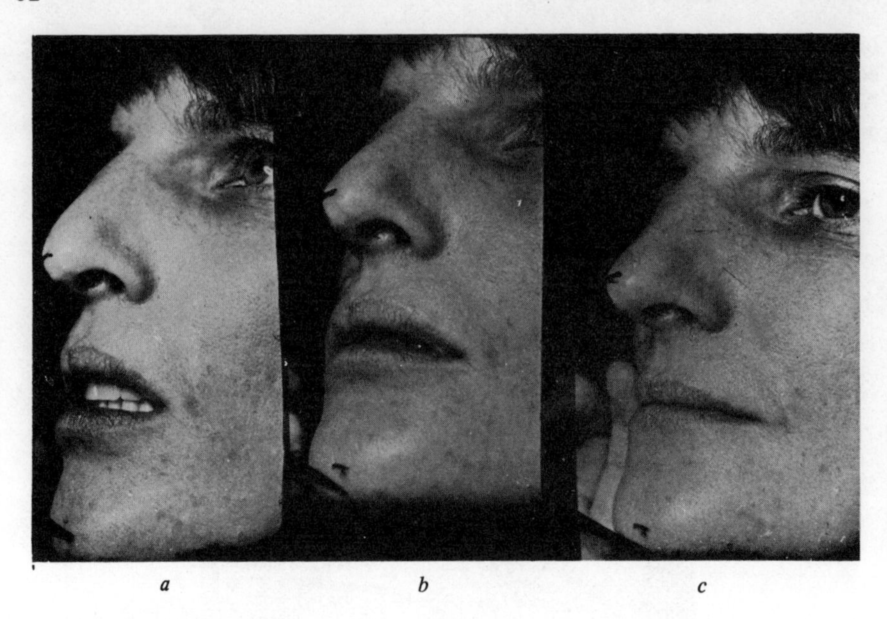

Fig. 8.2. *a*, Both dentures in the mouth, strained overopening. *b*, Upper denture only in the mouth. *c*, No dentures in the mouth, resting position.

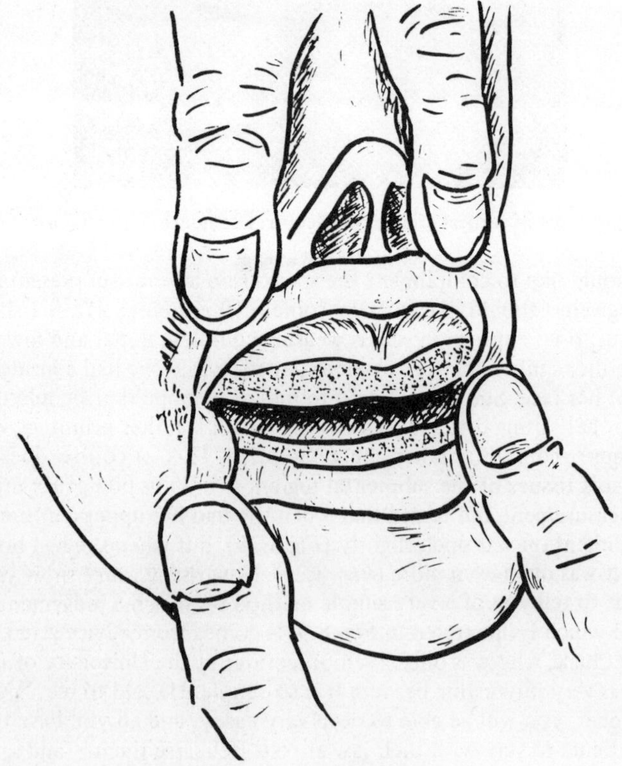

Fig. 8.3. Method of observing interalveolar vertical distance.

but please don't move your muscles" '. And the patient who is requested to do this, unless they are psychotic, usually obeys. In this particular case, you can see the very short distance between the upper and lower alveolus, but worse than that was the very restricted space between the tuberosity and the retromolar pad region behind. Now, the reason I show these pictures is because I feel that every prosthetist should do the simple thing first before he does anything else, and now, because Mr Nairn has not mentioned it, I just want very rapidly (and I will be brief Mr Chairman, because I can see that you are agitated) show the next one or two slides. *Fig.* 8.4 comes from the Royal Dental Hospital before Mr Nairn's time, and it is a lower tray with three pegs on it (669). The students lovingly called it 'the titty-peg tray'. The purpose of it was to take a lower impression and to apply a soft medium such as alginate to the upper surface. The patient was asked to

Fig. 8.4. The locating peg tray.

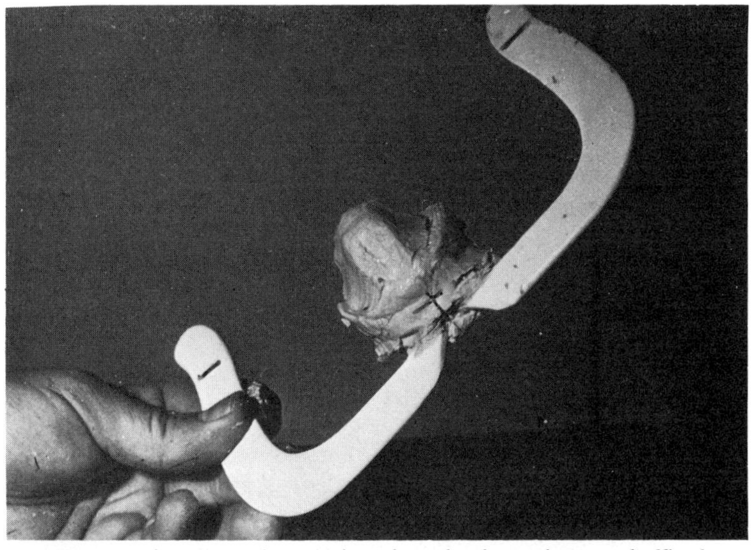

Fig. 8.5. Modified Fox's occlusal plane. Alginate interalveolar resting record affixed to intra-oral segment.

close to a comfortable position to record jaw relationships which at least were comfortable according to the patient. A more sophisticated method (670) is shown in *Fig.* 8.5. This shows a plastic Fox's occlusal plane, with an intra-oral segment which carries the alginate. The patient closes into that (*Fig.* 8.6), and later the upper and lower models are assembled quite easily into that resting record, as shown in *Fig.* 8.7.

This little contribution is meant to emphasize what I consider to be a provocative and beautifully presented paper by Mr Nairn.

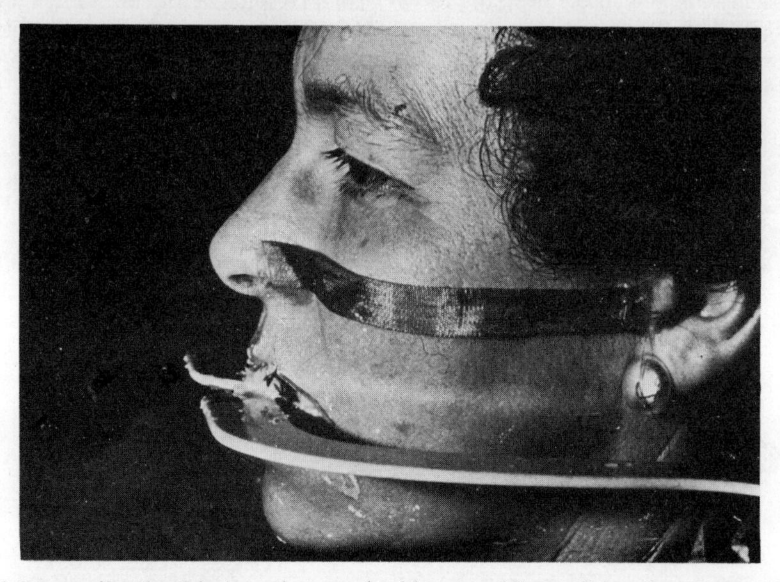

Fig. 8.6. Recording the alginate resting record, with external fin parallel to the ala–tragus line.

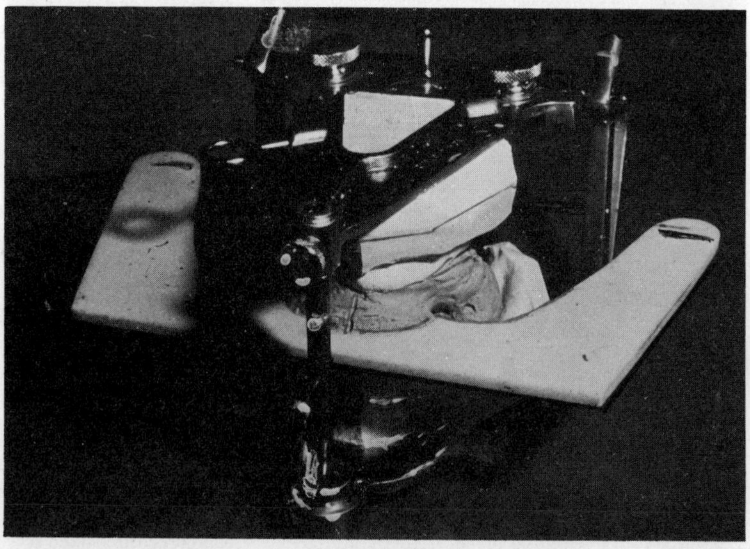

Fig. 8.7. Assembly of models with alginate resting record in articulator.

NAIRN: It is by no means uncommon to see patients who have had all their teeth out and who, when their mandible is in the endentulous resting position, have the maxillary tuberosity in contact with the retromolar pad. When the tuberosity is reduced and you check it again, the jaws are much closer than they were before. I have seen patients who have had all their teeth out and had to have a series of alveolectomies in order, gradually and slowly, to get some room in which to put an artificial replacement. I think the discrepancy between the natural occlusal vertical dimension which some people have, and the situation they ultimately like to come to when the teeth are removed, is sometimes quite extreme.
CHAIRMAN: Would anybody like to comment on Mr Nairn's point that there seems to be a difference between the posture at rest with natural teeth as compared with that shortly after teeth have been extracted ?
SUSSMAN: Surely, in each of the cases in which an immediate denture was fitted to a previously established occlusal height, and the clinical manifestations of a raised bite occurred later, the denture had been left out for quite long periods, probably at night.
NAIRN: It may well have been left out but the symptoms do not always develop very rapidly. Are you suggesting that, if they had left the dentures in all the time, this change wouldn't have occurred ?
SUSSMAN: I am suggesting that the changes must have involved some degree of muscle shortening which could not have taken place whilst they had a vertical relationship which was maintained by the existing natural teeth.
NAIRN: I wouldn't hazard a guess as to whether the muscles, in the absence of the teeth to hold the jaws apart, maintain a constant length, or whether there is any shortening of the muscles.

Paper No. 9

The Mechanisms of Postural Control in the Limbs and Trunk

A. J. Buller

My task today is to provide a lightning coverage of the physiology of the spinal reflexes concerned with the control of length of skeletal muscle. These reflexes—the stretch or myotatic reflexes—have been most extensively studied in the hind limb muscles of laboratory animals, especially the cat. Whether the principles of control which have been demonstrated for the limb muscles of laboratory mammals apply to human limb muscles is not yet clear, and whether the same ground rules also apply to those skeletal muscles that are innervated by cranial nerves, such as the jaw muscles, is completely unknown. However before conceding what I believe to be the central theme of Mr Nairn's talk this morning, namely that the mechanisms which are so important for the maintenance of posture in the mammalian hind limb play no part in determining the rest position of the jaw, it is important that we agree a common usage for some of the jargon terms that have been used in our first session. I shall therefore concentrate on the physiologist's definition of certain phrases and emphasize my personal view that if several of us use a single phrase to mean quite different things confusion must be the inevitable result.

The motor elements of the stretch reflexes are the skeletal muscles. In mammals all the limb musculature consists of twitch fibres. As Dr Yemm told us this morning these twitch fibres are organized into motor units, each motor unit consisting of a number of skeletal muscle fibres and the single motoneuron which innervates them. The number of muscle fibres innervated by a single motoneuron varies considerably, as does the distribution of the muscle fibres belonging to a single motor unit within the anatomical muscle, but these are matters that need not concern us today. Whilst it is tacitly assumed that all the muscle fibres innervated by a single motoneuron have the same contractile properties, several of this morning's speakers, including Professor Taylor and Dr Yemm, pointed out that different motor units within the same animal, and even within the same muscle, may differ markedly in their contractile characteristics. These contractile differences have given rise to the phrases 'fast-twitch motor unit' and 'slow-twitch motor unit'. It seems probable that such a clear-cut and unambiguous division of motor units into only two types is an over-simplification, since some intermediate types undoubtedly occur, but it is also equally apparent that the majority of motor units in limb muscles are either of the fast-twitch or slow-twitch type, and that these two variants of skeletal muscle are used for functionally distinct activities.

When a physiologist wishes to examine the contractile behaviour of muscle he typically uses one, or both, of two methods of recording. These methods are known as isometric recording and isotonic recording. Using the former method one end of the muscle is attached to a force transducer and the other end held rigidly, either by its attachment to bone or in a clamp. When the muscle is activated, either by direct stimulation or by indirect stimulation via the motor nerve, the muscle is prevented from shortening and the force developed by the muscle at a known length is measured by the transducer. You will recall, from the happy days you spent engaged in physiology practicals, that the isometric force generated depends upon the number and frequency of adequate stimuli reaching the muscle. If a single stimulus is used a twitch response is produced, if stimuli at a sufficiently high repetition rate are used a fused tetanus is produced. The force produced during the plateau of a tetanic contraction of mammalian skeletal muscle is typically some three to four times the maximum force produced during a twitch. This ratio of 3–4 is known as the tetanus/twitch ratio. When isotonic recording is used known loads are attached to one end of the muscle, the other end of which is held rigidly as during isometric recording. Under these conditions, when the muscle is activated by a tetanic train of stimuli the muscle will shorten and move the load, providing that the load on the muscle is less than the maximum isometric force that the muscle can develop. If, as is common, the load is suspended over a pulley, it will be raised during muscle contraction, and it has been found that the initial speed with which the load is raised, and which equals the speed of shortening of the muscle, is related to the size of the load. The heavier the load the slower it is raised. If, as has been pointed out above, the load either equals or is greater than the maximum isometric force which the muscle can generate the load will not be moved by the contracting muscle.

It is apparent that the two recording methods give different information about the contractile machinery. Isometric recording gives information about the force a muscle, or motor unit, can develop, and from measurements of the maximum tetanic force developed at a variety of different initial lengths a force/length diagram may be constructed. Isotonic recording gives information about the speed with which a muscle can shorten, and from measurements of the initial speed of shortening of the muscle fibres when lifting different loads a load/velocity diagram may be constructed. If the axes of force/length diagrams and the axes of load/velocity diagrams obtained from different muscles and different species are normalized it is possible to make certain important generalizations. Since skeletal muscles from different mammals differ greatly in total size but all have similar sarcomere lengths it is conventional to scale the vertical axis of the tension/length diagram (tension) in Newton's $\times 10^5/m^2$ cross-sectional area of muscle, and the horizontal axis (length) in sarcomere length (μm). Similarly, it is conventional to scale the vertical axis of the load/velocity curve (velocity) in μm/second/sarcomere and the horizontal axis (load) in load expressed as a percentage of the maximum isometric force which the muscle can develop. (Could I here enter a strong plea that in this day and age that we all use SI units?)

Plotted in this manner all tension/length relationships produce dome-shaped curves. The peaks of the curves occur (for mammalian skeletal muscles) at very similar sarcomere lengths and, more importantly, all skeletal muscles, whether fast-twitch or slow-twitch and whether from mice, rats, rabbits, cats or even humans, produce surprisingly similar maximum tetanic tension per unit cross-sectional area amounting to approximately 3×10^5 Nm^{-2}.

On the other hand, load/velocity curves from different muscles, whilst all showing a hyperbolic relationship, exhibit enormous variability in their maximum speed of shortening. Thus the sarcomeres of a mouse muscle can shorten more rapidly than the analogous muscle of a rat, and the sarcomeres of the rat muscle shorten more rapidly than those of a cat. Also there is, in any one species, a clear distinction between the maximum speed of shortening of fast-twitch and slow-twitch muscle. As the name suggests fast-twitch muscle has a greater speed of shortening for a given load (expressed as a fraction of maximum isometric force) than slow-twitch muscle. There is also a shape difference in the load/velocity curves of fast-twitch and slow-twitch muscle from the same animal which indicates that the latter type of muscle is metabolically more efficient during contraction, that is, it can sustain comparable contractions with a smaller usage of fuel.

The many measurements that have been made on mammalian skeletal muscle permit us, following the more generalized observations by that doyen of muscle physiologists A. V. Hill, to interpret the intention of nature in the development of two types of twitch skeletal muscle. It would appear that fast-twitch motor units are capable of developing, within the muscle of which they form part, a speed of shortening which, at maximum, is just short of a velocity which would cause the animal injury. It is these motor units which provide an animal with its maximum speed of movement. If the fast-twitch motor units within any mammal consisted of fibres having an appreciably faster maximum shortening velocity than normal the inertial forces generated by that animal when maximally activating those motor units would result in damage to the muscle, tearing of the attached tendons or even fracture of bone. One has only to consider how frequently injuries occur in racing animals like greyhounds, racehorses or sprinters in order to appreciate how near to self-inflicted injury nature has placed us. However the cost paid for high velocity is metabolic inefficiency. If animals used the same fast-twitch muscles to maintain stationary antigravity postures their energy requirements would be greater than they are. In fact, when an animal is stationary its fast-twitch motor units are quiescent and posture is maintained by the metabolically more efficient slow-twitch motor units. Thus, nature appears to have given fast-twitch and slow-twitch muscle approximately the same force-generating properties, but uses the former muscle type to provide the animal with its maximum (safe) speed of movement, and the latter muscle type to provide a more efficient mechanism for maintaining posture.

To revert from the general to the special. The distribution of fast-twitch and slow-twitch motor units in the muscles of man is not yet known. We were this morning shown a number of isometric twitch responses and at times we might have been led to think that these responses could be identified as fast-twitch or slow-twitch. It is my belief that we were given insufficient evidence on which to make such a decision. I must stress that the two muscle types are biochemically distinct, and that identification cannot be made simply by a measurement of the time to peak of a twitch response (the time from the start of the contraction to the instant of peak tension development). If the investigator is compelled to make only isometric recordings—and this is often preferable since the isometric technique is considerably simpler—a battery of tests including temperature sensitivity and response to repetitive stimulation are necessary in order to establish whether an isometric twitch response is originating from fast-twitch or slow-twitch muscle.

So much for the engines that are at the end of the large efferent fibres of the stretch reflexes. The sensory endings subserving the stretch reflexes are the muscle spindles. We were shown some of the structural complexity of these sensory endings this morning, suffice it for our purposes to state that, functionally, the muscle spindles lie in parallel with the striated (extrafusal) muscle fibres and consist of a number of intrafusal muscle fibres surrounded by a capsule. We have already seen that the extrafusal fibres are not identical (being a mixture of fast-twitch and slow-twitch fibres) so it comes as no surprise to learn that the intrafusal fibres also fall into two groups, the nuclear bag fibres and the nuclear chain fibres. Both types of intrafusal fibres are innervated by fusimotor nerve fibres which originate from the small anterior horn cells (or their counterparts in the motor nuclei of the cranial nerves). These small anterior horn cells are clearly separable from the large anterior horn cells which innervate the extrafusal muscle fibres.There is now good evidence that the two types of intrafusal muscle fibres receive functionally distinct motor innervation and in this respect also they resemble fast-twitch and slow-twitch extrafusal fibres. From the muscle spindles arise two types of afferent fibres. The larger diameter, faster conducting type 1A afferents take origin from the primary sensory end organ which is arranged as a spiral around the middle of a nuclear bag intrafusal muscle fibre. These 1A afferents typically make monosynaptic excitatory synapses on the large anterior horn cells innervating the extrafusal muscle within which the spindle lies, but the afferents also connect (monosynaptically) with the large anterior horn cells innervating synergistic muscles. Recently it has been shown that a few monosynaptic 1A afferents may produce inhibitory effects (478) but the significance of this finding is not yet fully understood. The slower conducting type II afferents arise from more diffuse sensory endings within the muscle spindle and make contact with large anterior horn cells only via interneurons within the grey matter of the spinal cord. These secondary endings also tend to increase the excitability of large anterior horn cells (323) but may also lead to inhibitory effects. Finally, to complete the traditional picture of the 'wiring' of the spinal stretch reflex mention must be made of the Renshaw inhibitory feedback loop. From a large proportion of the axons of large anterior horn cells collateral branches arise before the axons leave the spinal cord which pass back and synapse with Renshaw cells lying within the grey matter of the anterior horn. These connections are excitatory, but impulses passing along the axons of the Renshaw cells exert an inhibitory influence on the large anterior horn cells on which they terminate. Thus, when large anterior horn cells discharge impulses along their axons Renshaw cells are also excited which in turn leads to a temporary inhibition of neighbouring large anterior horn cells.

The integrated functioning of these interlocked circuits is still not fully understood, but a picture is beginning to emerge. One point now seems certain. The afferent input from the muscle spindles is not equally distributed to all large anterior horn cells, but is preferentially distributed to the anterior horn cells innervating slow-twitch muscle fibres. This suggests that the stretch reflex is primarily concerned with the maintenance of posture, which, as indicated above, is executed through slow-twitch muscle. Similarly the distribution of Renshaw inhibition may be primarily from the large anterior horn cells innervating fast-twitch muscle to the large anterior horn cells innervating slow-twitch muscle. This would allow a standing animal (using the stretch reflex afferents terminating on slow-twitch motoneurons to maintain posture) to rapidly activate fast-twitch motoneurons (or coactivate fast-twitch motoneurons and small anterior horn cells) via the corticospinal pathway thereby producing a rapid movement and simultaneously ensuring a 'turn-off' of the posture maintaining reflex via the inhibitory Renshaw loop. Such schemes must, for the present, remain speculative, but give added need for a full understanding of the different types of muscle involved in willed movement and postural activity.

In conclusion I reiterate my belief that if we are to unravel the many fascinating problems that remain in biology—including the problems of mastication—we must all adhere to the ground rules that have already been set. Information theory tells us that information transfer can only occur using an agreed set of symbols. We cannot suddenly change the meaning of some technical term to suit ourselves, neither can we extend the meaning of some phrase without prior universal agreement. Of course, we *can* do these wicked things, but if we do it will only be to the detriment of better understanding, and to the betterment of confusion and acrimonious dissension. Let us build on the shoulders of those that have gone before, using their definitions where appropriate or making it unambiguously plain that we find it necessary to introduce new, and clearly defined, jargon.

Discussion CHAIRMAN: B. MATTHEWS

GOLDSPINK: Recently we have looked at the efficiency of fast and slow muscles in isotonic movement and we have now got good evidence that the slow muscle fibres are the more efficient under these conditions. So we must think of them as being involved not only in maintaining posture but also in carrying out slow isotonic movements. When the cat is prowling very slowly, or when we are chewing in a very slow and routine way, it is quite likely that the slow muscle fibres are being used. At their optimum rate of shortening they are in fact more efficient than the fast muscle fibres at their optimum.

BULLER: I have intentionally drawn in black and white and I wouldn't disagree with what you have said but if I had designed the system I would only call in the slow-twitch muscle to perform the specific postural function and I would give the animal fast-twitch muscle to use relatively infrequently. I'm sure that there is an overlap as you say and it is not black and white.

DARLING: A very short question—it surprised me very much not to hear you mention the term 'resting position' which was used very greatly before. Does a physiologist believe in the resting position, or in a series of resting positions ? For instance as I was listening to you there were times when my teeth were tight together and times when they relaxed and I think if you had not been so interesting my jaw would have dropped open when I fell asleep.

BULLER: It probably did. I was talking only of the cat and the hind limb. We have a situation where if the cat is in a standing posture or if we adopt a sitting posture in a chair, my working hypothesis is that the slow-twitch muscles, which are economically the most efficient, are employed and that they are controlled by a length servo. A posture is maintained by all the muscles put at a fixed length. I know nothing about the jaws, but in the cat standing, or ourselves sitting, the muscles are not truly resting. We are only resting when we let everything go completely. Now what I do not know and would be interested to hear is whether the jaw muscles are truly at rest in what you call the resting position, or whether there is continuous activity to maintain the jaw in a 'rest' position.

CHAIRMAN: I think that Dr Yemm might have something to say about that later.

GOLDBERG: I can answer that question directly. We have recorded in the masseter muscle of humans and in the rest position you find very small motor units firing consistently at high frequencies. We do not have twitch times for these units but, in view of the long contraction times of the units that Dr Yemm (p. 29) found to be recruited at very low force thresholds, I would expect these units to be able to maintain a very steady level of tension.

BULLER: The dispersion of times-to-peak that Dr Yemm showed us did not cover the three-fold range that I would expect between fast-twitch and slow-twitch motor units in the hind limb of a mammal. The other thing is, and I do want to stress this, that simply by measuring a time-to-peak, one cannot tell whether the units are fast-twitch or slow-twitch.

SESSLE: I think some mention should be made of the slow and fast types of motor units in relation to recent work (651) which indicates that there are also differences in the effects of cutaneous afferents on slow and fast units in the limbs. Perhaps in the jaw muscles, where recurrent collaterals and Renshaw inhibition may be lacking, cutaneous and musosal afferents may be more involved.

BULLER: I agree that there is enormous convergence on to motoneurons, including joint receptors, Golgi tendon organs, and skin receptors. But the point I wanted to make about fast- and slow-twitch units is that the inputs from the muscles themselves appear to be differentially arranged.

SESSLE: Yes, but the point I am trying to make is that the cutaneous inputs might be differentially arranged as well.

Paper No. 10

Evidence that the Rest Position is Subject to Servo-control

E. Møller

INTRODUCTION

In the discussion of forces contributing to mandibular posture the main emphasis has been placed on the role of the elevators in maintaining vertical dimension. A simple mechanical analogue of muscular support consists of: (1) a contractile element of sliding filaments, (2) a stiff series-elastic element of filaments, z-lines and tendons, and (3) a parallel-elastic element of sarcolemma and connective tissue. Both contractile and parallel-elastic elements contribute to passive support, the task of active control being to adjust stiffness of the first.

When electromyography was applied to the study of postural activity in the muscles of mastication, it was expected that it would clarify and solve clinical problems in assessing a physiological rest position. Instead, the method added confusion by presenting conflicting evidence as to whether or not there is muscle activity when the mandible is at rest. Earlier disagreement on this subject may, to a large extent, be ascribed to differences in the equipment used to pick up the electrical activity and the procedures used in recording (404). However, in spite of technical improvements and quantitative evaluation, opinions have polarized. Is mandibular rest position controlled actively (361, 372, 404, 406) or determined passively (374, 641, 642)? This report aims to reveal that the apparent divergence may hold features of clinical importance.

REQUIREMENTS FOR SERVO-CONTROL

In the following, the term 'servo-control' is used according to Matthews (390) and Stein (540) i.e. as a servo-assistance to adjust or adapt contractions initiated by α-motoneurons by concomitant fusimotor innervation. Taylor and Davey (568) observed signs of fusimotor modulation during active contraction of the muscles of mastication in the cat, but no evidence exists of its importance in mandibular posture. Development of considerable force at the occlusal surfaces without damage to surrounding soft tissue demands precise nervous control of movements of the mandible comparable to that of finger movements. Delicate movements of the fingers depend strongly on the integrity of pathways assumed to serve α- and γ-activation (634). The ability to maintain a steady holding response in the absence of direct attention also depends on the integrity of these pathways (343). The requirements for servo-control in the muscles of mastication may therefore be determined from studies of fusimotor activity to the muscles of the hand.

At first sight, the positional servo of human fingers does not seem powerful: a sudden load applied to a finger extended with minimum effort results in obvious displacement (562). Recordings of impulses from single spindle afferents indicated insignificant fusimotor output to relaxed muscles of the hand, and their response to passive joint movement gave no evidence of fusimotor adjustment as means of maintaining position (593). However, during very weak contraction sufficient to hold the finger against gravity a steep rise of fusimotor outflow adequate to allow modulation was indicated (594). Taking into account the high sensitivity of spindles to small changes in muscle length (389) and the relationships between impulse frequency and torque in primary endings, Vallbo (594) calculated that 100 g weight at the tip of the finger would cause a movement of no more than 1°.

The following considerations of mandibular rest position are limited to the temporal and masseter muscles. Based on experience from muscles of the human hand, the idea of servo-control requires: (1) that these muscles are slightly active with the mandible at rest, and (2) that this activity has relevance to the position of the mandible.

POSTURAL ACTIVITY IN THE MUSCLES OF MASTICATION

During attempts to record activity in the muscles of mastication with the mandible at rest, it must be borne in mind that action potentials are absent in relaxed muscle. If subjects are instructed to relax, the experimental situation becomes irrational unless the position of the mandible is assessed simultaneously. Recordings taken from subjects selected as having minimum activity with the mandible at rest (102), or put in front of an oscilloscope to achieve movements of the mandible with the least possible muscle participation (641), have little chance of demonstrating activity. Efforts to obtain a relaxed experimental situation are not criticized but they bias the decision of whether a particular position is controlled actively or determined passively.

The postural activity in the temporal and masseter muscles to which I shall refer was recorded with surface electrodes from alert subjects sitting upright with their head unsupported and their eyes open. Since surface electrodes record from a larger area than do needle electrodes, they pick up more potentials during weak effort due to the amplitude—distance relationship (93) the amplitude of these potentials varies less with position than in recordings with needle electrodes. Hence surface electromyograms during weak effort consist of numerous small potentials, which can most suitably be evaluated by the level of their mean voltage. Large discrete action potentials were disregarded in the assessment of the mean voltage (404), since their amplitude reflects the distance to the active fibres more than the level of effort. Recordings made at rest and evaluated in this way have indicated a level of activity which is 2–5 per cent of that observed during chewing and swallowing (404). That activity was present at rest, was also established by comparing recordings obtained with the subject (1) sitting upright, head unsupported, (2) inclined backwards 45°, and (3) supine (372). Without instructions to relax, activity in the anterior temporal muscle decreased when the subject was moved from upright to supine position. Further evidence was produced by inducing deliberate relaxation. In subjects without symptoms or signs of functional disorders of the chewing apparatus, the activity in the anterior temporal muscle was systematically reduced during relaxation in the supine position. Electromyograms from the masseter muscles varied at random (406).

In subjects with functional disorders the level of postural activity is increased especially in the anterior temporal muscle (361). In such patients deliberate relaxation in upright position with headrest had little effect (406). However, in the supine position, the activity in the anterior temporal muscle was reduced virtually to the same level as in relaxed normal subjects.

The typical pattern of coordination of the anterior temporal and masseter muscles in a normal subject sitting upright with the head unsupported is characterized by slight but distinct activity in the temporal muscle and near absence of activity in the masseter (*Fig.* 10.1A). Since the recordings were obtained without instructions, to the subject who was alert with eyes open, they demonstrate muscle activity in what may be termed the 'alert rest position' of the mandible. If the subject is brought into supine position and instructed to relax, reduction of activity takes place in the temporal muscle with little or no change in the masseters (*Fig.* 10.1B).

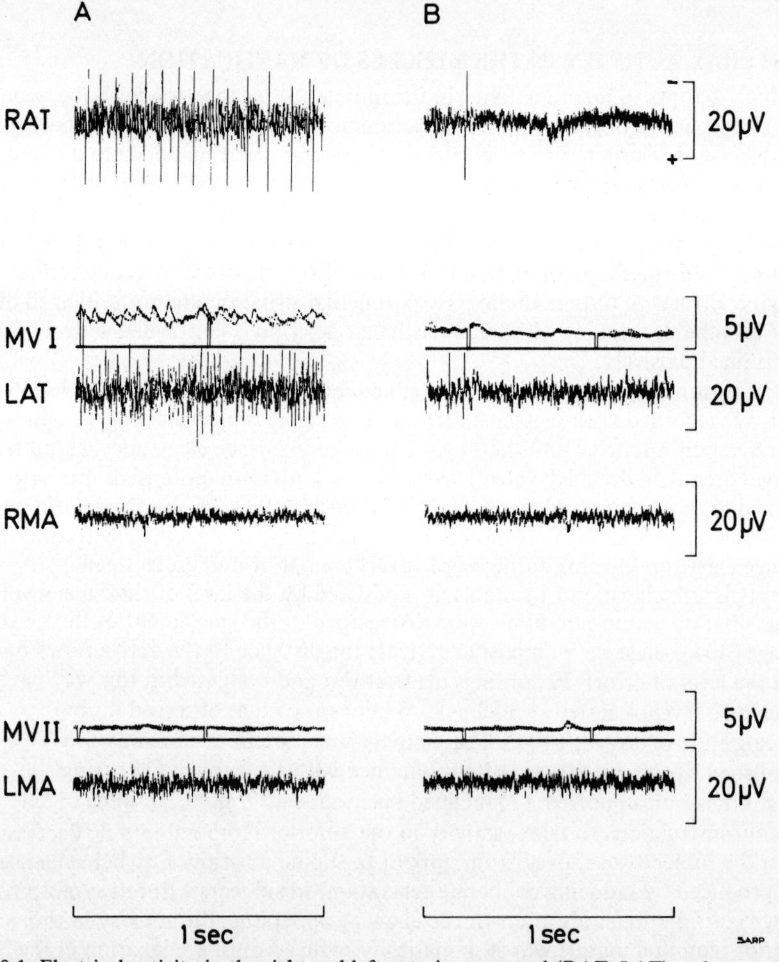

Fig. 10.1. Electrical activity in the right and left anterior temporal (RAT, LAT) and masseter muscles (RMA, LMA) and the corresponding mean voltages (MVI and II, heavy trace: right muscle, thin trace: left muscle, straight horizontal line: base line). Surface recordings from bipolar electrodes with the mandible at rest, the head unsupported and open eyes (A) and during deliberate relaxation in supine position (B). Male 20 years old.

Reduction of the activity in the anterior temporal muscle obtained merely by changing to supine position (361) was possibly acquired by putting the load of the mandible at a right angle to the direction of its fibres. This would reduce the activation due to a servo-assistance mechanism. The further reduction during deliberate relaxation (406) is assumed to be due to voluntary cessation of α-motoneuron activity.

Yemm and Nordstrom (642) suggested that the activity observed in the alert rest position might be the response to the stress of the experimental environment. It is unlikely that such stress would vanish immediately by changing to a supine position, which may seem even more strange to the subject than sitting upright. If it was an important feature, a general increase of postural activity could be expected in subjects with functional disorders. The differentiated patterns observed in such subjects, and the effect of changing to supine position (406) indicate that the experimental environment is of little importance.

POSTURAL ACTIVITY AND POSTURE

Garnick and Ramfjord (182) demonstrated that lowering the mandible beyond a clinically established rest position reduced the postural activity in the temporal muscle. Inhibition of the same muscle due to electrical stimulation of the oral mucous membrane (640) or the infra-orbital nerve in the cat (576), may result in mouth opening without contraction of the digastric muscle. However, these studies do not fully establish that the temporal muscles support the mandible in its alert posture. At present two series of experiments are being conducted to answer this question, and some preliminary results are presented here.

Kawamura and Fujimoto (315) demonstrated that eye closure reduced postural activity, especially in the anterior temporal muscle. This observation has been confirmed (*Fig.* 10.2). Typically, the activity in the anterior part of the temporal muscle decreased by 50 per cent during eye-closure, and reduction of activity also took place in the posterior part, while the masseter activity was unchanged. Simultaneous assessment of mandibular position showed that, if the activity in the anterior temporal muscle decreased, the mandible tended to drop between 1 and 2 mm.

In the other experiments, 1 ml of 3 per cent Carbocain® (BOFORS) was injected into the anterior part of each temporal muscle. Changes in the vertical position of the mandible were assessed from photographs taken in profile. The injection was followed by an obvious reduction of both postural (*Fig.* 10.3) and maximal activity. Postural activity was also reduced in the posterior temporal muscle, but usually unchanged in the masseters. The anaesthetic effect was maximal 10–30 min after injection and during this period the mandible dropped 2–4 mm. Therefore, the degree of postural activity in the anterior temporal muscle did influence the mandibular position. The variability of this activity may account for the postural oscillations demonstrated by Schwindling and Stark (512). The gap between active control and passive determination of mandibular posture seemed to be of the order of 1–4 mm.

SIGNIFICANCE OF DIFFERENCE BETWEEN ALERT AND RELAXED POSTURE

The importance of muscle activity in mandibular posture derives from the fact that it represents the fraction of tissue elasticity subject to nervous control. Whether the

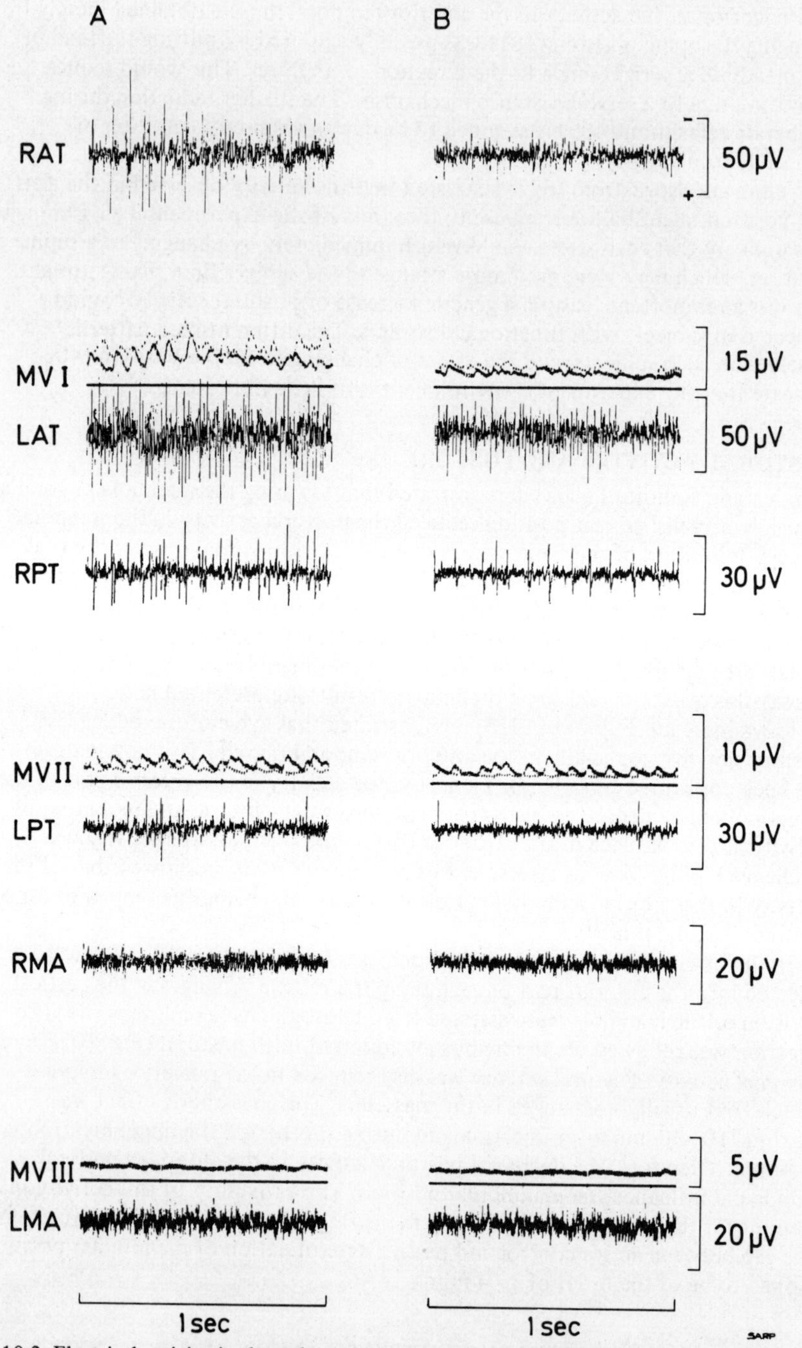

Fig. 10.2. Electrical activity in the right and left anterior temporal (RAT, LAT), posterior temporal (RPT, LPT) and masseter muscles (RMA, LMA) and the corresponding mean voltages (MVI–III, heavy trace: right muscle, thin trace: left muscle, straight horizontal line: base line). Surface recordings from bipolar electrodes with the mandible at rest, the head unsupported and open (A) and closed eyes (B). Male 23 years old.

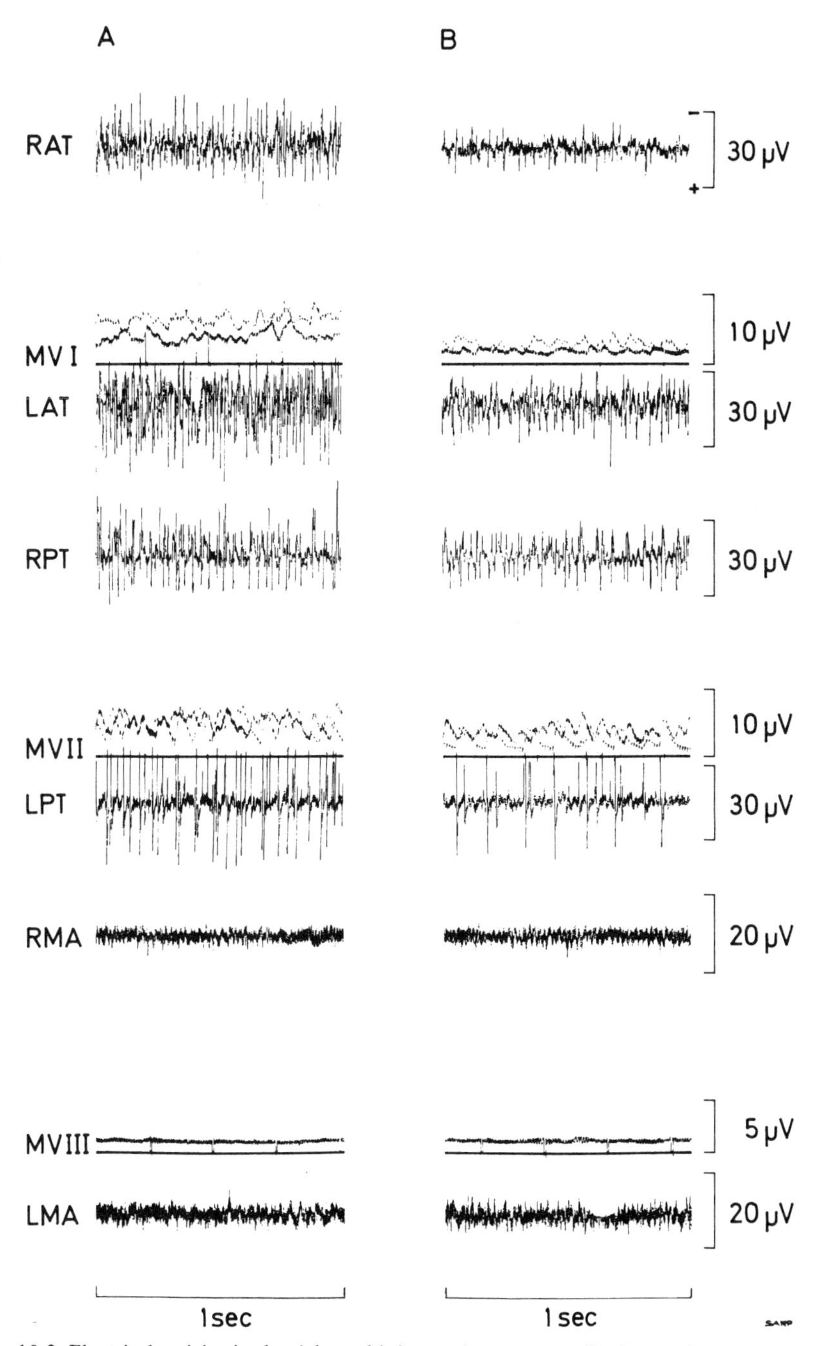

Fig. 10.3. Electrical activity in the right and left anterior temporal (RAT, LAT), posterior temporal (RPT, LPT) and masseter muscles (RMA, LMA) and the corresponding mean voltages (MVI–III, heavy trace: right muscle, thin trace: left muscle, straight horizontal line: base line). Surface recordings from bipolar electrodes with the mandible at rest, the head unsupported and open eyes obtained 10 min before (A) and 15 min after (B) injection of 1 ml 3 per cent Carbocain® (BOFORS) into RAT and LAT. Male 23 years old.

activity can be considered to be evidence of servo-control is not of overwhelming importance, especially since this concept is far from being clarified itself.

Radiographic studies have shown resting face height to follow changes in morphological face height caused by insertion of dentures, subsequent resorption of the alveolar processes (560, 561) or occlusion-raising procedures (108). Such adjustments call for control of tissue elasticity.

Asymmetry of postural activity in subjects with unilateral cross-bite may throw light on the mechanism of adjustment. This type of malocclusion is a morphological trait observed in the intercuspal position. Yet it influences postural activity and mandibular posture at rest, i.e. without tooth contact. In subjects with unilateral cross-bite there is on the average predominance of postural activity in the posterior temporal muscle on the side of the cross-bite (588) and a corresponding displacement of the mandible at rest (407). This situation is interesting because it does not involve the interplay between tissue elasticity and gravity present in the vertical dimension. Asymmetry of postural activity and displacement were not consistently related. In some subjects, displacement existed with little or no difference between the postural activity in the posterior temporal muscle on the side of the cross-bite and that on the side of normal occlusion.

The following series of events are suggested to have taken place:

First, through the activity of periodontal receptors, cuspal interferences of the unilateral cross-bite guide the α-motoneurons in their search for optimal intercuspidation during mastication. *Secondly,* the so-called servo-assistance mechanism is set for adjustment in posture. *Finally,* nervous adjustment becomes structural by a change in the number of sarcomeres in series, and thus the elastic properties of the muscles, related to a given position. Such structural changes accompanied by mechanical adjustment have been demonstrated to occur within a short interval of time (207, 452, 558).

The merit of active control is to put the muscles of mastication near their functional length and permit structural adaptation. Such adaptation may or may not be desirable. A large vertical, sagittal or transverse discrepancy between alert and relaxed posture of the mandible may indicate insufficient structural adaptation of the muscles of mastication and call for adjustment, e.g. of the occlusal position.

Discussion CHAIRMAN: B. MATTHEWS

BRATZLAVSKY: Do you have objective information indicating to what extent vestibular and tonic neck reflexes contribute to the resting position ?

MØLLER: No. My evidence is based on a clinical study of postural activity in the temporal and masseter muscles. The hypothesis I have put forward is based on such studies in which the activity was recorded with the subjects in different positions (372). It is not my object today to prove the existence of a servo-mechanism but to find what requirements one should look for if such a mechanism is to work. One requirement is the presence of some activity in the elevators of the mandible, which I have shown in the case of the anterior temporal muscle.

YEMM: I am not entirely happy that you have actually shown anything more than that there is often activity of the jaw muscles in a human subject in an experimental situation. I am not sure that you have made any particular case that the activity which you have demonstrated is in fact inherent in the mandibular rest position and in any way servo-controlled. You mentioned, for instance, that you got small

changes in mandibular position following various procedures such as after eye closure and injection of local anaesthetic. There is an enormous amount of literature on the effect of environmental stress or experimental stress on subjects, and if a subject is about to be injected with a local anaesthetic he is not going to be as relaxed as he otherwise would be. This is one of the great difficulties of working on human subjects; they will behave in an abnormal way under experimental conditions.
MØLLER: It would be impossible to prove that there was no anxiety, but would you feel that the anxiety would disappear if the patient is put in the supine position ?
YEMM: This is a possibility, not that it would disappear but that it would diminish.
MØLLER: It is difficult for me to understand that. In subjects with functional disorders in whom we have demonstrated an increase in postural activity (361) you also have the problem of anxiety. Yet, in such subjects you find a differentiated pattern of hyperactivity, not a generalized one as could be expected if anxiety was the only cause. Of course you need some anxiety tests to settle this problem, but I am not at all convinced that you are right in what you say about the importance of anxiety in this situation.
CHAIRMAN: There seems to be a problem in the definition of the term 'rest'.
MØLLER: I stated that quite clearly. If you try to relax the subjects before you record, or teach them to perform movements with the slightest possible muscle activity, then you will probably find no activity.
BULLER: Can you do what you said is done for other muscles, in the arm, for example; get them completely at rest, remove the effect of gravity and get to silence. Can you do this by supporting the jaw and achieve electrical silence ?
MØLLER: Yes, of course, this is a routine procedure when you use electromyography for neurological differential diagnosis; it is one sign of a normal muscle and is done very easily.
BULLER: And you can do that with these particular muscles, the jaw muscles ?
MØLLER: It can be done, in spite of what Dr Yemm says, with a needle electrode in the muscle to record action potentials for the purpose of measuring amplitude, duration, etc. But it is difficult with the subject in the sitting position; we do them in the supine position.
AHLGREN: Have you noticed individual variation in resting posture associated with differences in morphology or extra-oral muscles ? I think I have noticed that in some patients it is possible to establish a resting posture but in others not.
MØLLER: I think I tried to reach a compromise with Dr Yemm in my report, because I believe there is some importance in trying to find the difference between the relaxed and the alert posture. I exemplified it with the unilateral cross-bite in which displacement of the mandible and predominance of postural activity in the posterior temporal muscle coincide on the side of the cross-bite (this also in part answers Professor Ahlgren). In this situation there is no interplay between gravity and tissue elasticity. The association between mandibular posture and postural activity in the posterior temporal muscle in subjects with unilateral cross-bite I take as good evidence that the alert mandibular posture is controlled actively. At first the alert posture with displacement will probably differ more from the relaxed posture (without displacement) than at a later stage in which structural changes have occurred.
LANDGREN: Have you tried to apply vibration to the lower jaw, thereby vibrating the masseter, which would give you an increased inflow from muscle spindles and

possibly would affect your EMG, thereby showing the servo-mechanism in operation ?

MØLLER: I have not tried that and it is not very relevant to the question we are discussing here because I am not proving the existence of the servo-mechanism. I am trying to present some evidence to show that it may affect mandibular posture.

Paper No. 11

The Role of Tissue Elasticity in the Control of Mandibular Resting Posture

R. Yemm

For many years it has been considered that in a relaxed subject a small number of motor units is active in postural muscles, and produces muscle tone. The object of the present paper is to examine this concept and its application to the posture of the mandible.

THE CONCEPT OF MUSCLE TONE

The hypothesis that there is always a small proportion of motor units active in postural muscles originated from experiments on decerebrate animals. In such preparations, even though the reflex pathways associated with the muscles are undamaged, their control from higher centres is abnormal or abolished. The state of decerebrate rigidity in which there is pronounced activity of limb extensor muscles, is an extreme example of muscle tone. Such continuous muscle tension is clearly abnormal, and appears to be due to an increased excitability of the stretch reflex mechanism, as a result of the elimination of normal inhibition from higher levels of the central nervous system.

More recently it has been recognized that although some body postures are maintained by muscle activity, with servo-control involving the muscle spindles and the stretch reflex pathway, other postures appear to be maintained without muscle activity (41, 271, 297, 357, 464). In a relaxed conscious human subject the sensitivity of the stretch reflex system is not high enough to initiate a reflex contraction in muscles when they are stretched. Resistance to movement of a limb is generated, not by muscle contraction, but by tissue elasticity and friction (60, 357). It is only in response to extremely rapid muscle stretch, such as that used in evoking the knee jerk or ankle jerk, that a reflex muscle contraction is induced. This is in contrast to the decerebrate preparation in which even a slow muscle stretch will evoke a contraction.

The body posture of a relaxed subject is determined by forces other than those due to muscle contraction, predominantly gravitational forces and forces developed by tissue elasticity, particularly the elasticity of muscles. The elastic nature of skeletal muscle depends on both the contractile components and upon the connective tissue elements. It has been shown that, at their normal length in situ, muscles are stretched beyond the length to which they recoil when detached. Clemmesen (111) has described how the posture of a relaxed subject can be stabilized by the elastic forces developed by permanently stretched muscles, independent of active muscle contraction. Investigation of a number of body postures has such a mechanism. For instance, it has been shown that in the human subject standing erect many leg muscles are inactive at times. Some postures maintained in this way are unstable, and it seems that as deviations from equilibrium occur, muscle activity is initiated which provides correction (296).The stimulus for such activity may be vestibular or visual as well as from local proprioceptors. Other relaxed postures which are more stable do not exhibit controlled corrective activity and the body posture adopted is simply an equilibrium between internal and external forces (e.g. finger posture (357)).

Changes in posture occur following denervation of skeletal muscle. Clearly any posture normally maintained against external forces will be unattainable if relevant muscles are inoperative. However, in addition, normal relaxed postures will be affected by the changes in the physical properties of the muscle due to progressive atrophy.

MANDIBULAR POSTURE AND JAW MUSCLE ELASTICITY

Tissue elasticity, and especially muscle elasticity, must be a factor in determining mandibular resting posture. In the absence of muscle contraction, the mandibular rest position will be the equilibrium position determined by gravity and tissue elasticity.

If the observed rest position differs from the equilibrium position, the difference must be contraction of the elevator or depressor muscles, and perhaps lip muscles. An additional factor might be a subatmospheric intra-oral pressure maintained by anterior and posterior oral seals (166).

The role of tissue properties in determining mandibular rest position has received little attention. That forces may be developed by elasticity of jaw muscles when they are stretched is incontrovertible. The elastic properties of the muscles and other tissues of the mouth have been studies in this context (642). The experiments were performed on rats. Observation of the relationship of the upper and lower incisor teeth prior to administration of anaesthetic and during the period of induction showed that the unrestrained position of the mandible when fully anaesthetized was indistinguishable from the normal rest position of the conscious animal. In this position, the teeth were about 3 mm apart, measured at the incisors. In the anaesthetized animal, forces of about 60 g were necessary to open the mouth wide, and forces of 20 g were needed to close the mouth almost to occlusion of the teeth (*Fig.* 11.1).

In the conditions of this experiment the mandibular posture appeared to be determined by tissue elasticity. The considerable forces required to close or open the mouth were not due to muscle activity because death of the animal did not affect the results (*Fig.* 11.1). Experiments of this type on human subjects would also have to be performed under general anaesthesia with muscle relaxant agents to exclude the possibility of either voluntary or reflex muscle activity. This has not been attempted. In conscious relaxed subjects, using slow movements such as have been shown elsewhere in the body not to evoke a stretch reflex response, it has been shown that external forces of 500 g or more are needed to open the mouth wide (374).

Carlsöo (103) demonstrated a progressive increase in the activity of the digastric muscle during mouth opening. The resistance which this activity is overcoming could either be generated actively by elevator muscles or by elasticity. It is generally accepted that elevator muscles are not active during mouth opening.

Carlsöo's observations are compatible with the idea that the resting posture is maintained, at least partially, by elasticity. A more recent study supports this

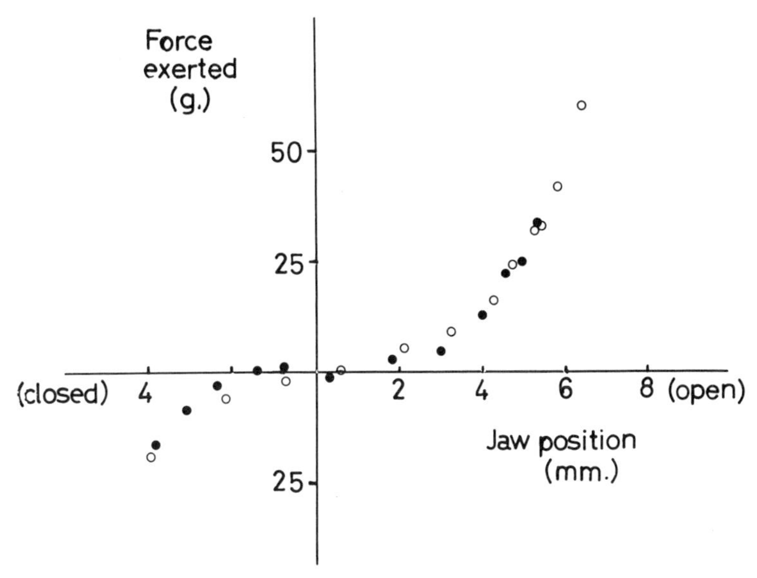

Fig. 11.1. Forces exerted upon a transducer incorporated in apparatus applied to the lower incisors and used to impose a series of jaw positions upon an anaesthetized rat (filled circles) and, on the same animal, shortly after death. The forces were measured along an axis at right angles to the occlusal plane of the upper posterior teeth. The intersection of the axes corresponds to the position adopted by the mandible when unrestrained. *(From Yemm and Nordstrom, 1974, Arch. Oral Biol.* **19**, *347–351.)*

contention (641). Human subjects were asked to open their mouth wide and to close slowly into centric occlusion. The activity of the digastric muscle and of the jaw elevators were studied during the movement. With the mouth wide open, the digastric muscle was active. A detectable but low level of activity of jaw elevator muscles in the wide open position diminished as closure started, and no further change in masseter, temporal or medial pterygoid activity was recorded until closure was almost completed. During the early phase of the movement, the active muscle component was a reducing level of activity in the digastric and not until near tooth contact was activity seen in the elevator muscles. The activity of the digastric muscle was minimal at about the position at which elevator activity was first detected (*Fig.* 11.2). The closing movement thus appeared to have two phases; the first phase being caused by elastic recoil of the tissues (including the jaw elevator muscles) and controlled by the digastric muscle. The active contraction of the elevator muscles in the second phase overcomes the elasticity of the submandibular tissues and gravity (the experiments were performed with the head erect). Estimates by the subject of the moment of reaching rest position corresponded closely with the cessation of digastric activity and the start of elevator muscle activity. Subsequent experiments using a movement transducer have confirmed this, showing that during closure elevator muscle activity starts at the rest position (*Fig.* 11.3).

The results of these experiments suggest that the passive properties of tissues are of primary importance in determining mandibular resting posture, and that a posture similar to mandibular rest position would be adopted in the absence of muscle activity. Experiments on human mandibular rest position indicate that it represents the equilibrium between tissue elasticity and external forces. The free-

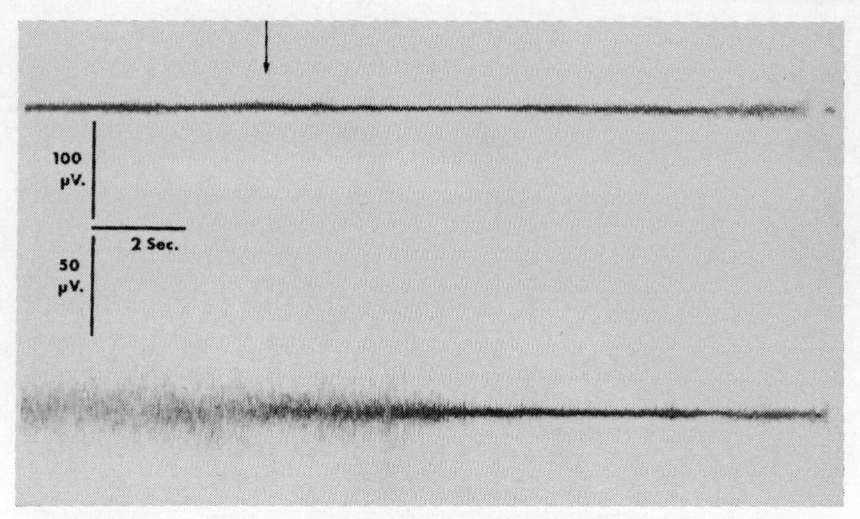

Fig. 11.2. Electromyograms of the masseter (upper trace) and anterior belly of the digastric muscles during slow closure of the mouth from wide open to tooth contact. The closing movement started at the point indicated by the arrow. Tooth contact was reached at the right side of the trace. (*From Yemm, 1975, J. dent. Ass. S. Afr.* **30,** *203–208.*)

way space, or interocclusal clearance, varies both with changes in head posture in relation to the rest of the body, and with changes in the orientation of the body in space (462).Not only are there changes in free-way space, but also there is evidence of variation in anteroposterior relationship, perhaps due to gravity (379).

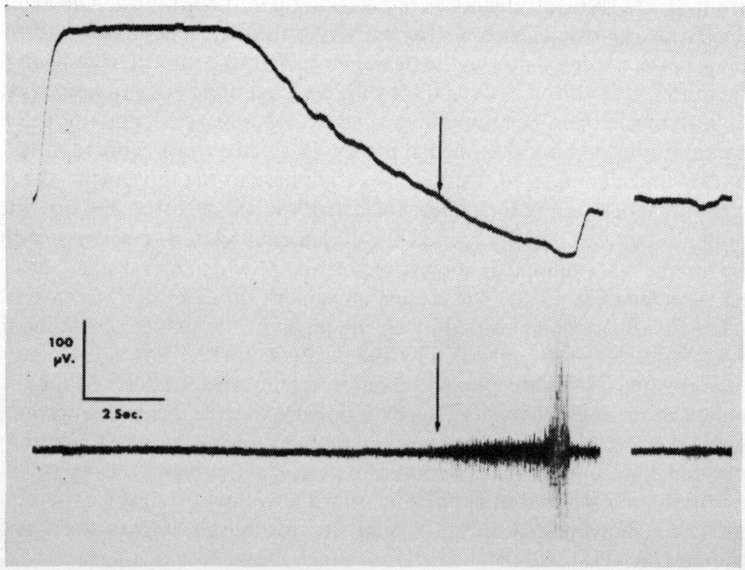

Fig. 11.3. Electromyogram of the temporal muscle with the mouth wide open, and during slow closure to tooth contact. The upper trace is a movement record. Tooth contact is indicated by the abrupt increase in amplitude of the EMG as the subject voluntarily clenched his teeth. A sample of a record obtained with the mandible in its rest position is given on the right and the position of the resting posture is indicated by the arrow on the main trace. (*From Yemm, 1975, J. dent. Ass. S. Afr.* **30,** *203–208.)*

Such instability is not consistent with a mechanism for precise control of resting posture by servo-controlled muscle activity, with feedback from receptors such as the muscle spindles.

MANDIBULAR POSTURE AND TONIC MUSCLE ACTIVITY

The question remaining is—do the jaw muscles play any part, either continuously or intermittently, in determining mandibular rest position ? Numerous attempts have been made to record activity from the jaw muscles of relaxed human subjects, the majority using surface EMG electrodes. The results are not conclusive, nor likely to be. Some reports (222, 404) claim to demonstrate continuous activity, particulary of the temporal muscle; others report intermittent activity or no activity (182, 288). However, in general the methods employed provided an inadequate basis for objective conclusions.

Surface electrodes enable a considerable volume of muscle to be sampled, but the necessary amplification introduces electronic noise which renders it difficult (at best, a subjective matter) to distinguish biological activity. In addition, the electrodes may sample too widely, even to include the brain since electrodes may be placed close to the temporal muscle when recording an EEG. Furthermore, relaxation may be difficult to achieve. It has been shown that environmental stress, such as might be engendered by an experimental situation, can initiate activity of the jaw closing muscles (451, 638). The use of needle electrodes to sample activity of the jaw muscles is also unsuitable for the determination of the presence or absence of tonic activity, because only small volumes of muscle can be sampled at one time, and also there is an increase in the stress upon the subject. However, in some current experiments I am studying the activity of single motor units of the masseter and temporal muscles using bipolar needle electrodes. Over 100 units from 4 subjects have been studied and none of these units has been active when the subject was relaxed. About 20 have become active during a movement from rest position to occlusion, the others firing at various levels of voluntary clenching. Since units firing at low levels of voluntary activity are the most easily discriminated with the needle electrode (by slowly changing its position) it is surprising that tonically active units have not been discovered, if they exist. Wyke (636) has postulated the existence of large units responsible for posture maintenance, especially in the temporal muscle, but I have found no evidence for these. Large tonically active units should be discriminated very readily, if present.

During these experiments, the EMGs of the muscles have been sampled with surface as well as needle electrodes. By averaging the surface EMG, it has been found that every unit isolated with the needle electrode has contributed to the surface EMG. However, while the contribution of some units to the surface EMG would be detectable above the noise level (*Fig.* 11.4), and may even be visible as a single event, other units, especially in the masseter, were represented by surface potentials down to or below 4μV. Such units would not be discernible above the noise level of the amplifier (*Fig.* 11.5, upper record) without averaging (*Fig.* 11.5, lower record). Thus surface EMGs may not be appropriate for detecting low levels of muscle activity. It is clear therefore that there is no simple and objective method available for determining whether continuous (tonic) muscle activity contributes to the maintenance of the mandibular resting posture. All that can be claimed at present is that if there is any activity at all it is at a very low level, with at most only a very few units active.

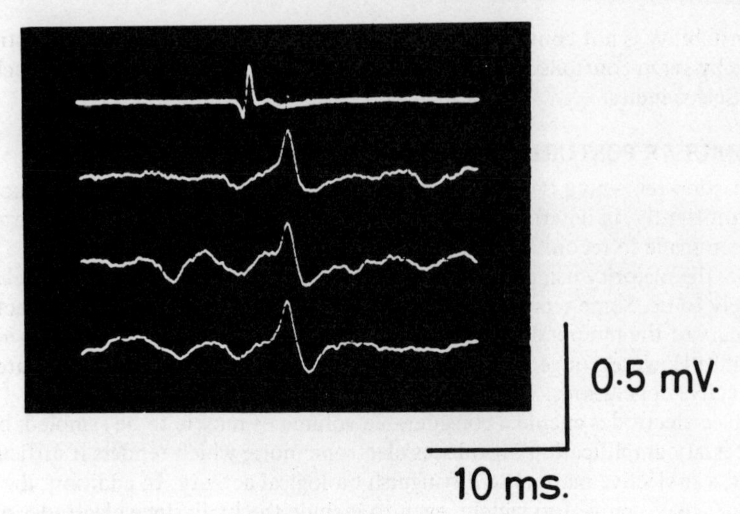

Fig. 11.4. Electromyograms from the temporal muscle of a human subject during voluntary clenching. The upper trace is 3 superimposed traces of the activity detected with a miniature bipolar needle electrode. The three lower traces are three simultaneously recorded surface electromyograms, displayed separately and not superimposed.

Since the position of the mandible changes when the orientation of the body in space is altered (379, 462) any simultaneous changes in the activity of muscles are not sufficent to prevent postural change. Nevertheless experiments reported by Lund et al. (372) provide some evidence that there may be some response by the muscles. It was shown that the electrical activity, particularly of the temporal muscle, was greater in an erect subject than when supine. The application of EMG techniques in this type of study, in which two or more experimental situations are compared in the same subject, is more justifiable than their use as a means of detecting the presence or absence of activity under similar experimental conditions but between subjects. Nevertheless even in these experiments some doubt must remain about the conclusion since the subjects may have experienced greater difficulty relaxing when erect than when supine, or may have tended to retain the same mandibular posture following a change in body position without remaining relaxed. Even so, this study provides the best available evidence for active stabilization of the mandibular rest position.

The fact that the mouth sometimes appears to fall open during sleep particularly when sitting, could be due to a reduction in any jaw muscle activity present, or to changes in intra-oral pressure or the activity of lip and tongue muscles. For instance, the cessation of habitual lip activity could either create the appearance of the mouth opening, or could actually permit a small change in mandibular posture. Another possibility is that the open position may arise only if the head drops back in the chair, when the postural change would be expected from the work of Prieskel (462).

Evidence of changes in mandibular posture with age has been obtained (560, 561). Such changes may be due to changes in tonic activity of the muscles in which case the gradual increase in face height reported to occur in subjects retaining their natural dentition could be due to a reduction in tonic activity of elevator muscles. Furthermore, a gradual increase in 'tone' of these muscles would account for the reduction in resting face height observed in wearers of full dentures. Alternatively, the postural changes may occur through age changes in the elastic properties of the tissues and structural adaptation of the muscles to changing working length. The latter property of skeletal muscles has been demonstrated in cat limbs (558) and in the rabbit digastric (417).

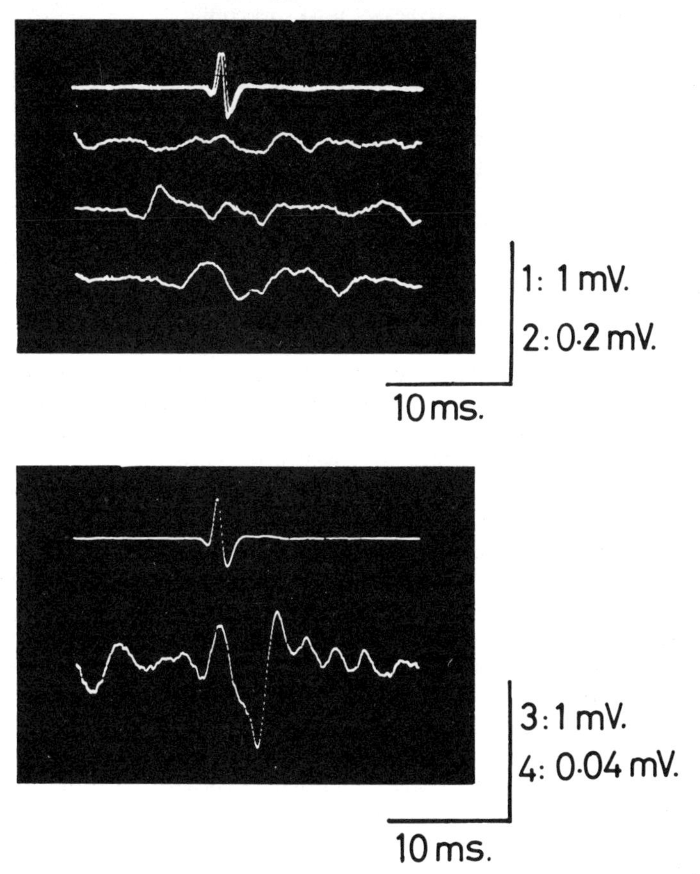

Fig. 11.5. *Upper record.* A unit from the same subject as shown in *Fig.* 11.4. The upper trace, from the needle electrode, is composed of three superimposed sweeps of the oscilloscope. The lower traces are three simultaneously recorded surface electromyograms displayed separately. The recording was made at the same session as the example in *Fig.* 11.4; the surface electrodes were undisturbed. *Lower record.* The average of 32 sweeps similar to those of the upper record, for the same unit.

CONCLUSIONS

At present, the evidence concerning the physiological mechanisms underlying the mandibular rest position is incomplete. It seems likely that the physical properties of the tissues are responsible for the basic position and that contraction of the jaw muscles is not a major factor. However, small changes in activity may occur in response to variations in body posture, indicating that some degree of control of posture through sensory feedback may occur. The parts played by other factors such as intra-oral pressure and activity of lip and tongue muscles remain to be investigated.

ACKNOWLEDGEMENTS
Thanks are due to Mr K. Robbins for his assistance in preparation of the illustrations.

Discussion

CHAIRMAN: B. MATTHEWS

CHAIRMAN: May I add a comment which I think is relevant here. Should the position sought by the hypothetical servo-mechanism happen to coincide with the position determined by passive forces alone, then one might have a servo-control system without any ongoing electrical activity recordable from the muscle in the 'resting posture'.
YEMM: I do not think that mandibular posture is a precise position and I agree with Bob Nairn, in his clinical assessment of the variability of the resting posture. I agree also with the things that Professor Brill said at the IADR meeting last week about there being a general area of facial dimension which is 'right', rather than a specific position. This fits with the observations on the anaesthetized rat where variability is introduced by friction or viscous components in the system.
BOSMAN: I have two questions about your *Fig.* 11.1. First, was there any hysteresis in the force-position curve and second, how do you interpret that curve ? I should expect evidence of dry friction as well as pure elasticity.
YEMM: The points in *Fig.* 11.1 were obtained with a step-wise movement in one direction from wide open to closed, so one would not expect any hysteresis to be shown. On each animal, we also made a complete cycle of closure, step by step, and then open again, step by step. This revealed a hysteresis effect. As you would expect from this, the separation of the jaws at rest was different depending on whether you pushed the jaw shut and let it go, or pulled it open and let it go.
BOSMAN: I should expect some dry friction, in which case, force could be applied without movement occurring.
YEMM: Was this not represented by the hysteresis in the system ?
GREENFIELD: Have you recorded during movement ?
YEMM: No. Professor Ahlgren (3) has stated that he considers that the elastic recoil of the jaw-closing muscles is partly responsible for initiating the closing movement in chewing.
AHLGREN: I found that the activity of the elevators in man did not start simultaneously with the closing phase, but I only studied the temporal and masseter muscles. Have you in your studies looked at the pterygoid muscles; perhaps these muscles start the action ?
YEMM: The picture we obtained from a few subjects with needle electrodes inserted into the medial pterygoid muscle was essentially similar to what I have shown for the masseter and temporal muscles.
WEIJS: Perhaps you might be interested in the fact that we have done some electromyography in rats and we found no activity in any of the masticatory muscles in the rest position. The results do not fit with the observations of Dr Møller, but they agree with your observations on the rat.
MØLLER: As to the onset of the closing movement, the medial pterygoid starts the chewing stroke, being active prior to the temporal muscles. This could explain the closing movement starting before the activity of the temporal muscles as Dr Ahlgren found.

Also in your elastic system, you attribute muscle activity to environmental stress. How do you account for the asymmetry of muscle activity and posture at rest in patients with unilateral cross-bite ?

YEMM: I have made no observations on unilateral cross-bite myself. Maybe the unilateral activity is the subject's response to the skeletal abnormality, and not an inherent characteristic of the nervous system.

CHAIRMAN: You are saying that the muscle activity is the result of conscious effort ?

YEMM: I wouldn't say conscious, any more than the cat is always thinking about the fact that it is standing up.

MØLLER: The point is, you consider this activity is due to a decision made by the subject. I find it very difficult to understand that a subject with a unilateral cross-bite would consciously keep his jaw to one side.

MEYER: When you were talking about the physical properties of skeletal muscle were you concerned about both the elastic and the viscous properties?

YEMM: I think that, if a posture is maintained without muscle activity, then it will depend both on elastic and viscous properties.

Paper No. 12

The Adaptation of Muscle to a New Functional Length

G. Goldspink

There are certain situations, in addition to normal postnatal growth, when a muscle has to adapt to a new functional length. After the extraction of teeth the resting posture of the mandible changes and this necessitates a change in length of the muscles of the masticatory system. Also, surgical procedures involving transplantation of muscle tissue are now being used on quite a wide scale. It has, therefore, become very important that we know something about the way and the extent to which growing and mature muscles are able to adapt to a change in functional length.

MECHANISM OF INCREASE IN LENGTH DURING NORMAL GROWTH

During postnatal growth, the bones of most species increase considerably in length and this, of course, is accompanied by an increase in the length of the muscles. However, the muscle fibres may not increase in length to the same extent as the belly of the muscle. In some muscles, the fibres run obliquely to the long axis of the muscle, and therefore some of the increase in length of the muscle is attributable to the increase in the diameter of the fibres. Even in muscles where the fibres do run parallel to the long axis, there seems to be some rearrangement of the fibres during growth. In these muscles, the tendons extend relatively further into the muscle as growth proceeds, so that the myotendon junctions become more staggered. Nevertheless, the fibres of most muscles do lengthen considerably during postnatal growth. This increase in length is associated mainly with an increase in the number of sarcomeres in series along the myofibrils (206, 624), and hence in the number of sarcomeres in series along the length of the fibres. The number of sarcomeres in series has been counted in individual, teased fibres from the soleus and biceps brachii muscle of mice at different ages (624). During growth, the number of sarcomeres in the mouse soleus was found to increase from about 700 to 2200, with most of the increase occurring before the animals were 3 weeks old (*Fig.* 12.1).

The way in which the myofibrils increase in length is a fascinating problem. This must of course involve the addition of new sarcomeres to the existing myofibrils; however, the point or points at which the sarcomeres are added to the myofibril has been a matter for debate for some time. Some authors (490, 503), have suggested that the myofibrils grow interstitially; in other words, new sarcomeres are added to the myofibrils at points along their length. The evidence for this theory is based on the fact that the sarcomeres of adjacent myofibrils are often out of register because of slight differences in sarcomere length. Therefore, for a given length of muscle fibre, some of the myofibrils will have an additional sarcomere,

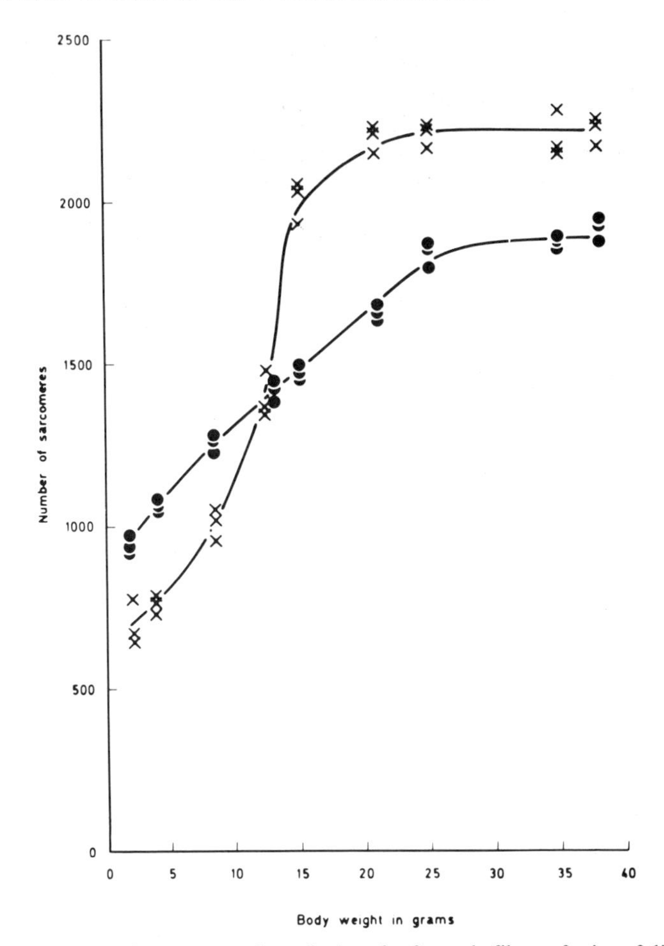

Fig. 12.1. The number of sarcomeres along the length of muscle fibres of mice of different ages (body weights) for the biceps brachii (•) and soleus muscles (×). (*From Williams and Goldspink (1971) J. Cell Science* **9**, *751–767.*)

and this is taken as evidence that a sarcomere has been inserted. However, in order to insert new sarcomeres in this way, it would be necessary, not only for the myofibril to divide transversely, but it would also involve considerable modification of the sarcoplasmic reticulum and transverse tubular system. Other workers (273, 282, 324, 378) have suggested that the lengthening process entails the serial addition of sarcomeres onto the ends of the existing myofibrils. Electron microscope studies of the end regions of growing muscle fibres show numerous ribosome formations (*Fig.* 12.2*a*) and myofilaments that are not properly organized into myofibrils (*Fig.* 12.2*b*), which suggest that this region is active in protein synthesis and myofibril assembly (282, 378, 624). The muscle–tendon junction is characterized by peripheral clefts and fingerlike invaginations into the ends of the muscle fibres (*Fig.* 12.2*c*). Muir (418) and Ishikawa (282) suggest hypothetical schemes whereby these sarcolemmal clefts provide regions where myofilaments might be added to the ends of myofibrils without the latter having to relinquish their attachment to the sarcolemma.

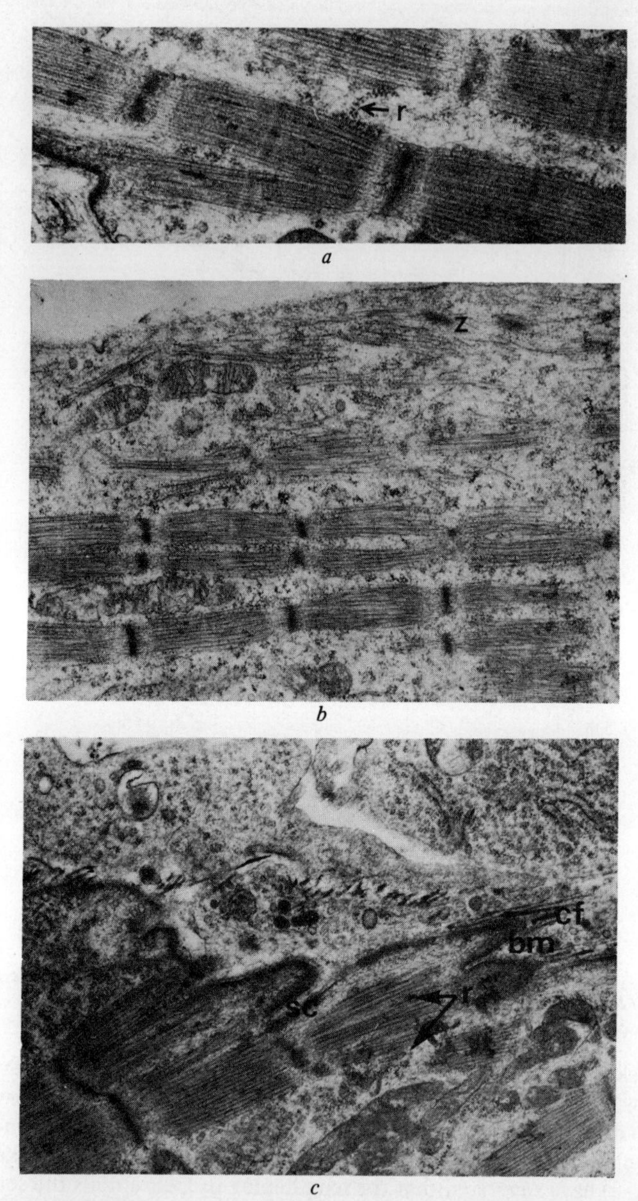

Fig. 12.2.*a.* Polyribosome chains at the myotendon region of a young mouse soleus muscle. The ribosomes (r) are numerous in this region of actively growing muscles and are often seen to be associated with the myosin filaments. *b.* An electron micrograph of the end region of a biceps brachii muscle fibre taken from a foetal mouse (18 days' gestation). At the periphery of the fibre the myofilaments are seen to be irregularly arranged. Dense bodies (z) can be seen which may be the precursors of Z-band material. Numerous ribosomes can be seen both round the periphery of the fibre and in the myofibrils. *c.* An electron micrograph showing the region of the muscle-tendon junction of a newborn soleus muscle fibre. In this region the sarcolemma forms many clefts (sc). The myofibrils tend to splay out near their points of attachment to the sarcolemma which appears denser and thicker in this region. Collagen fibrils (cf) seem to be connected to the outer surface of the basement membrane (bm). Again, groups of ribosomes (r) can be clearly seen. (*All from Williams and Goldspink (1971) J. Cell Science* 9, *751–767*)

More direct evidence for serial addition of sarcomeres has recently been presented (228, 624). This involved injecting tritiated adenosine into growing mice to label newly formed sarcomeres; adenosine is known to be incorporated into the structural ADP of the actin filaments. The soluble nucleotides were removed by glycerol extraction. Autoradiography was carried out on single teased fibres and longitudinal sections. Also, some muscles were sectioned transversely on a cryostat, and the amount of label in different parts of the muscle was determined by scintillation counting (*Table* 12.1). The data from the autoradiography (*e.g. Fig.* 12.3) and scintillation counting showed that most of the label was incorporated into the end regions of the muscle fibres, thus demonstrating that these regions are more active in the synthesis of actin and hence new sarcomeres. This, in turn, strongly suggests that the ends of the fibres are the regions of longitudinal growth and that new sarcomeres are most probably added serially to the ends of the existing myofibrils.

Table 12.1 ^{3}H adenosine incorporated into regions of a young and an adult biceps brachii muscle

Vial no.	*Distance along muscle mm*	*dpm/mm³ tissue*
Young muscle		
1	0–0·8	446
2	0·8–1·6	298
3	1·6–2·4	231
4	2·4–3·2	301
5	3·2–4·1	1067
Adult muscle		
1	0–0·8	7
2	0·8–1·6	1·3
3	1·6–2·4	5
4	2·4–3·2	5
5	3·2–4·0	17
6	4·0–4·8	42
7	4·8–5·2	20

Little is known about the physiological factors involved in controlling sarcomere number, and hence muscle length. However, Williams and Goldspink (624) found that, if the limbs of growing animals were immobilized by plaster casts (the casts were changed every few days so as not to interfere with bone growth), the addition of sarcomeres was very much suppressed (*Fig.* 12.4). When the plaster casts were removed, the muscles produced sarcomeres in series at a very rapid rate. They more than doubled their sarcomere number in one week and thus attained the correct sarcomere number and muscle fibre length for their size and age. This illustrates that length adaptations in the growing muscle can be very rapid indeed.

Not only does the number of sarcomeres have to increase during growth, but the other components of the muscle have also to increase in number or size. No information seems to be available on how some parts, for example the sarcolemma of the fibres, increase with growth; however something is known about the increase in the muscle fibre nuclei. Williams and Goldspink (624) have counted the total number of fibre nuclei in individual teased muscle fibres, while Moss (413) has counted nuclei in sections in muscle of different ages. They found that the fibre nuclei continue to increase in number beyond the stage at which there was no further increase in muscle fibre length. They also found that the larger diameter

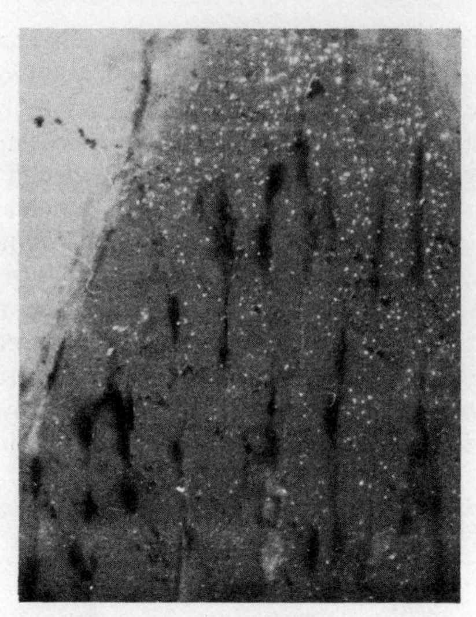

Fig. 12.3. An autoradiograph prepared from a longitudinal section of a biceps brachii muscle taken from a young mouse of 7 g body weight. (Initial body weight, 5 g.) The mouse had been injected with 75 μCi of [^{3}H] adenosine over a period of 3 days. (*From Williams and Goldspink (1971) J. Cell Science* **9**, *751–767*.)

fibres possessed more nuclei. Therefore, it must be concluded, the increase in the number of nuclei is associated with an increase in the girth of the fibres as well as the increase in length. The increase in the number of fibre nuclei does not apparently result from mitosis of the existing nuclei. Instead, certain cells known as satellite cells fuse with the fibres and thus donate nuclei to the growing fibres (391, 414, 519).These cells can be seen to be associated with the muscle fibres, particularly in young muscles, and they actually lie under the basal membrane of the fibres. The percentage of satellite cell nuclei with respect to muscle fibre nuclei has been shown, in the subclavius muscle of the rat, to decline by about 8 times between birth and maturity (12). Kitiyakara and Angevine (324) have carried out experiments using radioactively-labelled thymidine and found that most of the labelled nuclei were at the ends of the fibres. Aziz-Ullah and Goldspink (29) carried out a similar study on the mouse biceps brachii muscle and found that significantly more labelled nuclei were added to the proximal end of the muscle than to the distal end. This suggests that the satellite cells provide the necessary additional nuclei to allow the muscle fibres to increase in length as well as girth, and that muscle growth proceeds mainly at one end of the muscle rather than at both ends.

ABILITY OF MATURE MUSCLE TO ADAPT TO A CHANGE IN FUNCTIONAL LENGTH

The ability of mature muscle to adapt in length was first demonstrated by the experiments of Marey (385) in which he transplanted the distal end of the triceps surae tendon to a point farther down the calcaneum. In these experiments, the muscle adapted to its new functional length within a few weeks. More recently, Alder et al. (11) immobilized the rabbit tibialis anterior muscle in its shortened position by mechanically fixing the ankle in dorsiflexion. This resulted in a diminished muscle length and a change in the length/tension curve of the muscle in that the isometric tetanic tensions, equivalent in degree to those of the control muscle of the normal leg, were exerted at a shorter belly length and through a reduced range of movement of the foot. Similar results have been reported for cat muscle (563) in which it was also found that the length at which the muscle begins to develop contractile tension is increased if the muscle is immobilized in the lengthened position.

The mechanism of length adaptation has recently been studied at the cellular and ultrastructural levels (207, 558). Muscle fibre length, sarcomere length and the total number of sarcomeres along single teased fibres were determined for muscles

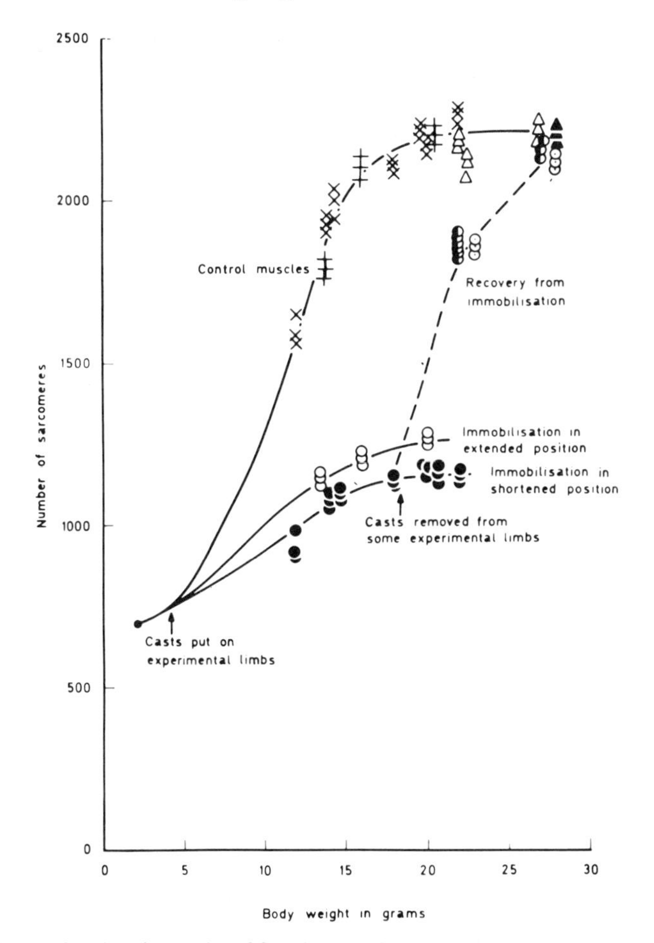

Fig. 12.4. Graph showing the results of four immobilization experiments with their controls. (1) The number of sarcomeres in fibres from soleus muscles which had been immobilized in the shortened position (●) and the number of sarcomeres in fibres from contralateral muscles (X). (2) The number of sarcomeres in fibres from soleus muscles which had been immobilized in the extended position (○) and the number of sarcomeres in fibres from contralateral muscles (+). (3) The number of sarcomeres in fibres from soleus muscles which had been immobilized in the shortened position until the mice weighed 18 g (◐) and the number of sarcomeres in fibres from contralateral muscles (△). (4) The number of sarcomeres in fibres from soleus muscles which had been immobilized in the extended position until the mice weighed 18 g (◎) and the number of sarcomeres in fibres from contralateral muscles (▲). (*From Williams and Goldspink (1971) J. Cell Science* 9, *751–767.*)

immobilized in different positions. Soleus muscles immobilized in the lengthened position were found to have 20 per cent more sarcomeres in series than normal muscles, whilst those immobilized in the shortened position had 40 per cent less than normal muscles. When the plaster casts were removed from muscles that had been immobilized in the shortened position, the length/tension curves and sarcomere number returned to normal within 4 weeks. Muscles that were immobilized in a shortened position and then immobilized in a second position were found to adjust rapidly to the new position with respect to their sarcomere number.

The effects of denervation on the response of the cat soleus muscle to immobilization at different lengths by plaster casts has been investigated for a period of 4 weeks. The denervated soleus muscle immobilized in the lengthened position was found to produce 25 per cent more sarcomeres in series, whilst those immobilized in the shortened position lost 35 per cent. This adaptation was essentially the same as in muscles that had been immobilized but not denervated. Denervation was found to have no effect on the recovery of muscles that had been subjected to 4 weeks' immobilization in the shortened position. In these muscles, the sarcomere number returned to normal within 4 weeks of removal of the plaster cast. The results for the normal and denervated muscles immobilized in both the extended and shortened positions are summarized in *Fig.* 12.5. The adjustment of sarcomere number to the functional length of the muscles does not therefore seem

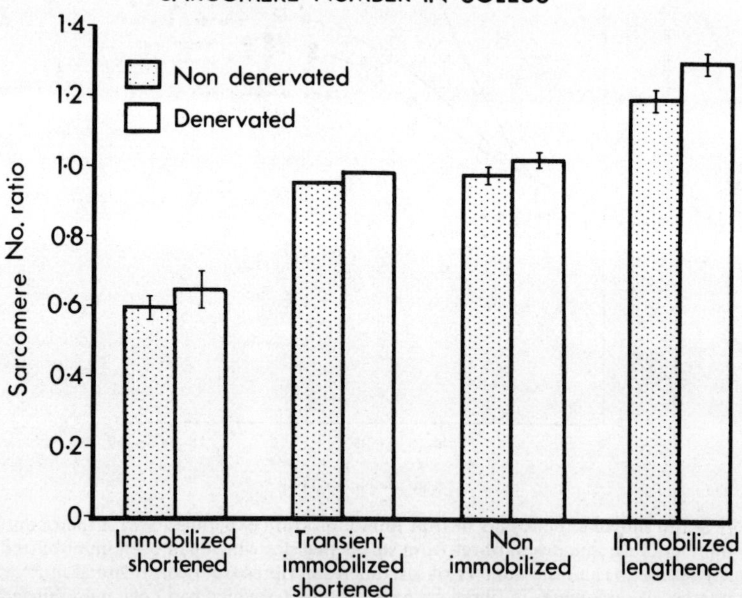

Fig. 12.5. Number of sarcomeres in a single fibre for non-denervated soleus muscles (hollow blocks), and soleus muscles denervated by resection of the nerve to soleus (stippled blocks). The blocks refer to the mean of the ratio of the sarcomere number for the experimental muscle to the contralateral muscle. The bars indicate the S.E. of the means. The various experimental situations were: immobilized in the shortened position, immobilized in the same position and then allowed to recover, non immobilized and immobilized in the lengthened position. (*From Goldspink et al. (1974) J. Physiol.* **236**, *733–742.*)

to be directly under neuronal control. It appears to be a myogenic response to the amount of passive tension to which the muscle is subjected.

Passive length/tension curves were established for cat soleus muscles that had been immobilized in different positions. Muscles that had been immobilized in the lengthened position showed no difference in their length/tension properties from those of normal muscles. However, those immobilized in the shortened position showed a considerable decrease in extensibility (*Fig.* 12.6). This was true whether the muscle was denervated or not. This decrease in extensibility (increase in passive resistance) appears to be a safety mechanism which prevents the muscle from being overstretched. This, of course, is important in muscles which have lost sarcomeres because stretching to their original length and beyond would result in the actin filaments being pulled out so far that there is no overlap between them and the myosin filaments and this would result in the muscle being damaged. At the present time we are uncertain about the reason for the decrease in extensibility of the shortened muscles. It could possibly be due to an increase, or rearrangement, of the connecting tissue or it could possibly be some change in the properties of the contractile apparatus itself.

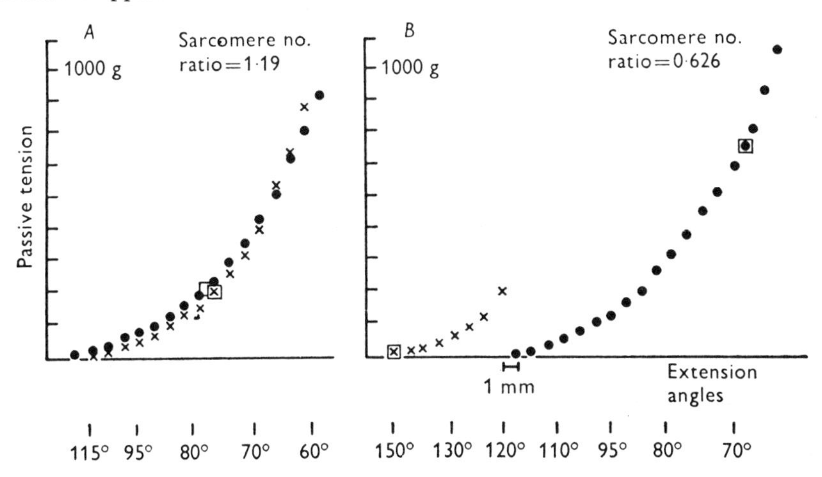

Fig. 12.6. Plots of passive tension against extension for muscles that have been immobilized in the lengthened position (*A*) and in the shortened position (*B*). The measured angles of the ankle during extension of the muscle are indicated on the abscissa together with a scale representing the change in muscle length. In each plot the length-tension relationship for the immobilized muscle (X) is given with that of its contralateral muscle (•). The sarcomere number ratio is given above each plot. (*From Tabary et al. (1972) J. Physiol.* **224,** *231–244.)*

The results presented here demonstrate that striated muscle is a very adaptable tissue and, in particular, they show that sarcomere number, and hence muscle fibre length, is adjusted to the functional length of the muscle. The physiological significance of this finding is apparent when one considers that the maximum contractile tension and the maximum rate of shortening of the muscle are obtained at the sarcomere length at which there is maximum interaction of the myosin cross-bridges with the actin filaments. Muscle is apparently able to adjust its fibre length and sarcomere length by producing more sarcomeres or by removing sarcomeres. In other words, it is able to adjust its sarcomere number to give the maximum functional overlap of the myosin cross-bridges and actin filaments.

When the limb was immobilized with the muscle at its maximum length, the muscle fibres were found to have produced 20 per cent more sarcomeres in series. On the other hand, when the limb was immobilized with the muscle in its shortened position the muscle fibres were found to have lost 40 per cent of the sarcomeres in series. These marked changes in sarcomere number were found to take place in a relatively short period of time. Also, the experiments in which the plaster cast was removed showed that, as well as adjusting to the new functional length, the muscle could readjust rapidly to its original length once the cast was removed. The length to which the muscle adjusts (its functional length) appears to be that which it occupies most of the time. When teeth are extracted, the position of the lower jaw changes and this is presumably accompanied by a change in the length of the muscles involved in mastication.

The link between the physical and molecular events involved in controlling the functional length of the muscle are not known; however, they do provide an extremely interesting subject for further study and it is especially important that this type of adaptation should be more fully understood because of its relevance to muscle physiology, physiotherapy, physiopathology and surgery, including, of course, dental surgery.

Discussion CHAIRMAN: B. MATTHEWS

LUND: Have you any information on intrafusal fibres ?

GOLDSPINK: No.

LUND: It would be interesting to know if, in a muscle which adapts to a change in length, there was any change in the receptors as well. One could postulate that, if the receptors did not change, there would be a very strange situation in the muscle control system.

GOLDSPINK: Yes, that is a very good point.

DARLING: How long did it take for this readaptation of length to take place?

GOLDSPINK: It is a very short time, a matter of 2–3 weeks in the cat. In the mouse it can take place in 3 or 4 days, but we have to be a little careful in extrapolating data from the mouse because its metabolic rate is very fast.

DARLING: Two or three weeks in man ?

GOLDSPINK: Yes.

HIIEMAE: You used the words 'lose sarcomeres' when the implications of the data were that you were dealing with growing animals. Could it not perfectly well be a case of failing to get sarcomeres, rather than positive removal ?

GOLDSPINK: Yes, in the growing animals that is certainly true, but not in the adult animals; they actually lose sarcomeres. I think that there can be no doubt in the case of mature muscles that sarcomeres are being removed.

HIIEMAE: Were your cats juvenile or adults ?

GOLDSPINK: They were adults.

KARLSSON: Henatsch and his group (251) have reported on the properties of spindles in different parts of a muscle and they found that there were very different responses to the same amount of stretch in different parts of the muscle. Furthermore, they found that there were different forces acting in these different parts. I think I am also correct that the sarcomere lengths in different parts of the muscle were different. I was interested to see that you counted sarcomeres all over the muscle and I wondered if you had any comment on differences in sarcomere lengths over a whole muscle.

GOLDSPINK: Sarcomeres at the ends of the muscle fibres are invariably shorter than those in the middle. Now this could be that these are the new sarcomeres and they are not yet quite fully functional, or it could be that, at the ends, there is more connective tissue which prevents the sarcomeres from being pulled out to the same extent as those in the middle. Now the problem I have glossed over is the really interesting part, that is, the link between the physical effect of a change in length of the muscle and the addition of sarcomeres. I do not know much about this as yet but I think that this is going to be an interesting problem. It seems, from our data, that the muscle adapts to the new length which it occupies most of the time and that muscle adaptation follows other things. It is probably not the primary cause of a change in posture, but it follows very quickly once there is a change in posture.

PICTON: Could I ask, what is your advice to the prosthetists ? Is it that their patients should wear their dentures all the time, so that there is no time during which the muscles should be allowed to develop this phenomenon you have just been describing ? Should dentures be fitted at once, as soon as the teeth have been extracted ?

GOLDSPINK: I do not think that is a fair question to ask a zoologist, I am not *au fait* with the clinical problems.

HANNAM: I think that your observations may have very profound implications for habitually altered jaw movement patterns, which we see commonly. The changes you describe could affect the movement pattern and the tensions developed by the muscles in fuction.

GOLDSPINK: I think that there are cases, obviously the spastic child is one, where some abnormality prevents normal adaptation from taking place.

AHLGREN: I wonder how long it takes for the shortened muscle to catch up with growth.

GOLDSPINK: In the growing animals it was only a matter of two weeks. It could even be shorter than that. My co-worker Mrs Williams, who really should take the credit for most of this work, tells me that she is of the opinion that it is only a matter of days rather than weeks.

DARLING: Is the amount of change half an inch or a foot ?

GOLDSPINK: It is 40 per cent of the muscle length; it is a very large change.

BULLER: Following Dr Hannam's comment, I do not think the adaptation will change the force. It should change the velocity of shortening, which is related to the number of sarcomeres in series, but not the force which depends on the number in parallel.

GOLDSPINK: Yes, I agree. In fact the sarcomere number is being changed to keep the force the same.

ORAMS: Does age affect adaptation ?

GOLDSPINK: That I do not know, but it is certainly something that we are going to look at.

HINRICHSEN: Regarding the small sarcomeres that you showed towards the end of the muscle. I have seen similar structures in the lips which surround secondary endings in muscle spindles. I thought that they would be of importance in perhaps deforming the secondary ending but I had not thought of them from the point of view of tacking on sarcomeres at the equatorial region of the spindle.

NAIRN: I would like to answer Professor Picton. If patients go on wearing their dentures to keep the muscle length, unfortunately the bone will go away and the jaws will get nearer together and the muscles will get shorter. There is really no way of stopping it.

Paper No. 13

Observations on the Resting Posture of the Mandible using Telemetry

G. H. Dibdin and M. J. Griffiths

We know that the concept of a fixed mandibular resting posture of a few decades ago is an oversimplification, and that it is influenced by many factors. Atwood (26), Thompson (582), and Brill et al. (88) have shown that all manner of short-term influences such as psychological state, posture, and fatigue, can change the resting interocclusal distance. This is in addition to long-term effects resulting from removal of occlusal contacts and ageing. The short-term influences make it desirable that methods for continuously monitoring mandibular posture should be as unobtrusive as possible. This is true both in clinical work and in experiments on the neuromuscular control of jaw posture. In defining jaw position, six measurements are needed: three linear (vertical, lateral and protrusive) and three rotational (in the occlusal, sagittal and coronal planes).

So, ideally, one should aim for a system to monitor all six of these movements continuously which is undetectable to the subject and allows him complete freedom of movement, without mechanical or psychological interference. Atwood (27) reviewed some of the methods available at the time. A fascinating array of electromechanical systems have been used; some monitoring most or all of the six degrees of freedom (45, 184), but some were extremely complex and unwieldy (187). Others were quite straightforward (603); even simple (463). All must have interfered considerably with function, however. Some measurements have been made using distance transducers in the mouth which required wires to project from between the lips (340, 341, 512). Cine film and tele-video methods have also been used, some making use of markers attached to the soft tissues of the chin and upper lip (30), and others (3, 150) using splints attached to the upper and lower teeth to avoid errors due to soft tissue movement. Ahlgren (3) estimated that errors due to soft tissue movement could be as much as 4 mm. Cine radiography has been used over short periods to demonstrate denture movement during mastication, but is impractible for the longer periods required in the study of mandibular resting posture. A disadvantage of many of these systems is that the subject's head has to be held in a cephalostat, or else allowance has to be made for movement of the head in relation to the camera.

Methods employing radio telemetry have also been applied to this field and, to date, have concentrated solely on the measurement of interocclusal distance. They have relied on two related ideas. In one (32), use is made of a phenomenon called the 'Hall effect'. In this, a voltage is developed across a conductor or semiconductor when placed in a magnetic field. This voltage is proportional to the current and to the applied field. Bando's (32) group made use of a permanent magnet mounted on one dental arch to cause a distance-dependent voltage to be generated in a 'Hall probe' mounted with a radio transmitter on the other arch. The Hall voltage caused the modulation of the transmitter to change, and this was converted to a proportional voltage for application to a chart recorder in the telemetry receiver. They mounted the components buccally on upper and lower premolars and molars. An advantage of their design was the presence of a small calibration switch (194) which provided a calibration signal when the teeth made contact in the centric position. This enabled errors due to slow baseline drift to be corrected. However, the assembly was rather large (32 X 14 X 8 mm excluding the magnet) and it weighed 7 g. It would be difficult to fit this into wax bite blocks or dentures without it being noticeable, let alone in the buccal sulcus of a subject with a full complement of teeth. In their study on a single subject the authors noticed that, when he spoke or swallowed without moving his head, the subject's interocclusal distance shifted from one apparently steady value to another of equal stability. This observation has also been made by some of the other groups. They also noted a mandibular vibration of amplitude varying from 0·1 to 0·5 mm.

The second idea makes use of the fact that if a metallic object is brought near to the coil of a tuned circuit, it shifts the resonant frequency. With the transmitter attached to one dental arch any metal crowns, fillings, or bands around the opposing teeth will increase the resonant frequency of the tuned circuit as the jaws move together; in contrast, a magnetic material like ferrite will decrease the frequency. In either case the frequency shift can be used as a measure of the interocclusal distance. A design produced by Thomson and MacDonald (586) used this system, with a coil wound on a 3-mm ferrite core and tuned to 10 MHz. This was the first attempt at radiotelemetry of mandibular posture, and the transmitter apparently fitted into an acrylic denture satisfactorily, although few details were given. The authors made recordings over periods of several minutes, and noted two effects; one was shifting of the rest position by up to 2·5 mm after swallowing, and the other was an increase in the interocclusal distance when the subject allowed his head to droop forward.

Joniot (295) reported another design, suitable only for studies of edentulous patients. The transmitter, tuned to about 2·5 MHz was fitted into the wax bite blocks used in the design of the dentures. With dimensions of 27 X 7·6 X 7·6 mm it was rather smaller than the design reported by Bando and co-workers (32). A ferrite plate on the opposing jaw gave a frequency shift as jaw separation changed. As with all these designs the response became increasingly non-linear, the greater the jaw separation, there was virtually no further change in output for separations above 7 or 8 mm. There seem to be some inconsistencies in the calibration Joniot (295) obtained. He reported that when the patient said the word 'Mississippi' the resultant interarch distance was 1·5–2 mm greater than his estimate of the true resting value, although this is not entirely clear from his published records. Twenty-three endentulous subjects were studied.

We chose a system somewhat similar to the last two I have described. However, we wanted equipment that we could put into mouths with fairly complete dentitions, so it was clearly desirable to keep the transmitter size as small as possible. The design details have been described elsewhere (145, 229). By attention to various points such as battery economy, techniques of moisture proofing, and choice of resonant frequency, we were able to produce transmitters with satisfactory stability. A typical transmitter weighing 750 mg is shown in black in *Fig.* 13.1. It measured approximately 10 X 7 X 5·5 mm and was held by a split retainer in the space left by the first molar. In this particular case a line of amalgam restorations in the opposing teeth produced the required frequency change. We investigated the error in measurement of interocclusal distance due to lateral and protrusive movement of the mandible (145) and found that for the smallest of the transmitters a lateral or protrusive movement of 2 mm changed the reading by about 10 per cent at a vertical separation of 2 mm. Although size needs to be kept small to avoid functional interference, the larger the transmitter coil, the smaller the error due to lateral and protrusive movement.

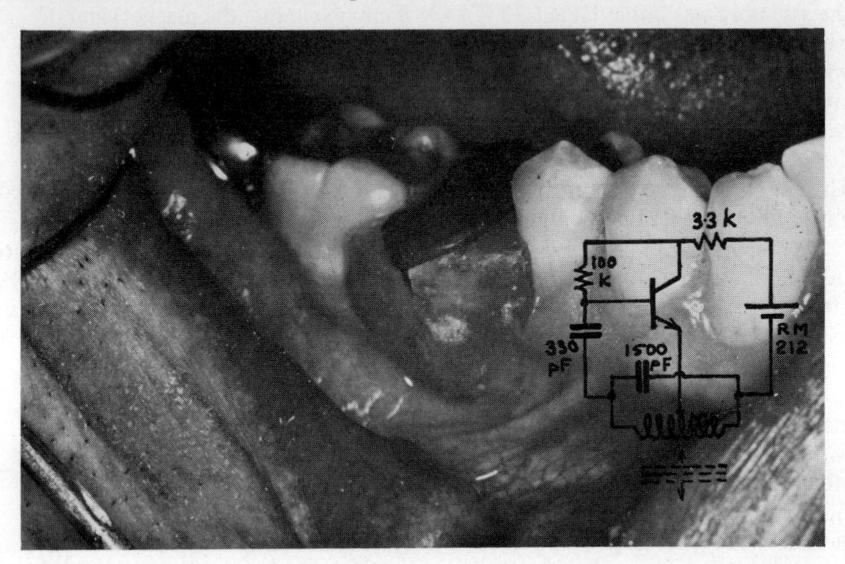

***Fig.* 13.1. The interocclusal distance sensor and transmitter fitted into a molar tooth gap by means of a split retainer.**

We have studied, amongst other things, the effect of speech, swallowing and posture on interocclusal distance and have also looked into the effects of stress. An example of the sort of trace obtained with the subject lying on his side is shown in *Fig.* 13.2. Note the change in interocclusal distance following speech; the same effect occurred with the subject in the vertical position. A similar shift occurred after swallowing. Also note in *Fig.* 13.2 the small regular mandibular movements, which proved to be associated with respiration. Dr Griffiths has also recorded interocclusal distance in one subject during overnight sleep with and without sedation, and has looked into the effect of attaching a bite raising appliance to the subject's dentition. He found that the subject very quickly adapted to the appliance with a correspondingly increased interocclusal distance and minimal increase in the number of occlusal contacts.

None of the designs devised so far permits studies to be carried out on subjects with a full complement of teeth unless the transmitter is placed in the buccal sulcus. From personal experience I know that even a small transmitter is to some extent obtrusive in this position.

In order to keep size to a minimum we have looked at the possibility of adapting the technique of passive telemetry for our purposes. With this technique, intraoral equipment which can interact with external signals is used to give the appropriate information, but without the need for any batteries in the mouth. The circuit we

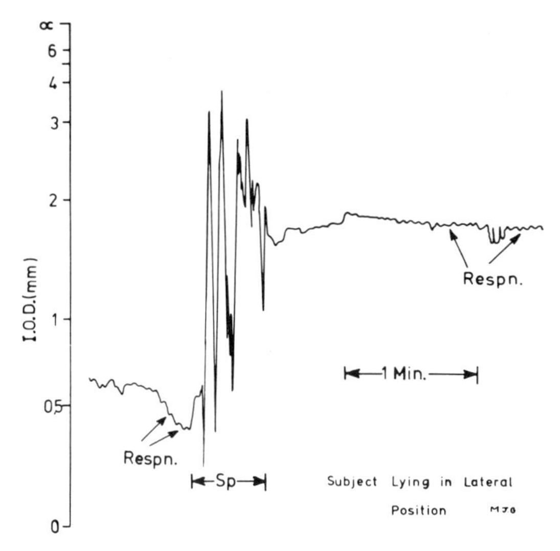

Fig. 13.2. Example of interocclusal distance measured on a subject lying on his side. The small rhythmic variations are due to respiration. There was a shift in the interocclusal distance after speech.

are using is shown in *Fig.* 13.3; but so far it has only been tested extra-orally. The passive components to be carried in the mouth consist merely of a coil, which can go around the teeth and an extremely small capacitor, which can be accommodated within an interdental embrasure. A pair of small coils is arranged on a pair of spectacles, one a transmitter and the other a receiver coil. By modulating the frequency of the transmitter signal it is possible to make use of the energy absorbed by the passive circuit as it comes into resonance, to lock the external transmitter

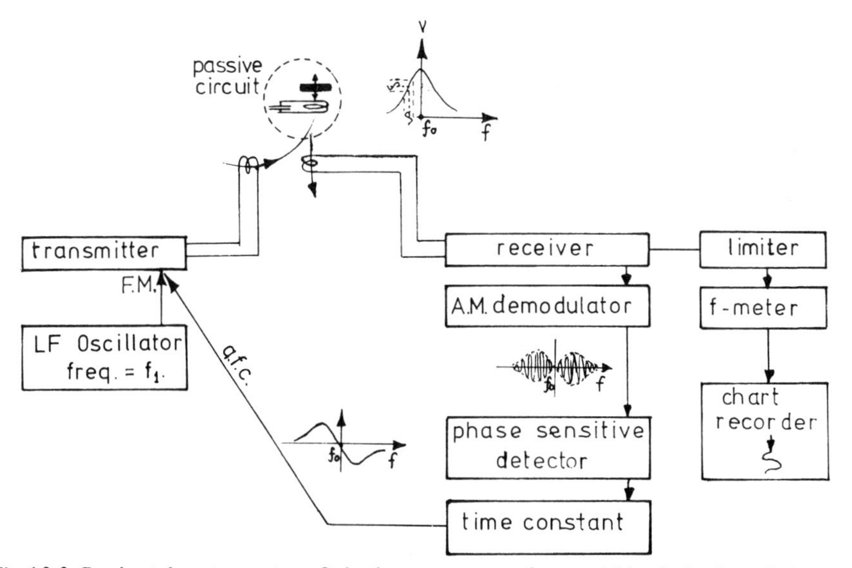

Fig. 13.3. Passive telemetry system. Only the components shown within the broken circle need to be in the mouth.

on to the natural frequency of the passive circuit. Since this natural frequency shifts with interocclusal distance, measurement of the frequency of the external transmitter gives an indication of jaw position. It is hoped that with this system we shall be able to make routine use of radiotelemetry for the study of mandibular posture.

Discussion CHAIRMAN: B. MATTHEWS

THEXTON: Is the frequency of your system affected by substances in the mouth other than metals, such as water ?

DIBDIN: That is one of the main points that concerned us to start with. Indeed, we may have gone unnecessarily far in trying to avoid the effects of soft tissues and water. What we have done is to choose a low frequency, where capacitance effects are minimal, and a circuit with a large tuning capacitance (1500 pF or 2000 pF). With this there is no detectable change due to soft tissue movement. However, as is commonly found, we did sometimes get trouble due to moisture creeping into the system but we have done trials over 10–12 hours in which drift has not upset the base line much. The few overnight measurements that Dr Griffiths has made have borne this out.

CHAIRMAN: I wonder if Dr Yemm would like to comment on whether those step-like shifts in jaw posture are likely to be within the range of the hysteresis he discussed.

YEMM: I can't answer that with respect to human subjects, but in the rat the hysteresis steps could be as much as 0·5 or 0·6 mm.

DIBDIN: We are talking about steps of 1–2·5 mm.

YEMM: This is clearly much larger than we found in the rat. However, the difference in size involved may mean that changes of the magnitude Dr Dibdin describes could be attributed to hysteresis.

CHAIRMAN: I thought Dr Yemm had estimated this in man also.

YEMM: Not in man.

DIBDIN: I think that perhaps some of the change is merely due to simple frictional effects and changes in the posture of the tongue. We did notice that there was a large change when the subject licked his lips when his mouth was dry. I wonder whether a shift in tongue posture or the way lips and cheeks stick to the teeth has some effect, in addition to the possible neurophysiological mechanisms regulating jaw posture.

GILL: Did temperature influence the transmitter ?

GRIFFITHS: Yes, it varied between 50 and 200 ppm per °C with different transmitters. We did not expect to get fluctuations of intraoral temperature beyond about 1°C. Within this range we get reasonable accuracy although we would like it to be better.

Paper No. 14

Masticatory Movements in Primitive Mammals*

Karen M. Hiiemae

The first mammals evolved from advanced mammal-like reptiles (the cynodont therapsids) about 200 million years ago. It is clear from the fossil evidence that some features of the jaw apparatus which are regarded as characteristically mammalian had in fact evolved earlier. The appearance of the squamo-dentary (as opposed to the quadrate-articular) jaw joint is, however, taken as the taxonomic watershed between reptiles and mammals and marks the point at which specifically mammalian evolution is considered to have begun. In some early transitional forms this joint functioned in tandem with the reduced reptilian joint before completely replacing it. When this occurred, the superseded quadrate and articular became incorporated into the middle ear as the incus and malleus. The postcanine teeth of the earliest mammals occluded (125, 128, 129), but did not have the complex arrangement of cusps, ridges and basins on a basically triangular crown which typifies the 'primitive mammalian' or tribosphenic molar (*Fig.* 14.1). This molar took a further 100 million years to evolve and persisted into the Eocene but not, unmodified, to the present. The tribosphenic molar is considered to be the type

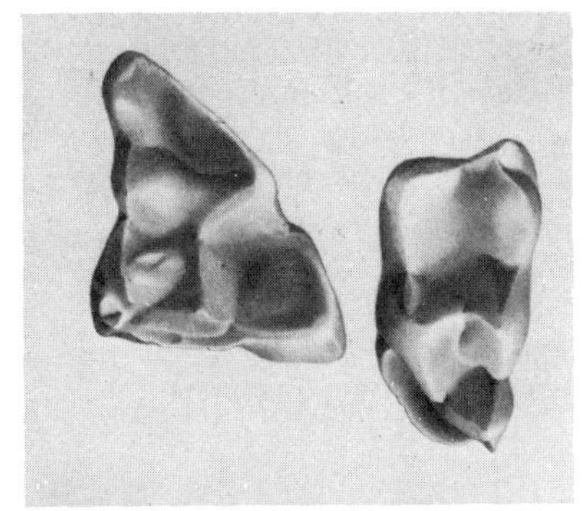

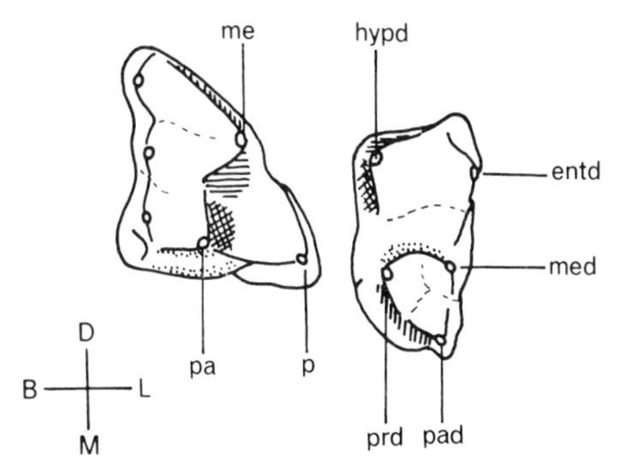

Fig. 14.1. Occlusal view of the upper (left in figure) and lower molars of the American opossum, *Didelphis marsupialis.* The major cusps on the upper molar are the protocone, p (which corresponds to the mesiolingual cusp on the human upper molar), the paracone, pa (the mesiobuccal cusp) and the metacone, me (the distobuccal cusp). The lower molar trigonid which occludes between the upper molar of its own number and the one in front has three cusps: protoconid, prd; paraconid, pad; and metaconid, med. The talonid has two cusps, human lower molar so the protoconid and metaconid correspond to the mesiobuccal and mesiolingual cusps respectively. The hatched areas indicate some, not all, of the matching wear facets, the edge of each ridge forming the cutting blade in each case.

*This paper was illustrated with a film.

from which all later placental and marsupial molars, however highly adapted, have evolved.

There are no extant 'early mammals' and very few living mammals have retained a relatively unmodified jaw apparatus. If, therefore, a primitive mammal is defined as one with pretribosphenic or basic tribosphenic molars, then its jaw movements in feeding have to be reconstructed from the wear facets on its teeth and data extrapolated from behavioural studies of a suitable living analogue such as the American opossum, *Didelphis marsupialis.* This primitive marsupial has slightly modified tribosphenic molars but is otherwise a good analogue (127), a view endorsed by Kallen and Gans (298). Various aspects of the structure of the jaw apparatus of the opossum as well as of its feeding behaviour have been, or are currently being, examined (126, 127, 130, 259, 261, 579, 580).

MOVEMENT RECORDING

There are a number of techniques currently available for recording jaw movements in mammals. Their applications and limitations have recently been reviewed (257). Cinefluorography, an interrupted technique, has been used in studies of *Didelphis* to record the movements of the cranium, lower jaw, tongue and the food in the lateral and dorsoventral projections. Most recordings have been made on 16 mm film at a camera speed of 60 fps which is sufficiently fast for the details of movement during the masticatory cycle to be analysed, particularly if markers are used (127, 259); although resolution is reduced by superimposition of the dental hard tissues when the teeth are in the intercuspal range. A combination of occlusal analysis (127) and cinefluorography can, however, provide the basis for a complete description of the movements of both jaws during the chewing cycle. Since masticatory movements are produced by cranial flexion and extension (*see Fig.* 14.9) coupled with mandibular elevation and depression (259), a distorted movement profile results from plotting successive jaw positions against a 'fixed' skull. Measuring the distance between two fixed points, such as the canines, as seen in frontal or lateral view (298) overcomes this difficulty but gives no information on the relative positions of the other teeth in the dental arch. This is a serious disadvantage when considering animals such as *Didelphis* where cuspal contact occurs from behind forwards. Measuring the 'gape', that is the angle between the upper and lower postcanine cuspal profiles, in successive frames and plotting this against time gives a record of the direction, amplitude and therefore rate of convergence or divergence of the tooth rows. In such plots (e.g. *Fig.* 14.2) full occlusion is conventionally presented as occurring at 1° gape. Interdigitation of the cusps of the most distal postcanines begins at 4° gape (measured both on cinefluorographs and on skeletal material) and is indicated by a fine line at this level in all the figures.

JAW MOVEMENTS IN FEEDING

Feeding is a progressive or 'ongoing' behaviour: the 'bite' (the particle of food in the mouth) is first collected or actively ingested ('bitten-off'), then progressively reduced and the resulting bolus finally swallowed. This series of activities is the 'masticatory sequence': all the movements involved, except possibly those of food collection, are cyclical. Within the sequence, the proportions of the basic cycle pattern are modified so that the initial 'handling' and 'transit' cycles associated with transfer of food from the incisors to the postcanines can be distinguished from those in which the food is puncture-crushed (teeth do not intercuspate) or chewed (teeth intercuspate)—*vide infra* and *Fig.* 14.4. A sequence may begin with puncture-crushing (PC) cycles if the 'bite' is separated from its matrix by the premolars ('ingestion by mastication') (259). Whatever the initial size or consistency of the food, the terminal swallow is always preceded by a characteristic 'preswallowing sequence' of two or three cycles. In the cat (263) these cycles show a steady diminution in amplitude. All masticatory (puncture-

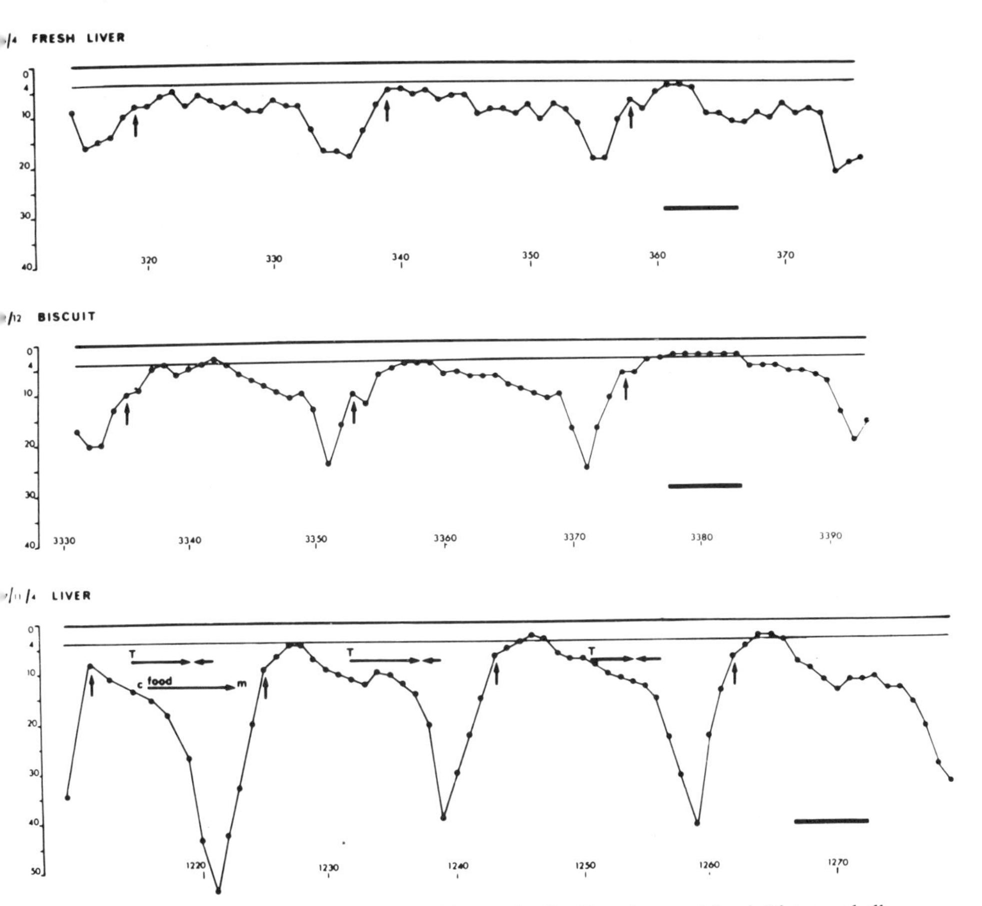

Fig. 14.2. Gape/time plots of puncture-crushing cycles for three types of food. These and all subsequent graphs are plotted in the same way. The vertical axis is degrees gape between upper and lower molar surfaces. The horizontal axis is time. Each dot marks the gape in a single frame of film taken at 60 fps in the lateral projection (every tenth frame is numbered below the plot). The time lapse between successive frames is 16·66 msec and the time bar is 100 msec. The fine horizontal line at 4° gape is the level at which intercuspation occurs. The vertical arrows indicate the point of tooth–food–tooth contact. The lumps of raw and (cooked) liver were approximately 20 mm³. Although the movement profile and the size of maximum gape differ, the total time taken for all cycles is comparable. The most regular movement is seen in PC cycles for liver (the first cycle in the sequence shown is a 'transit' cycle which is characterized by the absence of a power stroke (*see Fig.* 14.7)), the fast close, slow open and rapid opening stages of the chewing cycle are present. The food is moved from the canines (c) to the molars (m) by the tongue (T) during the fast opening stage of the transit cycle and the fast closing stage of the following PC cycle. The horizontal arrows show the protrusive movement of the anterior part of the tongue (arrowhead to right) in the slow opening stage and its rapid retrusion (arrowhead to left) in the fast opening stage of the opening stroke.

crushing or chewing) cycles have three strokes: *closing** in which the upper and lower teeth converge as the lower molars are moved rapidly upwards and laterally towards the uppers (fast close); *power* or *working* in which the food is triturated as the lower molars move anteromedially and upwards to the minimum vertical dimension for that cycle (slow close) and *opening* in which the teeth diverge. The opening stroke has two stages: an initial *slow opening* in which the gape slowly increases or is held within the range of 8–11° gape, followed by *fast opening* to maximum gape. Although the actual dimensions of the maximum gape change between successive cycles, this position forms a convenient and readily identifiable point in gape/time plots and is therefore used as the 'start-point' for each cycle.

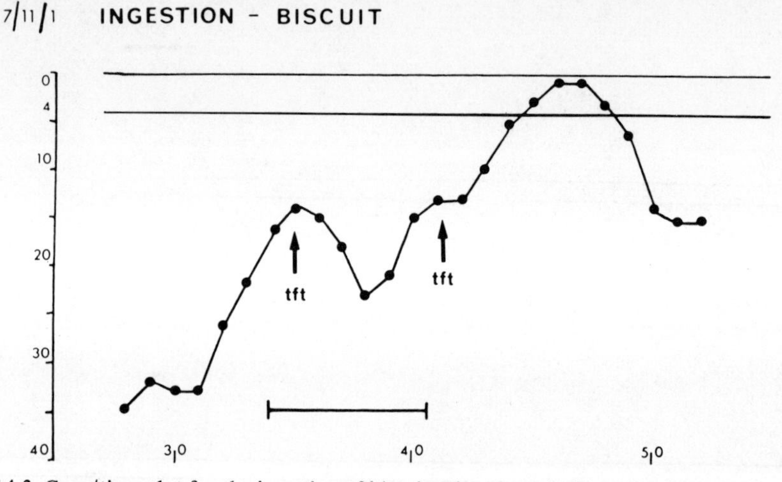

Fig. 14.3. Gape/time plot for the ingestion of biscuit. The arrows indicate the point of tooth–food–tooth contact (tft).

Movements in Ingestion

Some primitive mammals had very small incisors similar to those of *Didelphis.* In others these teeth were large and had complex crowns. The simple incisors of the opossum are used to collect and grip food and not to bite it (259). It is, therefore, a reasonable inference that the small subconical incisors of early mammals were used in a similar fashion. The spatulate incisors of the primates, capable of withstanding the occlusal loads of biting, were a later development and

---

*In a series of earlier papers (258, 259, 262) the terms *'preparatory stroke'* and *'recovery stroke'* were used to describe (*a*) the upwards movement of the teeth to tooth–food–tooth contact and (*b*) the downwards movement of the jaw to maximum gape after the power stroke. These terms are replaced by *'closing stroke'* and *'opening stroke'* respectively with the intention of facilitating comparisons with the literature on masticatory movements in man. (3, 404) However it must be emphasized that the *'power'* or *'working stroke'* as defined here for the opossum (and in papers on other mammals) is not strictly synonymous with the 'occlusal phase' described for man (5) since no clear distinction is normally drawn in the human literature between tooth–food–tooth and tooth–tooth contact (or the former is assumed to occur within the intercuspal range). The two can occur at different vertical dimensions. Further, the *'power stroke'* in the opossum is not synonymous with the 'isometric' phase described in man (and considered broadly equivalent to the occlusal phase by some authors) since there is some doubt about the existence of any period during the cycle in which conditions are truly isometric in respect of muscles acting on a stationary jaw (*see* text).

their appearance appears to be associated with the fusion of the mandibular symphysis (262).

The opossum collects small particles of food with the tongue and snout: softer mushy food is worked into a lump of suitable size by the tongue before being picked up by the incisors. The 'bite' is then transferred to the postcanines for triturition in one or more 'transit' cycles (*Fig.* 14.2) which have a distinctive profile and involve the highest rates of movement recorded (up to 0·8° arc/msec). If food has to be forcefully separated from its matrix, it is 'cut' or 'crushed' by the cheek teeth in a

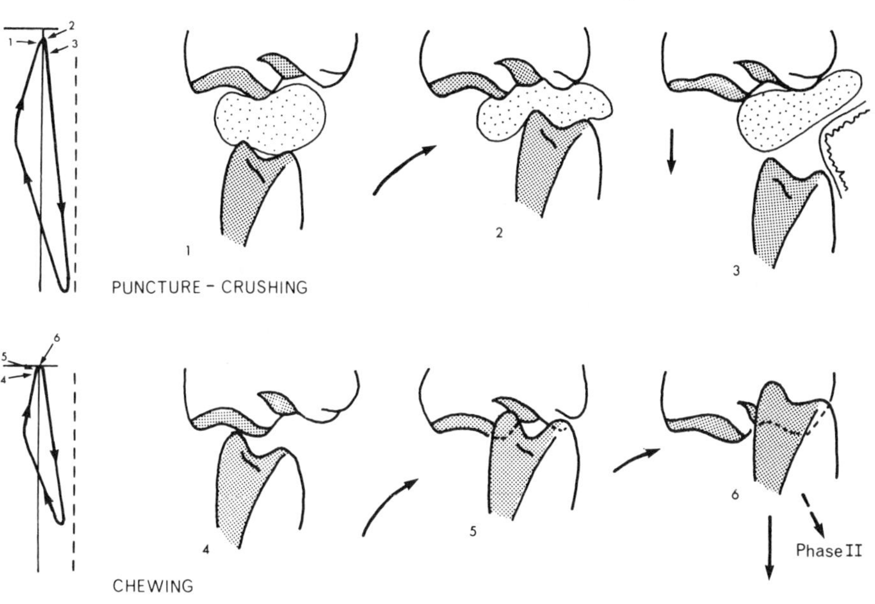

Fig. 14.4. To show the occlusal relations of the cheek teeth during puncture-crushing and chewing cycles in *Didelphis*. Reconstructed movement profiles for puncture-crushing and chewing with the working side on the animal's right are shown on the left of the figure. The horizontal bar marks the level of centric occlusion, the vertical a perpendicular to it. The broken vertical line presents the anatomical (midpalatal) midline. At maximum gape the mandibular symphysis (but not the cheek teeth) is in or very close to the midline position. The small arrows indicate the positions on the profile enlarged in the occlusal drawings. The teeth are shown in anterior (mesial) view. 1. Tooth–food–tooth contact on a large 'bite'. 2. Minimum vertical dimension on completion of the power stroke, the heavy arrow between 1 and 2 shows the direction and angulation (but without the anterior component) of movement of the protoconid. 3. Early in the opening stroke. (The heavy arrow indicates the degree of opening between 2 and 3.) The lump of food is being collected by the tongue. In drawings 4–6, the food has been omitted for clarity. 4. Tooth–food–tooth contact at the beginning of the power stroke, the reciprocally curved edges of the cutting blades will trap the food between them as the blades (hatched) shear past each other. 5. Halfway through the power stroke. The lower molar blade has sliced past the first shearing edge on the upper molar and is contacting the second. 6. Centric occlusion. The protocone (dashed line) is in the talonid basin of the lower molar which has now traversed both upper wear facets. The heavy arrows between 4 and 5 and 5 and 6 show the direction and angulation of the movement of the protoconid. The heavy arrow below 6 shows its movement through a position corresponding to 3. The broken arrow indicates the course of movement in a slightly more advanced tribosphenic molar which has developed a 'grinding' second phase to the power stroke. The buccal face of the protocone is then traversed by the lingual face of the hypoconid in a downwards and anteromedially directed loaded opening movement. The second phase is found in most primates, man and herbivores. The angulation of Phase I which corresponds to the whole power stroke in *Didelphis* and of Phase II depends on the relative heights of the cusps.

series of cycles indistinguishable from those occurring in the first stages of mastication. The number of PC cycles required depends on the size and consistency of the food. Current work has shown that given hard brittle food (biscuit) which breaks, a single cycle with an interrupted closing stroke may suffice (*Fig.* 14.3). In this typical case, the teeth were closed to tooth–food–tooth (tft) contact and the mouth slightly opened before being closed again; the biscuit broke on the second impact. After a short hesitation (17 msec) the closing movement was completed. Further experiments are required before any conclusions are drawn as to whether the first tft contact serves to provide a 'measure' of the size and consistency of the food and as to whether the degree of subsequent opening is correlated with the acceleration of the jaw required for the second and successful attempt at severance of the bite.

Movements in Mastication

Mastication in extant mammals with tribosphenic molars is unilateral, there is no functional or coincidental balancing side occlusion since the transverse width of the upper jaw is much greater than that of the lower jaw (the anisognathus condition) precluding simultaneous bilateral working contact (127). Centric occlusion (protocone in talonid basin) is also unilateral and is not coincident with 'centric relation' in the opossum. In *Didelphis* and other comparable mammals, the mandibular symphysis is mobile and forms an 'intramandibular' or 'third' joint (127, 498). The mandibular condyle is relatively free and there are no 'locking' preglenoid flanges on the articular fossa. The combination of these features means that all movements take place on or to the active side of the midline (unless working sides are being reversed), that each half of the mandible can rotate about its long axis and, therefore, that some independent movement of the working and balancing sides can occur. There is a regular divergent bilateral shift of both mandibular condyles in the closing stroke and a corresponding convergent medial shift in the power stroke (127).

When a portion of food has been ingested, it is first puncture-crushed and then chewed. These different modes of triturition rely on different parts of the cheek teeth for their effect and produce different patterns of wear (319). Puncture-crushing abrades the tips of the cusps leading to enamel loss and cavitation of the dentine, chewing produces attrition facets covered in fine parallel striations on the slopes of the ridges and cusps (*Figs.* 14.1 and 14.4). Both activities have the same path of movement within the envelope of motion but there are differences in the amplitude of the cycle in the vertical and, possibly, in the transverse planes (*Fig.* 14.4). The movement profile for a 'typical' chewing cycle in the opossum has been fully described elsewhere (127) and is illustrated in *Fig.* 14.4. (The profile in *Fig.* 14.4 differs in detail from that illustrated in Crompton and Hiiemae (126), this revision results from a more recent analysis of technically improved cinefluorographs.) It must be emphasized that the power stroke is not only upwards but also anteromedially directed.

Current work on feeding behaviour in the opossum is less concerned with descriptions of the movements in single chewing cycles than with the activity of the muscles producing it (130), with the modifications in the pattern of muscle activity and movement as the food is reduced during a single masticatory sequence and with differences in these patterns when different foods are handled (579). Earlier studies showed negligible variation in the behaviour of different animals feeding on the same food but considerable and consistent variation in their behaviour when

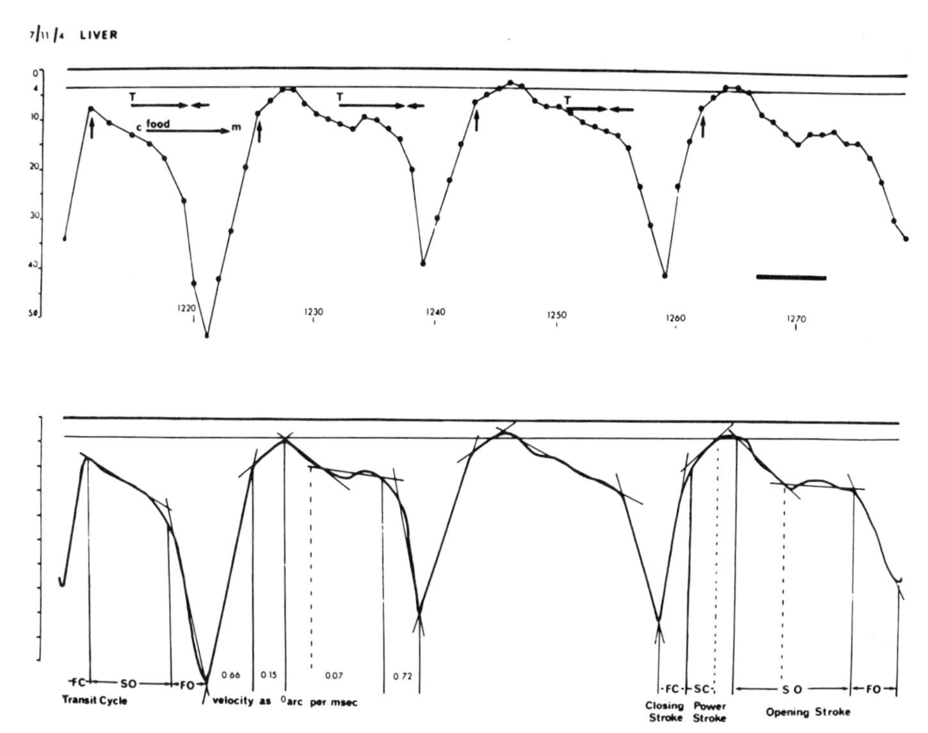

Fig. 14.5. The third trace from *Fig.* 14.2 with the profile smoothed to show the points at which significant velocity change occurs. In cycles where hard food is being puncture-crushed or chewed, the opening stroke usually begins, as in this example, with a definite opening movement to about 7° gape which is followed by a long period in which minimal movement occurs. When soft food is being chewed, or, as in *Fig.* 14.1, in cycles late in a sequence for hard food (i.e. when the food has been reduced and softened), this distinct initial opening does not occur, the 'slow open' movement begins from centric occlusion and continues to about 11° gape (*see also Fig.* 14.9). Note that the level at which tooth–food–tooth contact (heavy arrow top trace) is made and the minimum vertical dimension reached in each cycle diminishes as puncture-crushing proceeds.

supplied different types of food. These differences were reported but not quantitated (259). There is greater variation in the number of cycles and in their profiles early in the masticatory sequence than in the terminal stages before swallowing. Variation in the amplitude and duration of each movement is, in general, much less in chewing cycles than in puncture-crushing cycles (compare *Figs.* 14.2 and 14.5 with 14.6 and 14.7). On cinefluorographic criteria only, a puncture-crushing cycle is defined as one in which the teeth do not intercuspate, as the 'bite' is of sufficient bulk to prevent this (*Fig.* 14.4). In chewing cycles, the food has been reduced sufficiently to allow the teeth to intercuspate and therefore their relative movement to be governed by their shape.

Puncture-crushing

Forceful closures of the jaws on to a lump of food simultaneously squashes it between the opposing occlusal surfaces of the cheek teeth (crushing) and partially divides it by the penetration of the cusps (puncturing). All mammals so far examined including man ('chopping') (3) use this mode of trituration. In the pig it may be the only method (254) but in the opossum, the tree-shrew (262) and the little brown bat (298), all of which have basically tribosphenic molars, puncture-crushing is used to reduce the bulk of the food sufficiently so that intercuspation can occur, whereupon the cutting and, later (in both the behavioural and phylogenetic senses) grinding capacities of the teeth can be utilized.

It follows that the number of puncture-crushing cycles in a masticatory sequence depends on the size and consistency of the food. Soft food (semi-solid minced canned proprietary dogfood) is normally subjected to only one or two such cycles, really hard food (chicken gizzard or chicken leg) may be puncture-crushed for several minutes. Whatever the number of cycles, the overall duration of each is comparable (*Fig.* 14.2). There are marked differences in the maximum gape which may be independent of particle size (within certain limits) and relate only to food

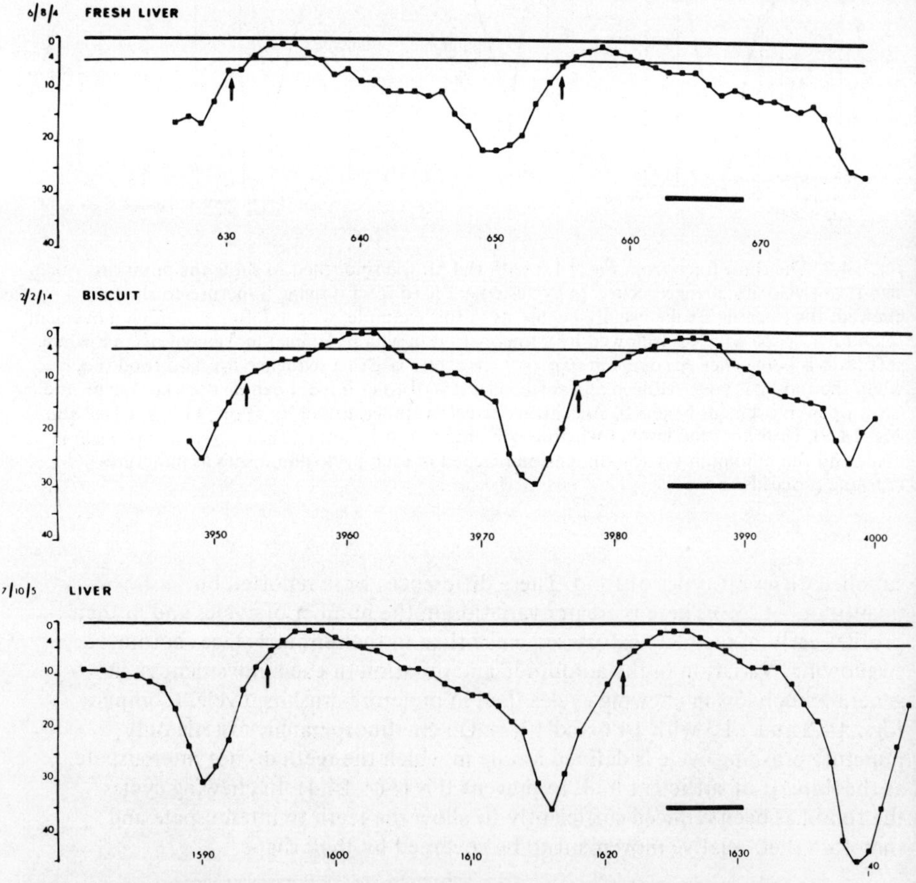

Fig. 14.6. Gape/time plots for the chewing cycles on the same foods as shown in *Fig.* 14.2. Although the amplitude of movement differs with the type of food, the durations and the profiles of the cycles are comparable. As in the puncture-crushing cycles there is a period in the opening stroke ('slow open') where downward movement is slowed or suspended.

consistency. (Work is currently in progress to investigate this possibility (263)). The only 'smooth' movement profiles were obtained when the animals were feeding on cooked liver, possibly due to the textural homogeneity of this material. Completion of puncture-crushing may be marked by an abrupt or gradual transition to chewing, depending on the nature of the food.

Chewing

The buccolingual width of upper tribosphenic molars is much greater than that of the lowers (*Figs.* 14.1 and 14.4). During the power stroke, a single cutting edge on the lower molar traverses at least one, and in some cases two, matching facets on the upper (*'en echelon'* shear (319)) before reaching centric occlusion. These edges are maintained throughout the life of the tooth although the height of the cusps is progressively reduced by wear.

All chewing cycles, irrespective of the food, have a comparable duration and general profile (*Fig.* 14.6). However, the amplitude of such cycles is related to the bite size and its consistency, so that large maximum gapes in puncture-crushing cycles are followed by smaller, but still large, gapes in chewing. Chewing cycles are always longer than puncture-crushing cycles. This increase in duration coupled with the reduction in amplitude leads to the lower angular velocities during the closing stroke and the fast open stage of the opening stroke shown in *Fig.* 14.7 (compare with *Fig.* 14.5). The nearest approximation to a 'steady state' in mastication is reached in the later stages of the masticatory sequence when, for

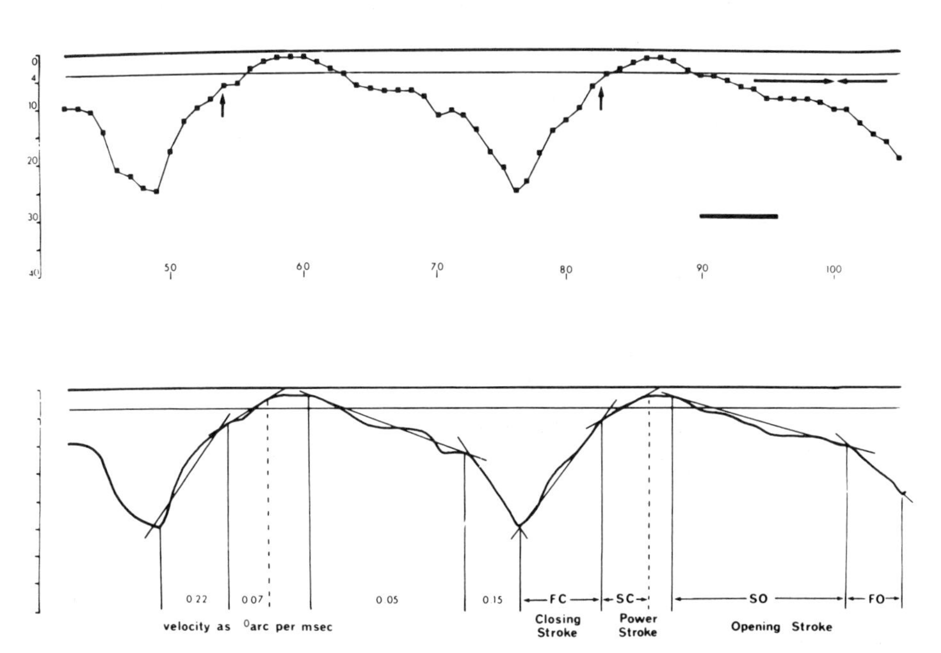

Fig. 14.7. Chewing of biscuit, to show the velocity changes and the phases of the chewing cycle. Comparison with *Figs.* 14.2 and 14.5 shows that the rates of movement are lower, there is a longer period in which no vertical movement can be seen in the lateral projection cinefluorographs and the total cycle time is relatively increased.

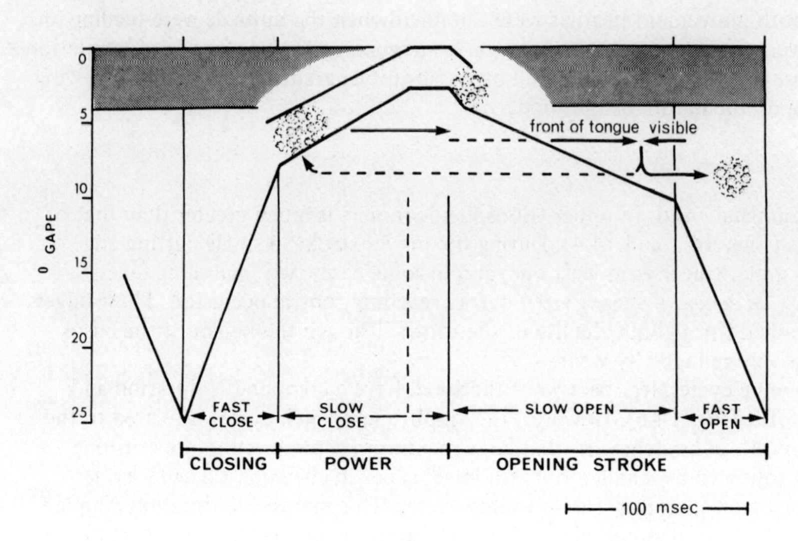

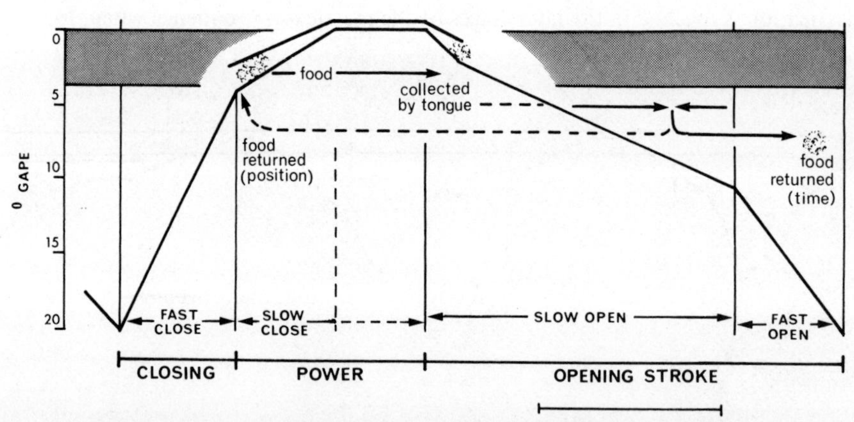

Fig. 14.8 'Averaged' puncture-crushing (n = 18) and chewing cycles (n = 24) with a total duration of 315 and 400 msec respectively, to illustrate the profile and approximate duration of the four main stages of the masticatory cycle and their variation as between puncture-crushing and chewing cycles. The duration of the power stroke is comparable in both cycles (92 msec and 104 msec respectively) but the time in which the jaw is in tooth–food–tooth contact and continuing to slowly close to the minimum vertical dimension is much longer in puncture-crushing. During the period (48 msec in chewing) in which no visible vertical movement is occurring the jaw is moving anteromedially (*see Fig.* 14.4). The total length of the 'slow open' stage of the opening stroke is much greater in chewing (170 msec as opposed to 102 msec in puncture-crushing cycles). If the duration of each stage of the cycle is expressed as a percentage of the total cycle time then these differences virtually disappear so the closing stroke takes 16 per cent or 17 per cent (figures for puncture-crushing cycles first) of the total time; the power stroke, 29 per cent or 26 per cent; the 'slow open' stage of the opening stroke 39 per cent or 43 per cent and the fast opening stage 16 per cent or 15 per cent. During the 'slow open' stage the hyoid bone is moving upwards and forwards so elevating and protruding the base of the tongue. The movements of the anterior part of the tongue as seen in the cinefluorographic records are shown.

a short period before the preswallowing sequence, the profile and duration of each chewing cycle is very similar to, or identical with, that of its predecessor. It is not yet clear whether the characteristics of such cycles are specific to the mastication of one type of food only or whether cycles late in the sequence are the same irrespective of the original nature of the food.

The Time Course of Chewing

It is clear from the results obtained on studies of *Didelphis* and from other work on the cat (263) that there is no 'standard' or 'typical' chewing cycle representative of all masticatory cycles. Puncture-crushing cycles are highly variable both within a single sequence and as between foods, chewing cycles show less variation within single sequences but also show differences with changes in the food. Pooling data merely generates an average value for any parameter with a large standard deviation. This is a recent finding: pooled data has previously been presented (259, 262). It is possible that work in progress will show that if, for a specified food, chewing cycles from late in the sequence are considered in isolation, then a behaviourally and statistically satisfactory 'standard' can be established for each such controlled situation. However, such 'regular' cycles can only be identified in the context of the whole sequence.

'Averaging' data can, however, provide a way of illustrating the essential similarities in the details of the movement profile in puncture-crushing and chewing cycles as seen in gape/time plots. *Fig.* 14.8 shows an 'averaged' puncture-crushing (n=18) and chewing cycle (n=24) solely to illustrate their components. The puncture-crushing cycle is shorter (315 msec) than the chewing cycle (400 msec). This difference is almost entirely attributable to a 35 per cent increase in the duration of the slow opening movement in chewing. During this stage of the cycle the jaw may move slowly and steadily downwards (the normal pattern for soft food) or be depressed to about 7° gape and then either be held, slowly opened, or even, occasionally, slightly closed (seen when the animal is feeding on hard food) before the beginning of rapid opening. It may prove significant that in the later stages of triturition of hard food, the profile of the slow opening stage conforms to that routinely seen in the chewing of soft food. In either case, the movement profile is attributable to changes in the rate of mandibular depression and not to both cranial and mandibular movement (*Fig.* 14.9). A similar movement profile is seen in gape/time plots for the tree-shrew and some primates (263) as well as in the cat. During this stage, the food is being moved within the mouth by the tongue; the heavy arrows in *Fig.* 14.8 show the movements of its anterior part. Recent work (130) in which the movements of both the jaw and hyoid have been plotted against a synchronized electromyographic recording of activity in both the elevator and hyoid musculature, has shown that the slow opening phase of the chewing cycle is coincident with an upward and forward movement of the hyoid which brings the base of the tongue upwards and forwards. In the chewing of soft food, the end of slow opening and the begining of the rapid downward movement of the jaw is coincident with the completion of this hyoid movement and with the initiation of its downward and backward movement towards its most retruded and depressed position. This is reached at the end of the power stroke. At the present time the relationship between the movements of the hyoid and jaw during other types of chewing and in lapping and swallowing is being investigated.

There is a short period, of the order of 17 msec in puncture-crushing, and of about 50 msec in normal chewing cycles (this figure is on occasion exceeded) in which vertical movement of the jaw is apparently suspended. This stage of the chewing cycle has, in man, been called 'isometric'. However, in the opossum and, arguably, in man, movement has not completely ceased. The power stroke in *Didelphis* is upwards and forwards and medially directed to bring the teeth into centric occlusion. The last stages of this movement occur within the intercuspal range of gape and cannot be clearly seen in lateral projection radiographs. Whilst electrical activity in the elevator muscles (temporalis, masseter, superficial masseter and medial pterygoid) ceases fairly abruptly at the end of the 'slow close' stage of the power stroke (261), the effect of their contraction persists for some time, given that their time to peak (twitch) tension is of the order of 15–18 msec with a half-relaxation time of about 20 msec (263, 580). The possibility that part of this 'occlusal' phase (4) does include a period in which the lower jaw is held in centric occlusion cannot, however, be excluded although such a delay is patently not a prerequisite for normal jaw function. No pause, or pause in excess of 17 msec, is required for the depressor muscles to initiate jaw opening; *Figs.* 14.2 and 14.3 show that this can follow immediately on rapid closure.

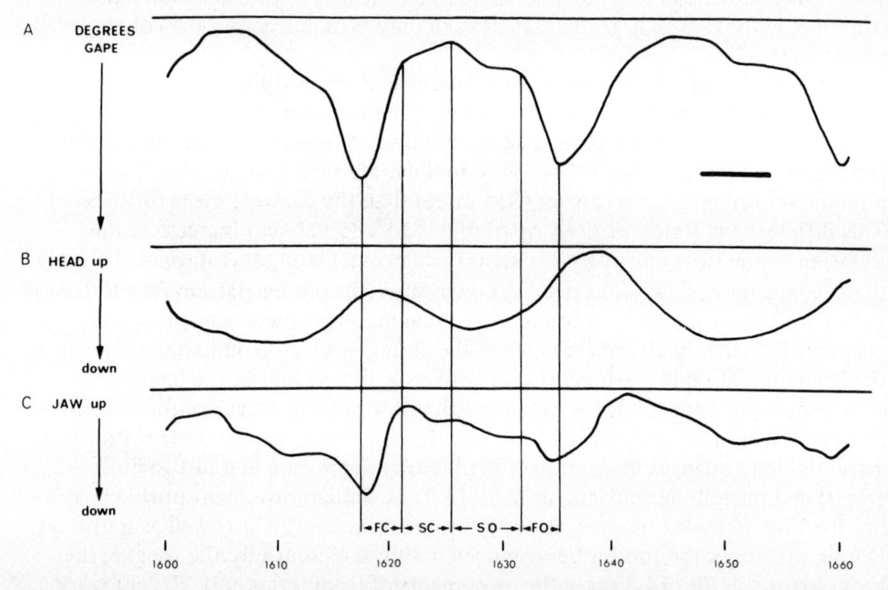

Fig. 14.9. Movement plots to show the contribution made by cranial flexion and extension (B) and mandibular elevation and depression (C) to the cycle profiles shown in A which is a conventional gape/time plot. Traces B and C have been plotted against external reference points. Trace C demonstrates that the change in velocity between the 'slow open' and 'fast open' stages of the opening stroke is attributable to an alteration in the rate of mandibular movement rather than that of both the head and lower jaw. (FC, Fast close; SC, Slow close; SO, Slow open; FO, Fast open.)

Occlusal Relations in the Power Stroke

The occlusal relations of upper and lower molars in the power stroke are illustrated in *Fig.* 14.4 for cycles early in the puncture-crushing stage of mastication and for late in the chewing sequence. All mammals so far studied have a medially and, to a variable extent, anteriorly directed power stroke (257). In primitive mammals with tribosphenic molars, and their predecessors (125), the power stroke was also upwards to the minimum vertical dimension or to centric occlusion (Phase I). The primate lineage is characterized by the development of a fourth cusp on the upper tribosphenic molar, the hypocone (distolingual cusp in man) which occluded into the trigonid basin of the lowers. This increased the 'cusp' into 'basin' occlusion of the cheek teeth and was associated with the beginnings of a second, grinding, component of the power stroke with an anteromedial but downward direction. This second movement, Phase II, became progressively more important in the primates (319). At the same time cusp heights were reduced and the movement into and out of centric occlusion became much flatter. With heavy occlusal wear in man (58) the vertical component in both phases may be almost or completely lost.

A detailed description of masticatory movements in feeding is possible if a combination of cinefluorographic and occlusal analytical techniques is used. The data obtained from studies of *Didelphis,* an evolutionarily significant mammal, have been used to explain the evolution of the tribosphenic molar (125) and its later modifications (99). There is accumulating evidence that, if these techniques are correlated with EMG studies by synchronizing the film and EMG recording (580), then, not only can the pattern of muscle activity related to movements of both the hyoid and jaw be demonstrated, but the system itself provides an opportunity for studying the control of mastication in the normal intact animal. Such experiments are in progress and the results obtained so far, coupled with the results obtained from studies of feeding in other mammals (254, 298, 600, 619) provide accumulating evidence that the control and execution of masticatory movements may be broadly similar in all mammals (257). The opossum may prove to be not only a basic structural but also a behavioural model for mammals in general.

Acknowledgements
The studies on *Didelphis* described here were undertaken first at Yale University and later at the Museum of Comparative Zoology, Harvard University with Professor A. W. Crompton and recently, Dr A. J. Thexton. I should like to thank them both and also the many people who have facilitated our work by providing technical and secretarial assistance. Throughout, support has been given by the U.S.P.H.S. (N.I.H.) with Grants Nos: DE 06248 and 03219.

Discussion CHAIRMAN: D. J. ANDERSON

HANNAM: We are beginning to see the same sort of patterns in man, but I am not sure what is happening in man during the opening stroke.
HIIEMAE: It was not until recently that I realized that if you put a large lump of apple in your mouth, the way you chew it is going to change from cycle to cycle, and yet we have all been busily (and I am as guilty as anybody) searching for a nice standard archetypal chewing cycle which I do not think exists.
HANNAM: The only thing that can be done is to use a substance like gum, but even then there is a fair amount of variability.
HIIEMAE: I think that is the only way to do it, but it does not produce normal behaviour; it is a laboratory test situation.
LUND: Do you feel that you are dealing with two different processes with your

puncture crush and mastication, or is it one basic process which is being modified by the physical properties of the food and sensory feedback ?

HIIEMAE: What we have is a repetitive behaviour which seems to show localized shifts and variations depending on the stage of the masticatory sequence and type of food. The other important factor, which is terribly difficult to quantitate, is the intrinsically changing nature of the food as it is reduced.

LUND: We have noticed in monkeys that, for our whole range of rhythmical movements, many of the differences in the movements can be explained by changes in the degree of jaw opening; the other parameters seem to follow. Thus, wide jaw opening is associated with a high velocity of opening, followed by rapid closure. These three parameters seem to go together. Similarly, with slow opening, the jaw does not open very far and it closes slowly. Do you find this ?

HIIEMAE: Something I ought to have said, which is extremely important, is that cranial flexion and extension are as much a part of mastication as are mandibular elevation and depression; the film showed it but I did not point it out. It is perfectly possible to close the mouth forcefully by moving the head only. There has been a paper recently which purports to describe mastication in the monkey where mastication, as described, was in no way normal because the head had been rigidly fixed. I believe you are quite right, but you have to be careful how you examine the rates of movement because they are in fact composite and not a simple mandibular oscillation.

WATT: Could you tell me how you measure the degrees of gape and what the error is in the measurement ?

HIIEMAE: The technique I use is to back project on to a perspex screen at a very small magnification to give optimal resolution. Then a lateral skull tracing is made and a line is drawn straight through all talonid basins of the lower molars and another straight through all the buccal cusps of the upper molars. This then gives you a degree of gape. The error on repeat sequence is not much more than one degree. I am quite proud of that but it starts to get tricky when there is intercuspation. There are, however, other reference points that can be used and in a lot of the earlier work we put metal markers in.

WATT: In the intercuspal position, how do you know within one degree of gape error, that the teeth do not stay in the same position.

HIIEMAE: I only showed you lateral projections. The profile of movement was established using dorsiventral projections as well. There was a long dorsiventral sequence on the film in which you could see movement in this plane too.

Paper No. 15

Masticatory Movements in Man

J. Ahlgren

Masticatory movements in man have always interested dentists, particularly those working in the field of orthodontics and prosthodontics, where treatment should secure, or at least not interfere with, masticatory function, i.e. that tooth contacts must be in harmony with mandibular movement.

Various methods have been used in the study of masticatory movements. At the beginning of the century graphic registrations from face bows were used, but more sophisticated methods, such as cinematography (3, 25, 58, 255, 265, 338, 511),

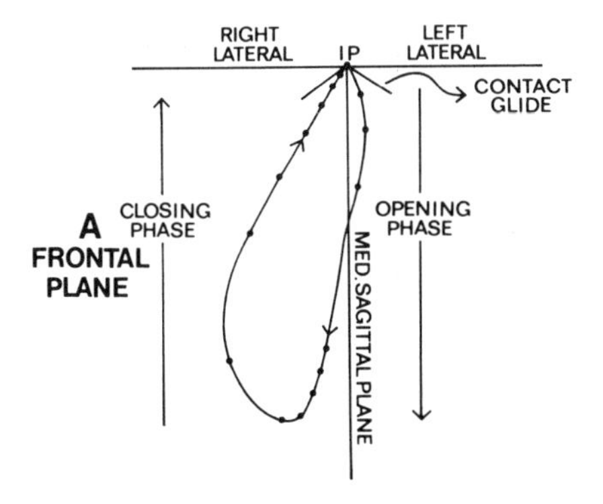

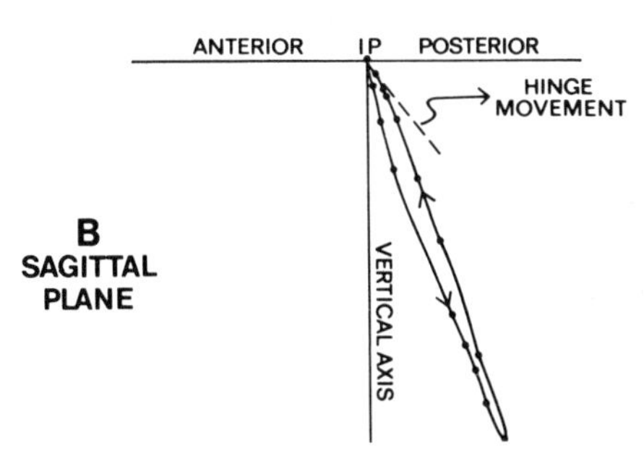

Fig. 15.1. The chewing cycle, illustrated with a cinematographic record of the movements of the lower incisors. The definitions used in this report are included. The time interval between two dots on the path of motion equals 1/24 sec. IP = intercuspal position. The plane of orientation both in the frontal and sagittal views is the occlusal plane. This applies also in subsequent figures.

cinefluorography (21, 325, 327, 523) and various electronic recording methods have since been devised (5, 8). Direct observation of the functioning mandible and attrition patterns of the teeth have also proved valuable since mastication can then be studied without any interference (4, 397).

This paper concerns the variation between individuals of the movements of the mandible in mastication, and their variation with the occlusion and kind of food being chewed. I will confine myself to the movements of the incisors and molars in the frontal and sagittal planes. The movements of the mandible in the transverse plane are of less interest. The movements of the condyle will not be considered.

Definitions: The masticatory cycle illustrated with a cinematographic record (*Fig.* 15.1) consists of three phases: (1) the opening phase—during which the mandible is depressed; (2) the closing phase—during which the mandible is elevated; and (3) the intercuspal (occlusal) phase—during which the teeth are in the intercuspal position. The intercuspal position (IP) corresponds to habitual occlusion and is characterized by maximal intercuspation of the occluding surfaces of the teeth. Terminal hinge movement is the most retruded opening of the mandible (centric relation).

Chopping movements are defined as crushing food with mainly vertical strokes, grinding movements are those that break food with a shearing force when the mandibular teeth move across the maxillary teeth.

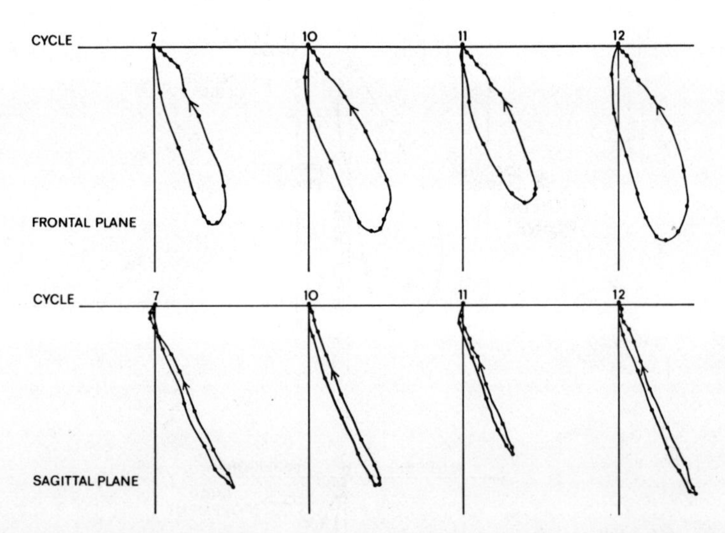

Fig. 15.2. Cinematographic records from a single subject of carrot-chewing, showing regular and characteristic chewing patterns.

PATTERN OF MOVEMENTS

Frontal Plane

Each individual has a rather characteristic basic pattern of movement. It appears mature and well coordinated around 4–5 years of age, by which time the primary teeth have erupted. But the pattern of consecutive chewing cycles is continuously changing and two consecutive cycles are never exactly alike (3, 5, 265, 327) (*Fig.* 15.2). The opening movement rarely goes straight down, but deviates to one side. In most cases it deviates to the chewing side, but frequently the opening phase starts with a contralateral deflexion (41 per cent of the cycles during carrot-chewing) and then swings over to the chewing side (*Fig.* 15.3). In grinding movements the contralateral phase during opening is particularly noticeable, and accentuates the impression of a wide masticatory stroke (75 per cent of the cycles during gum-chewing) (3, 58). Usually the closing phase is lateral to the opening phase although often this relationship is reversed, and the closing phase passes medial to the opening movement, i.e. a reversed masticatory stroke (*Fig.* 15.4). Sometimes the entire opening phase deviates towards the side opposite to that on which chewing is taking place. These abnormal masticatory movements

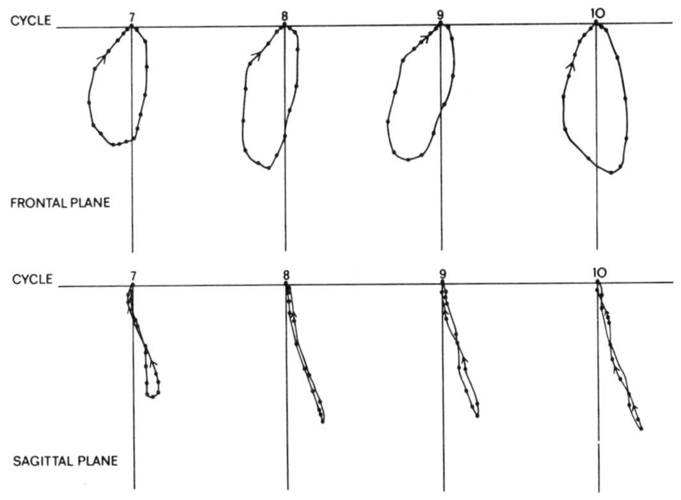

Fig. 15.3. Cinematographic records from a single subject of carrot-chewing, showing the contralateral phase during opening.

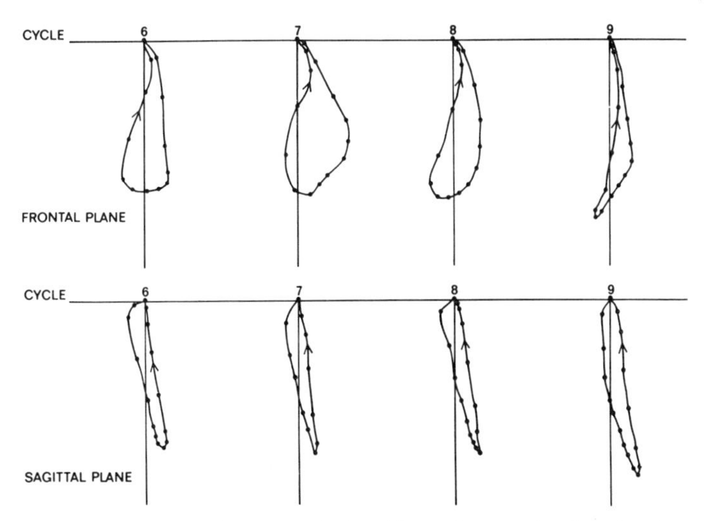

Fig. 15.4. Cinematographic records from a single subject of carrot-chewing, showing reversed masticatory cycles.

frequently occur in crossbite malocclusion. The most lateral point of the chewing cycle is situated about midway through the closing cycle for grinding movements, but is lower for chopping movements (*Fig.* 15.4). This further emphasizes the wider and more oval masticatory cycle during grinding movements.

The occlusion of the teeth is of significant importance for the development of masticatory movements. In general, individuals with normal occlusion have regular and coordinated masticatory movements (3, 58) (*Figs.* 15.2 and 15.3). Cases with malocclusions of the teeth have an irregular pattern with frequent crossings of the opening and closing movements (3,5,265) (*Fig.* 15.5). Cases with temporomandibular joint problems also show an irregular chewing pattern (25). A large vertical overlap of the incisors and cuspids (overbite) reduces the lateral deflexion of the masticatory stroke and the movements have a more vertical direction, i.e. chopping strokes. On the other hand, if the overbite is small and the lateral incisal guidance reduced, the movements assume a more oval and wide form with a more prominent lateral component, i.e. grinding strokes (3, 5, 327).

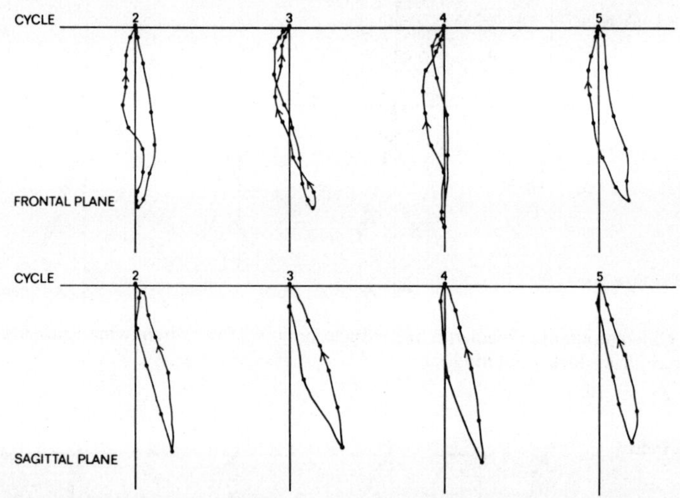

Fig. 15.5. Cinematographic records of carrot-chewing in a subject with malocclusion of the teeth showing irregular movements.

The wide variation within and between individuals of the masticatory movements is explained by the infinite variation of afferent inflow from the periodontium, tongue, gingiva and joint receptors during natural chewing and, consequently, the muscle activity. The irregularity and incoordination of the movements in patients with malocclusion and temporomandibular joint problems are probably due to impairment and/or incoordination of the proprioception from the periodontium and temporomandibular joints. The character of the food influences the chewing pattern (3, 58, 327, 523). Tough test foods, such as roast beef and toffee, require wide grinding cycles, whereas carrots and peanuts require vertical chopping masticatory strokes. The correlation between food, occlusion of the teeth and masticatory movements is demonstrated both by the chewing pattern of people living under primitive conditions (Australian aborigines) and by civilized people of European origin (*Fig.* 15.6).

The dimensions of the masticatory cycle are shown in *Table* 15.1. The means of the vertical dimension of the chewing cycle are between 16 and 20 mm. The means of the lateral deflexion lie between 3 and 5 mm. This dimension increases during grinding movements and decreases during chopping movements (3).

Unilateral chewing is the rule in most people of European origin, whereas in subjects of more primitive origin chewing occurs bilaterally, alternating between the left and right side with marked regularity. The prefered side for chewing is the side with the larger number of teeth (molars and premolars) in contact during lateral gliding (3, 58, 265).The aborigines thus have a bilaterally balanced occlusion.

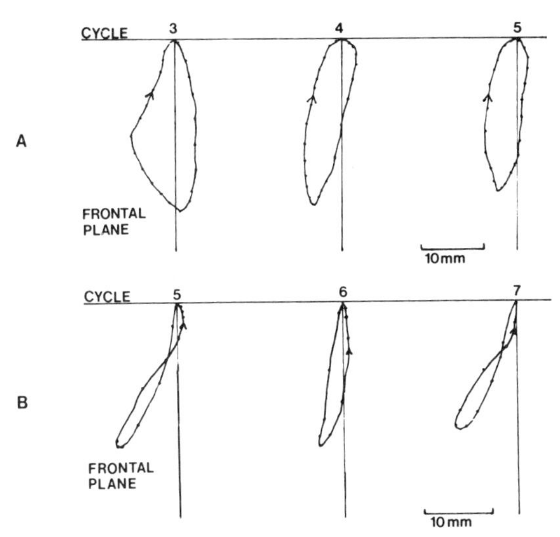

Fig. 15.6. Cinematographic records of an Australian aborigine chewing roast beef (redrawn from Beyron, 1964) (A) and a Swedish boy chewing carrot (B).

Table 15.1. Vertical and lateral dimensions (mm) of the masticatory movements in the frontal plane

Movement – dimension	*Hildebrand (265)**	*Beyron (58)†*		*Ahlgren (3)**	
	n=19	*n=26*		*n=34*	
		$\bar{x}$	SD	$\bar{x}$	SD
Vertical	16–17	17·1	2·19	19·8	3·1
Lateral	3–4	5·3	0·7	5·0	1·9

*European group.
†Australian aborigines.

Sagittal Plane

The movement pattern in the sagittal plane during mastication has not the same individual pattern as in the frontal plane. The opening and closing movements resemble quite well an ordinary empty open–close cycle. They do not differ greatly between individuals, types of food, or types of occlusion (*Figs.* 15.2, 15.3, 15.4 and 15.5). The closing movement has no definite positional relation to the opening movement (7, 327, 511).Sometimes the curves cross each other (52·3 per cent), sometimes the closing phase is situated more anterior (15·4 per cent) and sometimes more posterior (31·2 per cent) to the opening movement (7). The angle between the paths of movement and the occlusal plane has a mean value of 71·5° (SD=7·5) (7). The cycle is situated anterior to the path of terminal hinge closure, which has a mean angle of 54·4° (SD=7·1). The opening movement frequently starts with a protrusive phase which appears to be part of the contralateral deflexion of the mandible seen in the frontal plane (7, 511) (*Fig.* 15.4). This protrusive lateral movement makes possible a grinding action not only during closing, but also during opening.

Table 15.2. Duration (sec) of the masticatory movements

Test bolus	*Duration*				
	*Ahlgren (3)** *n=21*		*Beyron (58)†* *n=26*	*Hedegård et al. (250)** *n=14*	
	$\bar{x}$	SD	$\bar{x}$	$\bar{x}$	SD
Gum	0·77	0·15			
Carrot	0·58	0·09			
Roast beef			1·00		
Toffee				1·06	0·24
Bread				0·83	0·16

*European group.
† Australian aborigines.

DURATION AND SPEED OF MOVEMENTS

The duration of the masticatory cycle varies between 0·6 to 1 sec (*Table* 15.2). Different test materials have a significant influence on the duration, sticky and tough food substances increasing it.

The speed of the masticatory movement varies between subjects, within each cycle and between different test food (3, 511, 522). The mean speed of the movement during carrot chewing is 7·5 cm/sec (SD=1·37) (3). The speed appears to be slower for grinding movements than for chopping strokes. The opening speed is, on the average, higher than the speed during the closing phase. The opening phase is a 'ballistic' type of movement with relaxed antagonists, whereas the closing phase is a 'rapid tension' movement with antagonistic control guiding the mandible exactly into intercuspal position. The variation in speed within a masticatory cycle is demonstrated in *Fig.* 15.7. It shows that the mandible accelerates at the beginning of the opening and closing phases, but decelerates towards the turning points of the cycle. The highest speeds are usually found in the middle of the opening and closing phases. The deceleration was especially marked when the mandible approached the intercuspal position. In the intercuspal position the mandible comes to a standstill approximately 100 msec, ($\bar{x}$=102), before the next cycle begins (3, 5, 327).

TOOTH CONTACTS, MUSCLE ACTIVITY AND CHEWING FORCE

Tooth contact occurs in the intercuspal position in 84 per cent of all chewing cycles during natural chewing (carrot and peanuts) (3). Similar results have earlier been reported (1, 17, 217). Chewing cycles without tooth contact occur mainly at the beginning of a chewing sequence, i.e. the crushing strokes. As the bolus gets more and more triturated, occlusal contacts occur in every cycle. The occurrence of tooth contact glide during the closing phase depends on the type of food and type of occlusion. In primitive people chewing tough food with a well abraded dentition,

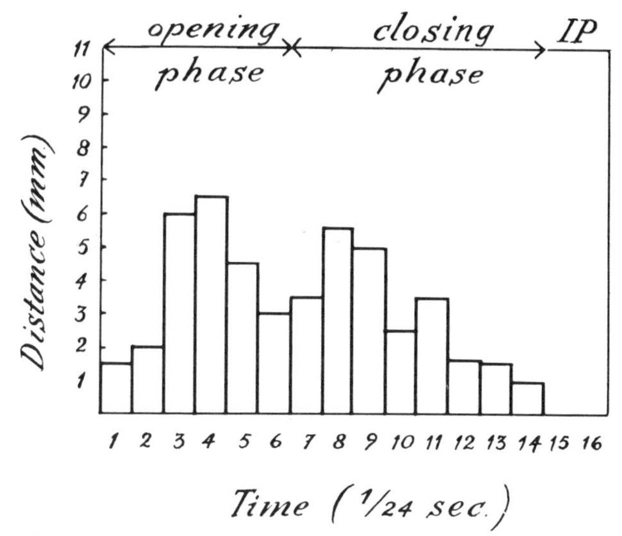

Fig. 15.7. Time-distance histogram of the chewing cycle (test-food carrot). In intercuspal position (IP) the mandible comes to a standstill for 2/24 sec.

the contact glide is large ($\bar{x}$=2·8 mm, SD=0·35) and occurs in nearly every chewing movement, whereas in Europeans living on a more easily-triturated food, contact glide is small ($\bar{x}$=0·90, SD=0·36) and occurs in only 54 per cent of the cycles (*Table* 15.3). The long glide of the aborigines clearly demonstrates the vigorous grinding movements of a primitive group of people.

Table 15.3. Amount of tooth contact glide during the closing phase of mastication

Test bolus	*Tooth contact glide (mm)*				
	*Ahlgren (3)**		*Beyron (58)†*		*Hildebrand (265)**
	n=19		*n=26*		*n=14*
	$\bar{x}$	SD	$\bar{x}$	SD	$\bar{x}$
Carrot	0·9	0·36			
Roast beef			2·8	0·35	
Meat					1·3

*European group.
†Australian aborigines.

The individual variation of cuspal glide is wide, and in some subjects there is no tooth contact glide at all, but merely a chopping stroke. This explains why there are so many different opinions about the importance of cuspal guidance during the closing masticatory stroke. The amount of glide is determined by the type of occlusion and the type of food being chewed. Hildebrand (265) found a contact glide of 1·3 mm for mastication of meat in people with malocclusion. This increased to 1·75 in people with normal occlusion. The angle of the contact glide to the occlusal plane during the closing phase also varies between occlusions. In people with advanced attrition (58) this angle is very small ($\bar{x}$=18·3°) whereas in Europeans without much cuspal abrasion (7) the angle is much larger ($\bar{x}$=36·9°).

Thus, as the attrition proceeds and the cuspal guidance is reduced, the closing stroke assumes a more horizontal direction when moving into the occlusal position. A summary of some of the parameters of a chewing cycle is shown in *Fig.* 15.8.

A contact glide of the occlusal surfaces during opening is also common. It is usually found in grinding movements with a wide contralateral deflexion at the beginning of the opening phase (*Fig.* 15.3) (3, 58, 265). The contact glide during opening is sometimes combined with a protrusive component, which can be seen on the sagittal recordings of the opening phase (7) (*Fig.* 15.4). This contralateral–protrusive movement has also been described by Mills (397) on the basis of studies of the attrition patterns of the teeth. He calls this movement the lingual phase of occlusion and it occurs when lower buccal cusps occlude with upper lingual cusps. It is a functional movement which increases the triturating effect of the chewing cycle. A similar mesiolingual movement has also been found in lower primates (98, 262).

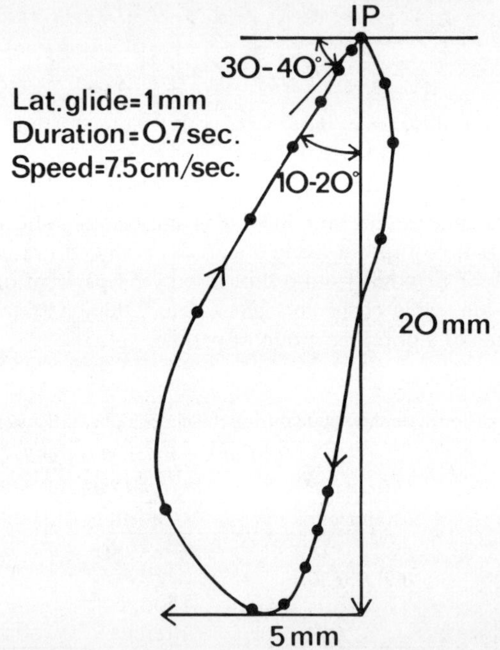

Fig. 15.8. Diagram of incisal movements in the frontal plane during mastication. The values are approximate.

At the end of the closing phase, the chewing force increases and reaches a maximum in the intercuspal position (*Fig.* 15.9). The peak chewing force occurs in the middle of the intercuspal phase, which lasts between 40 and 170 msec ($\bar{x}$=102·5 msec) (8). The muscle activity (recorded as peak integrated EMG) of the temporal and masseter muscles precedes maximal chewing force by a mean value of 41 msec (SD=26). The force gradually disappears but usually outlasts the intercuspal phase. This is because some chewing force has to be maintained to produce the gliding occlusal contacts during the opening phase.

Summary

Each individual has a characteristic pattern of mandibular movements in chewing.

But there is continuous variation between consecutive chewing cycles. The occlusion as well as the type of food influences masticatory movements. In general, subjects with a normal occlusion show simple and well coordinated masticatory movements, whereas subjects with a malocclusion have an irregular, complicated pattern of movements. Extreme vertical overlap (overbite) reduces the lateral component of the chewing cycle and produces vertical chewing strokes. Tough food usually is associated with grinding strokes with pronounced lateral deflexion of the mandible, whereas soft food produces more vertical, chopping movements. The speed and duration of the masticatory movements also vary with the type of occlusion and kind of food.

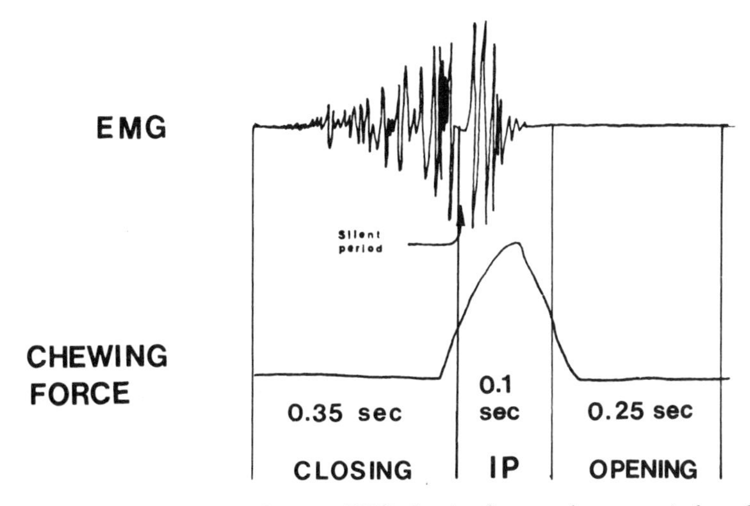

Fig. 15.9. Diagram of the relation between EMG, chewing force, and movement phase during mastication.

Tooth contact in the intercuspal position occurs in nearly every chewing cycle. The occurrence of tooth contact glide during the closing phase depends on the type of food and occlusion. In primitive people chewing tough food with a well-abraded dentition, the contact glide is large and occurs in nearly every chewing cycle, whereas in Europeans, living on a more easily triturated food, contact glide is small and occurs in approximately half the cycles. In the intercuspal position the mandible stops moving for approximately 100 msec before the next cycle begins. The chewing force is greatest during the short pause in the intercuspal position.

Discussion CHAIRMAN: D. J. ANDERSON

HANNAM: We are currently involved in measuring jaw movements in three planes and recording the associated electromyographic activity of six muscles during functional chewing movements. Briefly, our technique consists of recording displacement by means of a transducer which senses the field strength of a small magnet cemented to the labial surface of the lower incisor teeth (Kinesiograph, Myotronics Research Inc., Seattle, Washington), recording muscle activity by means of surface electrodes placed over the right and left anterior temporal, posterior temporal and masseter muscles, and using a digital computer to sample all the data recorded during right-sided gum chewing, left-sided gum chewing, and open–closing–clenching manoeuvres.

Our displacement data seems to fit very well with Professor Ahlgren's observations, but I have really mentioned the study because I would like to illustrate some of our problems, already familiar to Professor Ahlgren and several other members of this symposium, associated with managing the data that one accrues during experiments of this kind. Even though we have a 16K disc-based minicomputer to help us, we are still faced with handling 9 channels of input data, each of which must be sampled every millisecond so as to give us adequate resolution. The well-known variability of chewing strokes and muscle patterns from chew to chew means that we must sample at least 30 strokes for each sequence to establish a representative pattern of behaviour. A total input of this size places severe constraints upon any sampling system. As a result we have been obliged to limit our sampling time to include only the closing strokes, and to curtail each sampling period to one second. A diagrammatic example of this arrangement is shown in *Fig.* 15.10. Even these limitations mean that a complete run on a single subject represents 810K words of stored data.

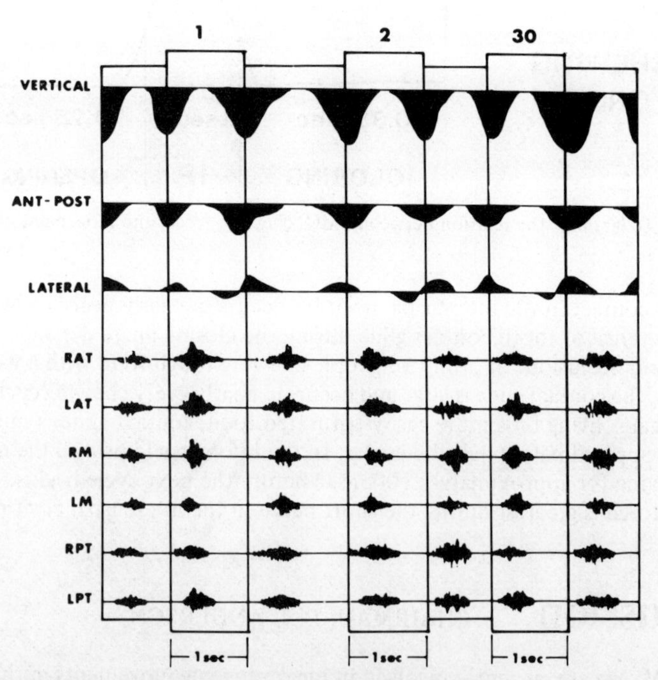

Fig. 15.10. Diagrammatic representation of the signals recorded during an experimental run. The top 3 traces represent displacement data, from above-down vertical movement, anteroposterior movement and lateral movement of the incisor point. The remainder of the traces represent electromyographic activity recorded from the right anterior temporal (RAT), left anterior temporal (LAT), right masseter (RM), left masseter (LM), right posterior temporal (RPT), and left posterior temporal (LPT) muscles respectively. The areas enclosed by the vertical boxes represent the periods during which data is sampled by the computer, 30 periods for each sequence of gum-chewing.

Moreover, our problem does not disappear once we have managed to cope with data acquisition. We still have the task of further data reduction to permit permanent, long-term storage, and then of statistical management, which has to be carried out so as to give us manageable figures without masking or losing important information. Professor Møller mentioned to me a couple of years ago that he had begun to believe that the expression of muscle activity in the form of a series of variances might almost be more useful than attempting to express it as a series of mean values, and we have to agree with him. At present, having assembled detailed measurements of displacement and muscle contraction for individual chews, some examples of which are presented in *Figs.* 15.11 and 15.12, we are expressing the performance of a subject as a series of variances and distributions for each of the many parameters measurable.

Distributions of labelled data are calculated according to set bin widths of, for example, 50 msec for time measurements or 2 mm for displacement measurements, and in this way we are attempting to compare performances, in a statistical sense, for the same individual, and between individuals. It would be inappropriate to

EVENT	msc
TRIG TO MAX OPEN	201
TRIG TO MAX LAT	232
TRIG TO MAX VERT	367
MAX OPEN TO CLOSE	281
DURATION CLOSE	134

POSITION IN mm	1 MAX OP	2 ⅔ OP	3 ⅓ OP	4 CLOSED	5 MAX LAT	6 5 (mm)	7 MIN OP	8 ½ OPEN
VERT	-18·5	-10·5	-3·6	0	-16·0	-5·0	-3·5	-12·6
LAT	2·7	0·9	0·3	0	2·8	0·5	0·7	2·5
ANT POST	-8·1	-18	-0·6	0	-5·8	-0·9	-0·3	-4·9

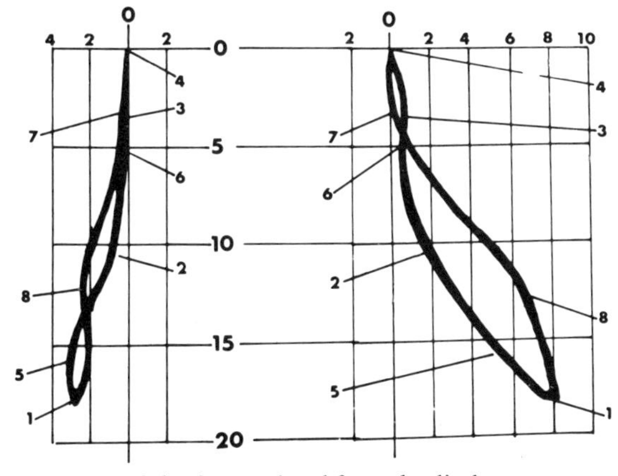

Fig. 15.11. Typical example of the data retrieved from the displacement records by the computer for each single chewing stroke. The lower left trace represents data recorded during a single right-sided chewing stroke, and displayed in the frontal plane. The lower right trace represents the same data displayed in the sagittal plane. All calibration figures are in mm. The centre table illustrates typical values sought by the computer from the lower traces, and can be referenced by means of the figures 1–8. The upper table illustrates examples of timing data obtained from the same chewing stroke. All chewing strokes are analysed in this manner.

0 1 2 3 4 5 600msc

SECTOR	1	2	3	4	5	6	7	8	9	10	11	12
RAT	0	3	6	10	16	46	68	61	27	13	0	0
LAT	0	0	3	4	24	47	33	33	12	0	0	0
RM	0	0	0	0	11	32	150	87	80	15	0	0
LM	0	0	0	0	0	6	18	24	10	4	0	0
RPT	0	0	7	8	13	18	31	23	7	0	0	0
LPT	0	0	0	0	4	10	23	22	24	9	0	0

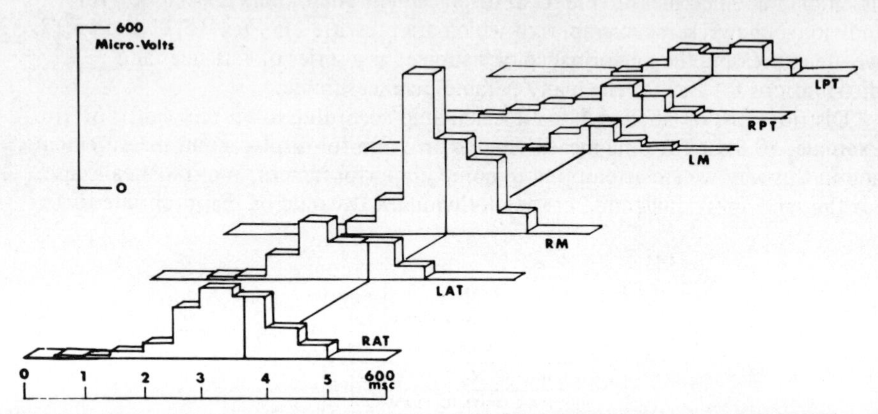

Fig. 15.12. Example of the way in which electromyographic data for each chewing stroke is collected. The lower figures represent the data for all six muscles (labelled as in *Fig.* 15.10) expressed in each case as the mean activity during each consecutive 50 msec time interval. In this example there are 12 such sectors, and the actual values derived by the computer for each sector are shown in the upper table. All chewing strokes are analysed in this manner.

describe the details of these experiments any further at the present time. It has simply been my intention to point out some of the factors which tend to impede the recording and reliable assessment of variables like mandibular movement and associated muscle contraction during normal function.

GREENFIELD: Would you like to speculate on the ideal occlusion and morphology of teeth for modern man with a modern diet.

AHLGREN: I do not think I can answer that question. Do you mean an ideal occlusion from an orthodontic or prosthodontic point of view ?

GREENFIELD: I was really wondering whether it is necessary to know how teeth come into occlusion when, I would think, they so rarely contact during mastication with a modern diet.

AHLGREN: Modern man has a contact glide of about 1 mm. I myself have one chopping side and one grinding side. It is practically impossible to chop tough steak; there must be some grinding action.

PICTON: You showed contact apparently in centric in each cycle. I take it that your examples were from a particular part of each chewing sequence. How did you choose these traces ?

AHLGREN: They were taken in the latter part of a chewing sequence. In the earlier part, contacts are rare but as the chewing sequence continues, there are more and more contacts.

PICTON: Where there was no contact, were there any other differences in the chewing cycle ?

AHLGREN: Not at all, only that the teeth did not come into occlusion.

Paper No. 16

Observations on Normal and Abnormal Temporomandibular Joint Function

G. Lindblom

The anatomy and function of the temporomandibular joint have been well described in detail, in many textbooks and papers. However, very little information can be found about the changes in the joints that occur during normal development or associated with abnormal function or pathological conditions.

In clinical prosthodontia, we dentists are daily concerned with the temporomandibular joint and its function. That there are normal and abnormal occlusions, and that the occlusal vertical dimension can and does influence the function of the lower jaw, are facts that can be attributed to the mechanics of this joint. In assessing the mechanics of the joint, we try to record the inclination of the condylar path indirectly from intra- or extra-oral records of the position of the mandible in protrusion. With the aid of these records we guide the movements of study models on an articulator to study the supposed condylar pathway.

In all clinical investigations—as distinguished from the theoretical ones—practical limitations necessitate compromise. The aim of this study was to measure the shape and size of the fossa, the height of the articular tubercle as well as the different positions and movements of the condyle. The study was based on my clinical material, collected during a period of 40 years, and which included more than 2000 radiographs of the temporomandibular joint.

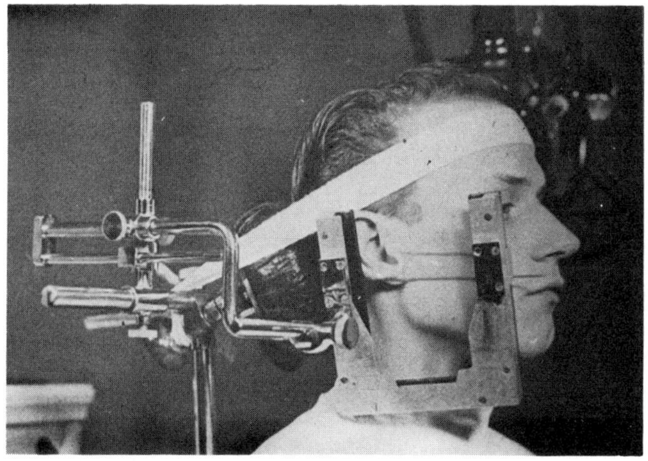

Fig. **16.1. Patient's head adjusted to the Camper line. (*From Lindblom, 1960.*)**

In each case about 30 variables were recorded during the examination. Many of these factors were later found to be of no value when statistically analysed, and therefore have been excluded. On the other hand, some new factors were identified and would have been valuable if only they had been included from the beginning. For example, information about right and left-handedness could have been of interest in the study of cranial asymmetry and, especially, temporomandibular joint asymmetry.

The material for this investigation consists of 318 cases requiring bite rehabilitation, some of whom had temporomandibular joint arthrosis, and 61 control cases. For each patient a complete bite analysis was performed using study models mounted on a Hanau articulator using my own technique. All 2000 radiographs of the temporomandibular joint were taken by the author using his own method and apparatus (351). With this technique the subject's head is kept vertical with the Camper line horizontal (*Figs.* 16.1 and 16.2). The X-ray projection is angled 15° to both frontal and the horizontal planes. The standard error for measurements made on duplicate exposures, with varying intervals between them, was 0·15 mm.

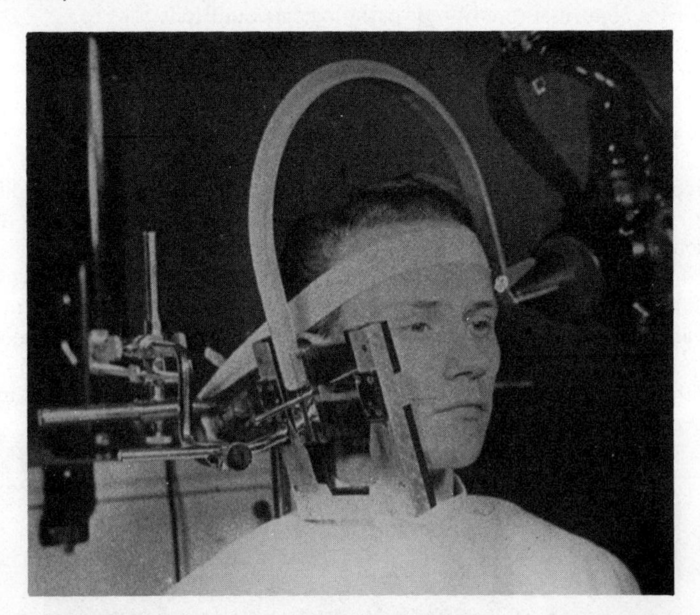

Fig. 16.2. Beam direction indicator. (*From Lindblom, 1960.*)

Examples of the pictures obtained by this method are shown in *Fig.* 16.3. They have been measured for evaluation of different fossal and condylar parameters: including depth and breadth of the fossa, height and angulation of the articular tubercle (*Figs.* 16.4 and 16.5), and condylar positions and displacements during different degrees of sagittal opening (*Figs.* 16.6 and 16.7). Furthermore, the effects of age, sex, the type of bite and bilateral asymmetry have also been investigated. The material has been subdivided according to side and the presence or absence of arthrosis.

Ninety-five of the patients (22 men and 73 women) suffered from temporomandibular joint arthrosis (arthrosis group) and a further 223 (100 men and 123

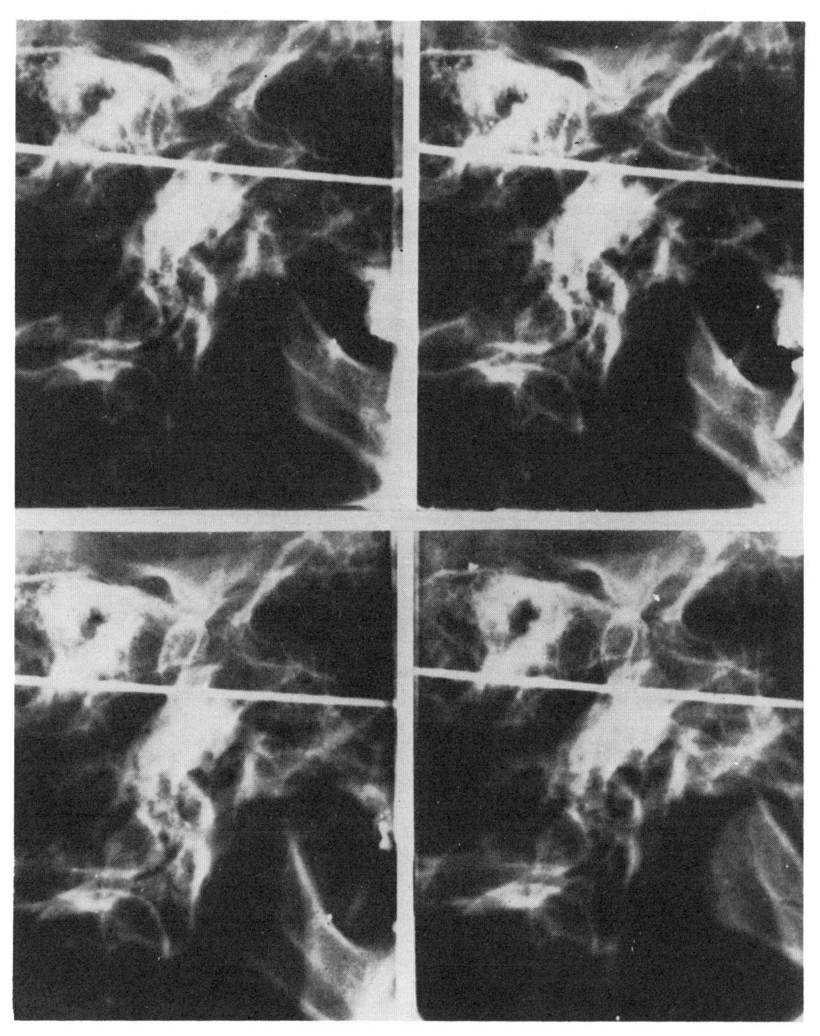

Fig. 16.3. Temporomandibular joint radiographs taken with the jaw in different positions. Upper left: Occlusion; Upper right: Rest position: Lower left: Open (2 cm): Lower right: Protrusion (edge-to-edge). (*From Lindblom, 1960.)*

women) required bite rehabilitation for other reasons (rehabilitation group). Thus, in the arthrosis group, women were in the majority in a ratio of about 3:1. This is in accordance with the findings of other authors. The joint disturbance may be localized on either side or both. For unknown reasons, pure left-sided arthrosis seems about twice as common as right-sided. Is there perhaps a strain on the left side in right-handed persons ?

A classification of the temporomandibular joint ought to be made according to the size of the fossa, e.g. its depth and breadth and the height of the articular tubercle, and not according to its shape (shallow, normal and deep), as has been most common up to now. The larger the fossa, the greater its depth, the more circular its contour and the steeper and more S-shaped its anterior wall (*Fig.* 16.8).

The radiograph gives a valuable guide to and confirmation of the articulator registration. This is particularly important with big fossae which seem to have thicker and more wedge-shaped discs, which will influence the inclination of the condylar path in protrusion. Thus in these cases the functional condylar path will always be less steep than the anatomy of the fossa would indicate. This is a most important fact, unfortunately often overlooked. In this connection, however, the vertical and horizontal overbite also play a very important part in determining the path of a protrusive movement of the mandible with the incisor teeth in contact. During a free opening movement, on the other hand, the neuromuscular influences predominate. A suitable formula for the size of a fossa could be the product of its depth and breadth. Left joints generally had higher values than right joints in

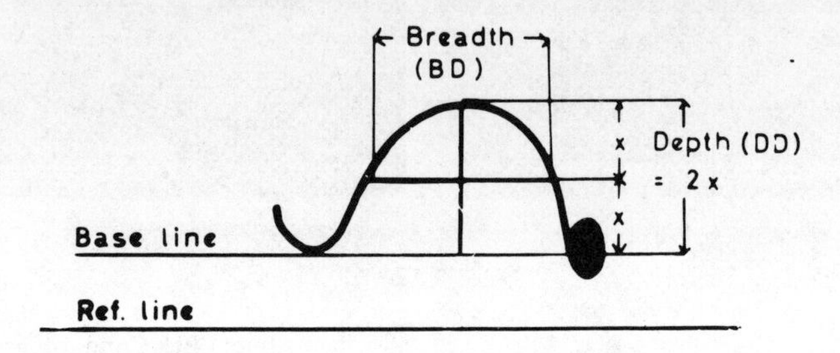

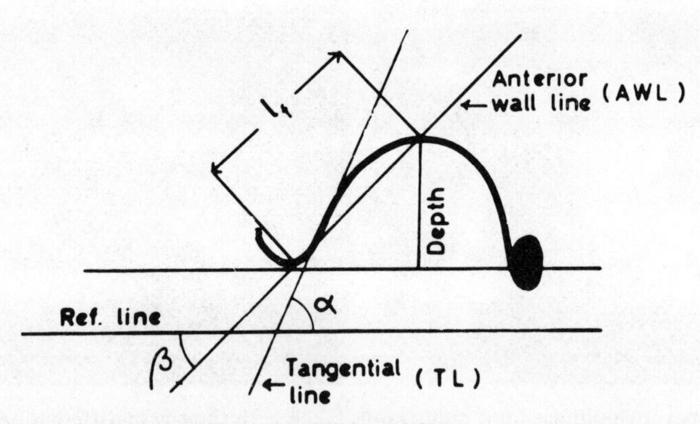

Fig. 16.4. Fossal measurements. (*From Lindblom, 1960.*)

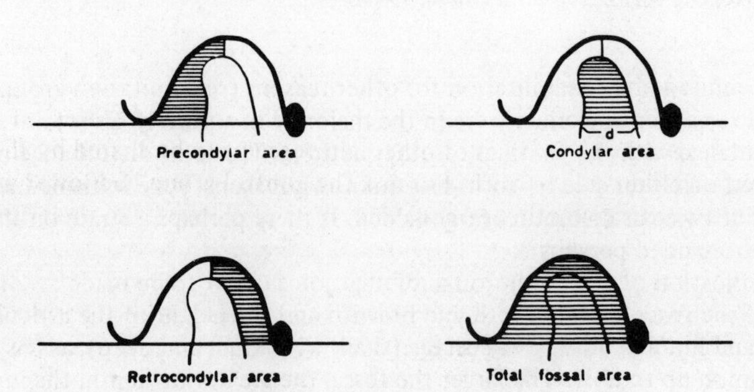

Fig. 16.5. Planimetric measurements. (*From Lindblom, 1960.*)

the control group as well as in the rehabilitation group. However, it is remarkable that this bilateral asymmetry could not be found in the arthrosis group and perhaps therefore the absence of bilateral asymmetry could in some way be a predisposing factor for the development of an arthrosis.

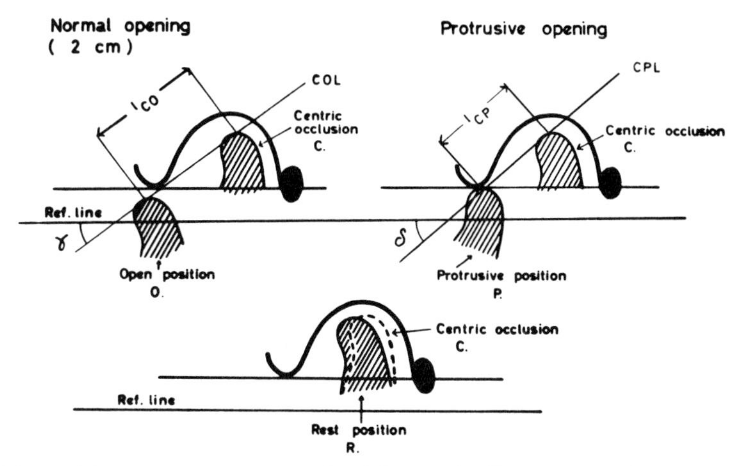

Fig. 16.6. Condylar measurements. COL: Condylar opening line; CPL: Condylar protrusive line; l_{CO}, Length of condylar opening path; l_{CP}, length of condylar protrusive path. (*From Lindblom, 1960.*)

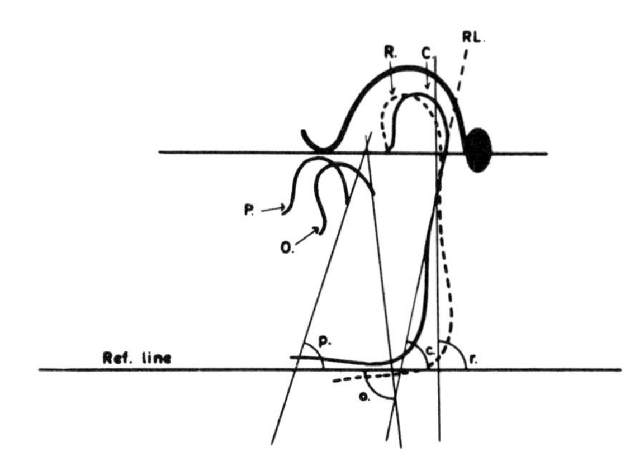

Fig. 16.7. Condylomandibular rotation axis in Centric occlusion (C), Rest position (R), Open position (O) and Protrusion (P). RL, Ramus line. (*From Lindblom, 1960.*)

The fossa appears to become somewhat shallower during the process of ageing (from about 40 years onwards) and this tendency seems to be shared by the control, rehabilitation and arthrosis groups. The control group had lower values for joint size than the two other groups. The size and shape of the fossa did not seem to be correlated with the type of bite and a deep vertical overbite was not correlated with the depth of the fossa. Men seemed to present somewhat bigger fossae than women. The sizes of the fossa and the condyle are well correlated.

The condylar movement from occlusion to the rest position is a combined rotation and translation, i.e. there is no pure rotation. The rotation has been found to be about 2° and the translation, which was not measured directly, was found

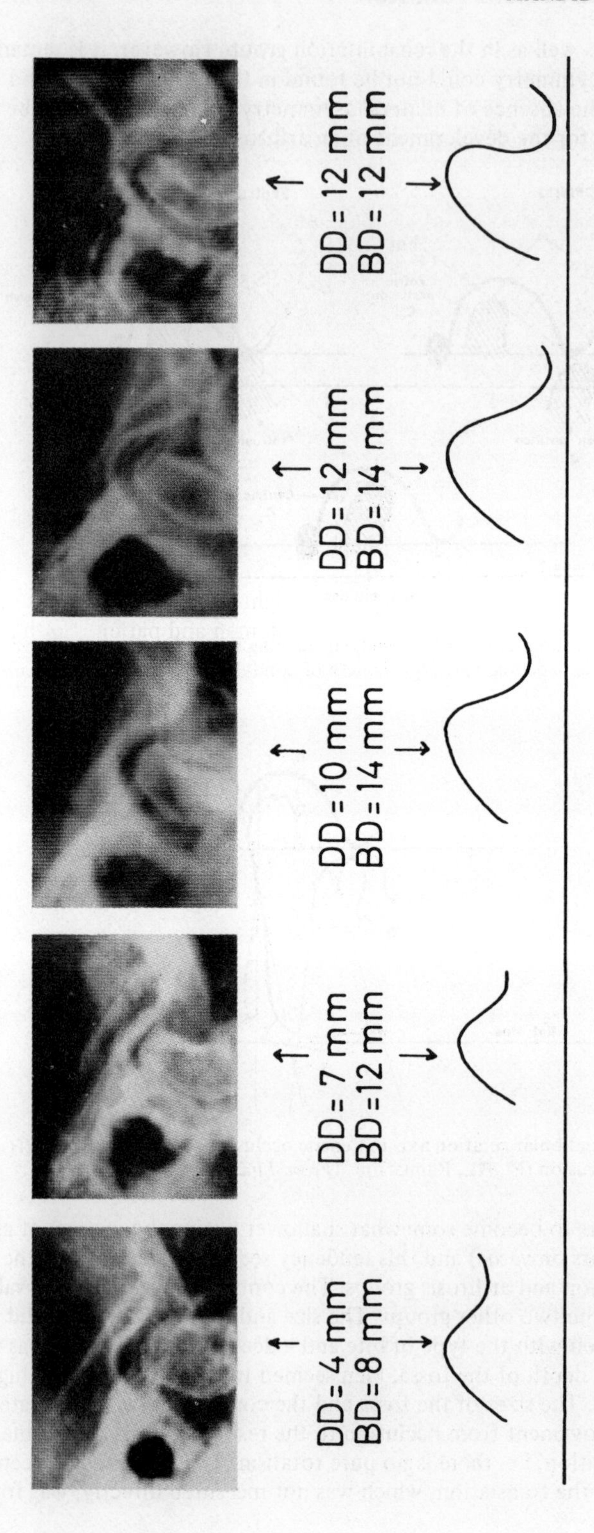

Fig. 16.8. Five different sizes of the TMJ with corresponding differences in shape, representing the minimum and maximum sizes from over 2000 pictures. DD, Depth diameter; BD, Breadth diameter (*see Fig.* 16.4). (*From Lindblom, 1960.*)

to take place mainly in the vertical direction. During sagittal movement from occlusion to 2 cm opening, a rotation takes place of about 10° and a translation of about 10 mm, along a path about 33° to the Camper line. The Camper line has been found to be more suitable than the Frankfort line as a reference when making measurements on the temporomandibular joint radiographs. A pure protrusive movement to the edge-to-edge position involves as a rotation of 0·7° and a translation and inclination of about the same magnitude as for free opening.

The condylar path determined on the articulator was at least 13° less steep than both the opening and protrusive paths measured on radiographs. This difference appears to be due partly to physiological factors and partly to deficiencies in the articulator and bite registration technique employed. No correlation seems to exist between condylar path inclination and rotation, nor between the articulator registration and the other parameters. On the other hand, as might be expected, a correlation is present between the length of the condylar path and the rotation. Normal and abnormal bites show no differences in the shape and size of the joint. An exception to this may possibly be found in postnormal bites with normal vertical overbite. These cases exhibit a tendency towards bigger fossae.

As to the very complicated problem of temporomandibular joint arthrosis, the following conclusions can be drawn from the present investigation: purely left-sided arthrosis seems to be about twice as common as right-sided. Arthrosis was observed to be about three times as common in women as in men and patients in the arthrosis group were somewhat younger than those in the rehabilitation group. The type of bite seemed not to be a predisposing factor in the development of an arthrosis. Left-sided arthrosis cases seemed to present somewhat bigger fossae than right-sided, and this might be explained by the fact that no bilateral asymmetry could be shown to exist in the arthrosis cases.

A longitudinal study has also been carried out on 85 patients with dysfunctional disturbances (arthrosis) and different kinds of discomfort from their temporomandibular joints (352).The duration of the study in each patient was from 2 to 30 years. A group of 68 patients included seven times as many women as men. In my two previous studies (351 and *Acta Odontol. Scand.* **11**, 1; 1953) the corresponding ratios were 3 : 1 and 4 : 1. The age range was 19–60 years for women and 20–60 years for men.

Treatment was completely successful in a high percentage of cases and in all 68 cases in which a direct dental origin could be shown, such as malpositioning of the lower jaw through faulty occlusion, i.e. where a deviation of the mid-line of the lower jaw occurred in habitual centric occlusion. Lateral displacement of the mandible was far more common than distal displacement.

After short-term observation (less than 2 years) of 68 treated cases, 59 had responded with complete disappearance of dysfunctional symptoms from their temporomandibular joints, while 9, although not completely cured, had definite improvement of their symptoms.

Sixty-one cases were observed for between 2 and 30 years, and of these, 59 showed a complete absence of any dysfunctional symptoms from their temporomandibular joints. One case had no improvement and there was one in whom some slight symptoms remained.

Discussion CHAIRMAN: D. J. ANDERSON

INGERVALL: In a group of children and young adults I have shown (659) that the morphology of the temporomandibular joint is related to the morphology of the face. You said that there was no relationship between the morphology of the joint and the occlusion, but there certainly is a relationship with the morphology of the face as a whole. *Figs.* 16.9 and 16.10 show one individual with a shallow fossa and one with a deep fossa. The subject with the deep fossa has a rectangular type of face, whereas the subject with a shallow fossa has a triangular face.

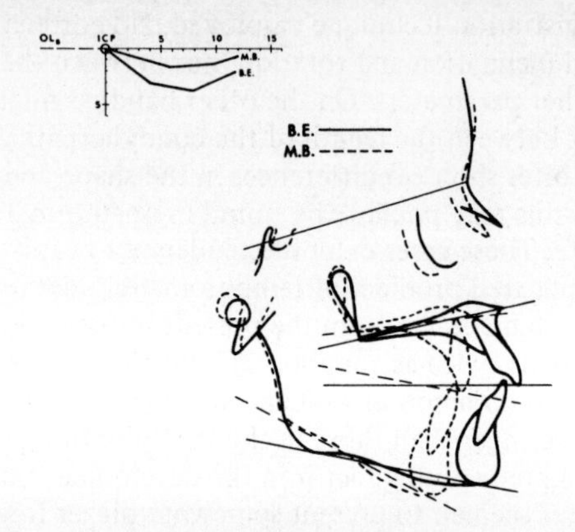

Fig. 16.9. **Facial morphology of two 8-year-old girls with large and small inclination of the condylar path and tubercular height. (*From Ingervall, 1974.*)**

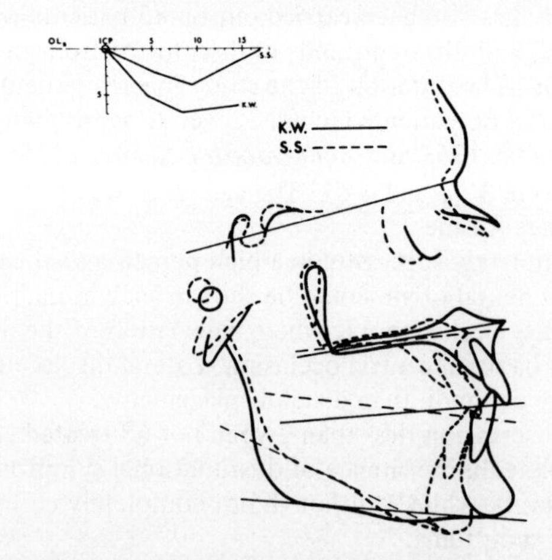

Fig. 16.10. **Facial morphology of two 23-year-old men with large and small inclination of the condylar path and tubercular height. (*From Ingervall, 1974.*)**

LINDBLOM: The question which has interested me is: what is guiding the mandible ? Is it only the joint or only the occlusion ? Both, of course. I have tried to take my X-rays in both a relaxed position and a partly opened position, in which the teeth do not exert any influence. As soon as any teeth come in contact there is a battle between the effects of the joint and the teeth and their intercuspidation.

NEURAL INPUTS TO THE MOTONEURONS OF THE MUSCLES OF MASTICATION

Paper No. 17

Reflexes Elicitable from the Jaw Muscles in Man

B. Matthews

The reflexes which can be elicited from the jaw muscles provide information about the connections which exist between afferent and efferent nerves involved in the control of the muscles, and something of the time course of the effects produced by stimulating different groups of afferent nerves. They provide very little direct information, however, on the mechanisms of control of the muscles in normal function.

The reflexes I shall describe are illustrated with a series of records which, with the exception of *Fig.* 17.2, were all obtained from the same subject (Dr Yemm) in a single session during which the recording electrodes were not moved and the amplifications and the time scales used were kept constant. In each record, five successive responses were superimposed when the photograph was taken. These records are reproduced from a more extensive review (388).

JAW JERK

The jaw jerk is elicited by tapping on the chin. *Fig.* 17.1 shows EMGs recorded from the skin over the masseter and anterior digastric muscles when this was done. The records were synchronized to the stimulus by tapping the chin with a hammer which gave a signal at the moment of contact and this signal was used to trigger the oscilloscope sweep.

The principal response to a sudden downward movement of the mandible is a reflex contraction of the masseter muscle which has a latency of about 8 msec (*Fig.* 17.1A). With the subject relaxed, there is no evidence of any response in the digastric muscle; the slight disturbance in the base-line in *Fig.* 17.1A is due to movement of the skin. A similar reflex can be demonstrated in the temporalis muscle, but no observations seem to have been made on the medial pterygoid.

If the jaw jerk is repeated while the subject contracts his elevator muscles, the response of the masseter is greatly increased in amplitude (*Fig.* 17.1B). Furthermore, after this initial response, the muscle does not return immediately to its previous level of activity, but passes through a period of decreased activity (the so-called 'silent period') during which the muscle appears to be totally inactive. With records that extend over a longer period it is sometimes possible to see evidence of further cycles of increase and decrease in amplitude of the signal before it stabilizes to the level present before stimulation.

The record in *Fig.* 17.1C was obtained with the subject contracting his depressor muscles, with a thumb placed under his chin to prevent his mouth opening. The digastric activity present prior to stimulation was reduced in amplitude after a

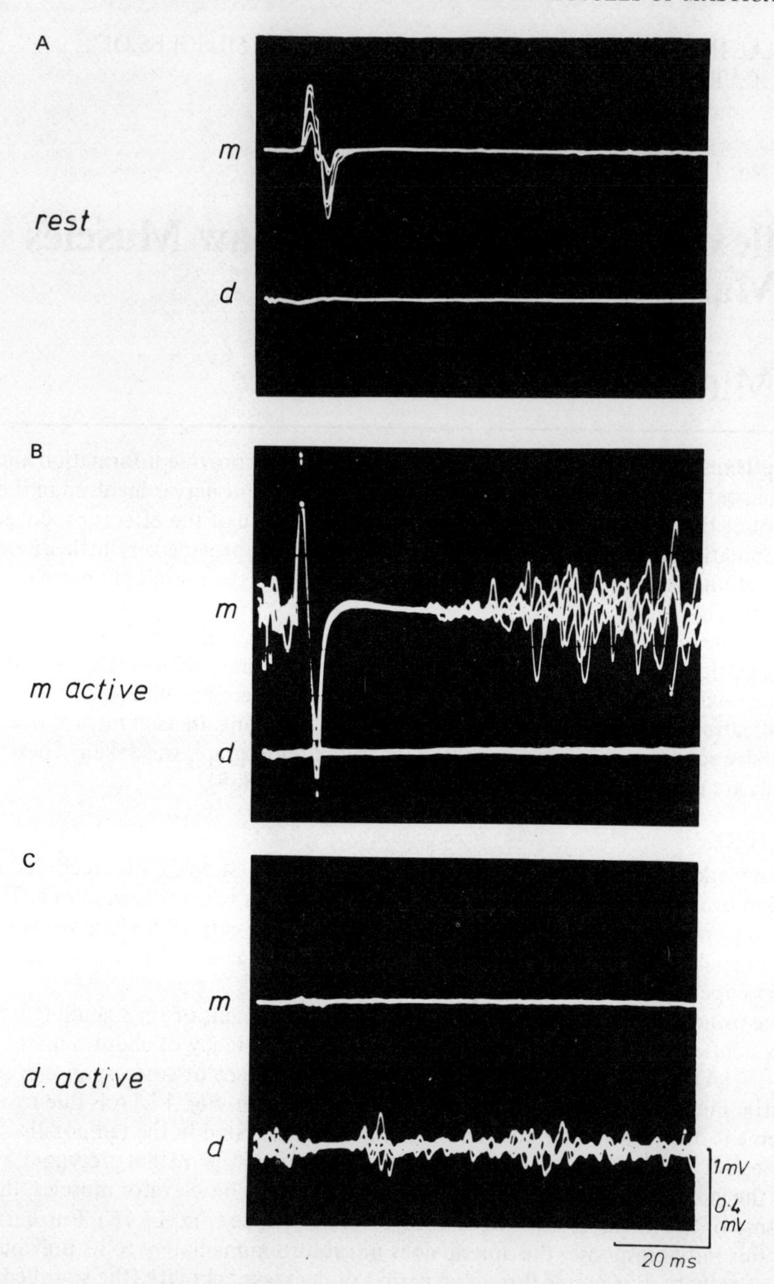

Fig. 17.1. Jaw jerk. EMGs recorded with surface electrodes from the right masseter (*m*) and the anterior belly of the right digastric (*d*) muscles. The jaw jerk was elicited by tapping on the subject's chin with a special patella hammer which incorporated a switch to indicate the moment at which the stimulus was applied. The signal from the hammer was used to trigger the oscilloscope beam so that each sweep began, on the left, at the time the chin was tapped. The jaw jerk was elicited while the subject relaxed, bit on a rubber bung between his right premolars, and contracted his depressor muscles against the resistance provided by a thumb under his chin. (*From Matthews, B. 1975.*)

latency of 12 msec for a period lasting about 10 msec. Also, the response of the masseter was smaller than that produced when the muscle was relaxed (*Fig.* 17.1A).

In common with tendon jerks in the limbs, the jaw jerk is often reinforced during voluntary contraction of other muscle groups (238).

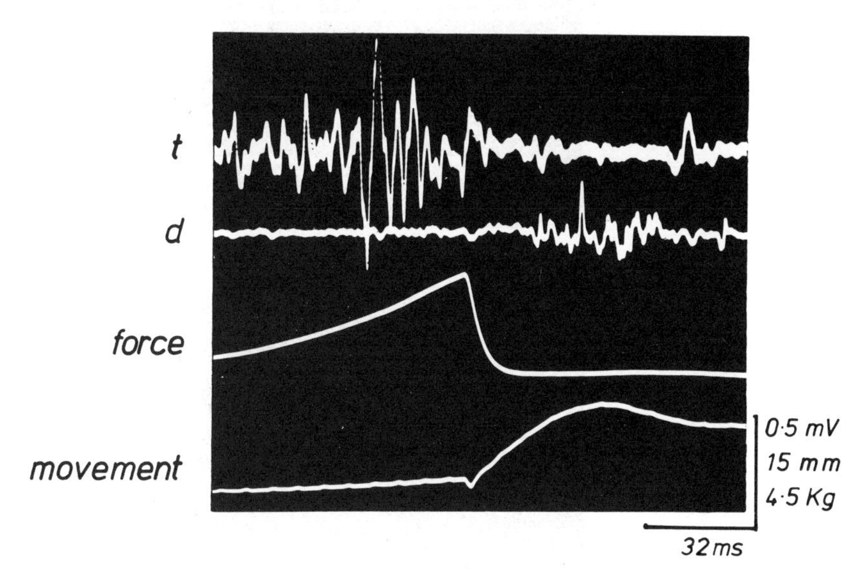

Fig. 17.2. Unloading reflex. The upper two traces are EMGs recorded with surface electrodes from the right temporalis and the anterior belly of the right digastric muscle while the subject bit on a piece of brittle plastic and broke it. The piece of plastic was incorporated into a device which was placed between his right premolar teeth and which provided the record of the biting force. The movement of the mandible was also recorded. The oscilloscope beam was triggered automatically when the force reached about 50 g. The initial opening was 23 mm measured at the premolars, and during the record the minimum opening was 11 mm. The small downward deflexion of the movement record at the time of fracture was due to vibration of the movement recording system. (*From Hannam, Matthews and Yemm (1968), Arch. Oral Biol. 13, 362.*)

THE UNLOADING REFLEX

There is no detectable response in either the depressor or elevator muscles of a relaxed subject if the chin is tapped from underneath, nor is there any response if the stimulus is applied while the subject contracts his depressor muscles against resistance.

However, a sudden closing movement during contraction of the elevator muscles produces an unloading reflex (241).To obtain the records shown in *Fig.* 17.2, the subject bit hard on a piece of brittle plastic tube which eventually broke. The piece of tube was held in a device which gave a record of the biting force, and a separate instrument was used to record jaw movement. During the first part of the record, the masseter activity and biting force increased and, as the tube bent under the strain, the jaws closed together a little. The moment at which the fracture occurred is indicated by the sudden closing movement and the decrease in force. After a latency of about 20 msec there was an almost simultaneous decrease in masseter activity and increase in digastric activity. The closing movement was arrested before tooth contact occurred.

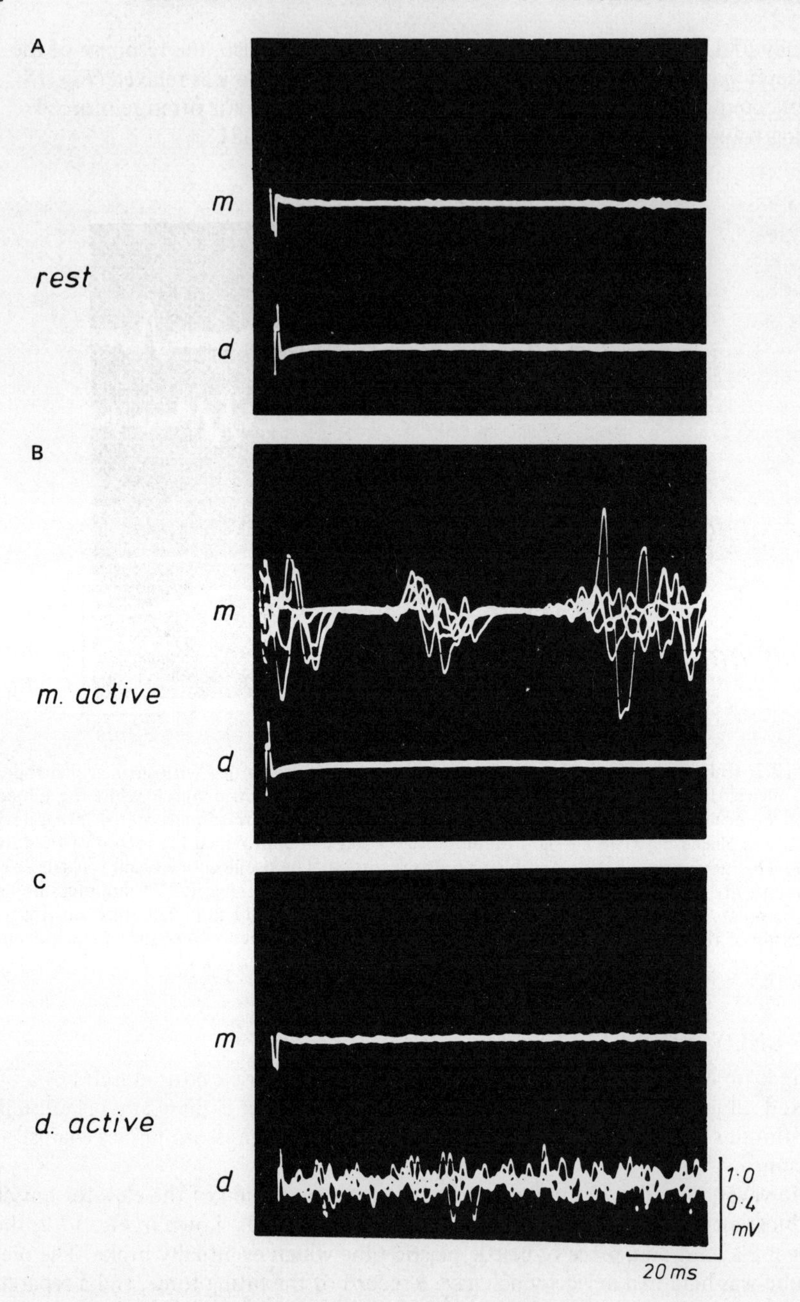

Fig. 17.3. Effects of electrical stimulation. Electrical stimuli of an intensity that caused pain (15 V, 1 msec duration) were applied twice a second with bipolar electrodes to the palatal mucous membrane just behind the incisors. The electrical stimuli were applied while the subject relaxed, bit on a rubber bung between his right premolars, and contracted his depressor muscles against the resistance provided by a thumb under his chin. The recording conditions were the same as for *Fig.* 17.1. (*From Matthews, B., 1975.*)

RESPONSES TO ELECTRICAL STIMULATION OF MUCOUS MEMBRANE

Electrical stimulation of the mucous membrane while the elevator and depressor muscles are relaxed produces no recordable response in the jaw muscles, even when the stimulus causes considerable pain. In *Fig.* 17.3A stimuli were applied to the incisive papilla and the stimulus intensity was as high as the subject could tolerate. A similar result is obtained with stimulation of the mucous membrane over the root of a tooth (199). When the stimuli are applied during voluntary contraction of the elevator muscles, they cause a reduction in the background muscle activity. In *Fig.* 17.3B there are two periods of reduced activity in the masseter EMG, with latencies of approximately 13 and 45 msec, which are separated by a brief return of activity. Similar responses in the elevator muscles have been reported following electrical stimulation of several different intra-oral sites (79, 272, 639, 640, 643).

Even when the subject contracts his depressor muscles, there is no evidence of a response in the digastric muscle (*Fig.* 17.3C). This is in contrast to the situation in experimental animals where a similar stimulus produces a brisk digastric contraction.

Goldberg (199) has shown that, prior to the inhibitory effects in the masseter muscle, electrical stimulation of the mucous membrane over the root of an upper incisor produces a transitory activation of the masseter with a latency of about 7 msec. No such excitatory effect has been reported with other sites of stimulation and none is apparent in *Fig.* 17.3.

RESPONSES TO MECHANICAL STIMULATION OF TEETH

In many respects, mechanical stimulation of teeth causes similar effects to those produced by electrical stimuli. When the muscles are relaxed there is no detectable response in either the elevator or depressor muscles if a single upper tooth is tapped (*Fig.* 17.4A), even when the intensity of the stimulus is as hard as one can reasonably apply without risk of damaging the tooth. If the elevator muscles are contracted when the stimulus is applied, a reduction in elevator muscle activity is produced, but there is no response in the digastric (*Fig.* 17.4B). The initial reduction in amplitude of the masseter EMG is followed by an abrupt return of activity and a second, less pronounced phase of depression. With a smaller stimulus than that in *Fig.* 17.4, the most obvious effect is the initial decrease in activity.

It has been shown that the silent period in the masseter muscle is often preceded by a brief increase in activity (199, 243, 518). This is sometimes difficult to detect in the raw data (*Fig.* 17.4B) but can be found reliably using averaging techniques.

EFFECTS OF TOOTH CONTACT

When the jaws are snapped together and the upper and lower teeth come into sudden contact, the reflex changes which follow are very similar to those produced by mechanical stimulation of a single upper tooth. There is a transient activation of the elevator muscles which is followed by a silent period and, often, later phases of increased and decreased activity (*Fig.* 17.5). There are no significant effects on depressor muscle activity. The silent period in the elevator muscle EMG produced by tapping the teeth together in this way has been described often (84, 242, 419), and there is also evidence that it is present after tooth contacts during mastication (5, 6, 17, 242). The origins of these effects have not been established with certainty. They are, no doubt, partly due to periodontal mechanoreceptors but receptors in the muscles and joints could be involved too.

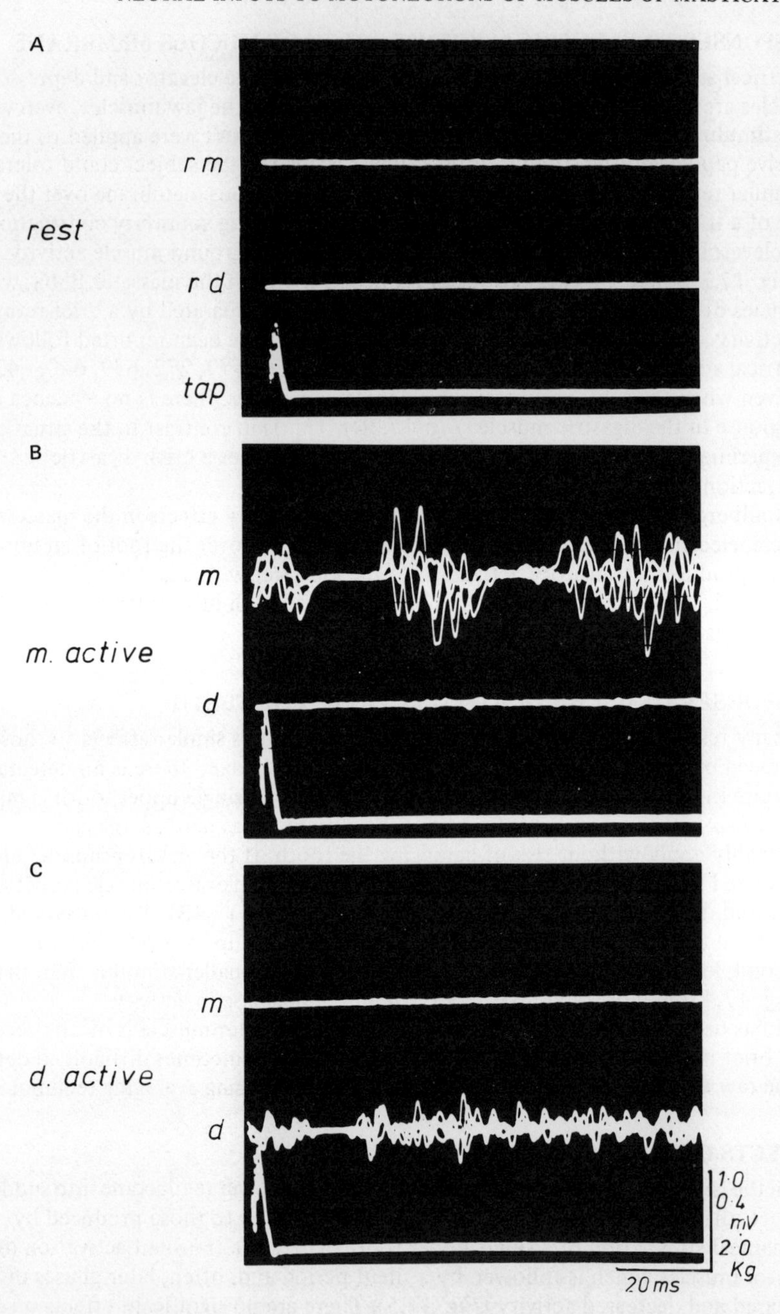

Fig. 17.4. Effects of tapping a tooth. The subject's upper right central incisor was tapped on its labial surface with a small metal bar which incorporated strain gauges. The oscilloscope was triggered from the strain gauge record at the beginning of each stimulus. The stimuli were applied while the subject relaxed, bit on a rubber bung between his right premolar teeth and contracted his depressor muscles against the resistance offered by a thumb under his chin. The recording conditions were the same as for *Fig.* 17.1. (*From Matthews, B., 1975.*)

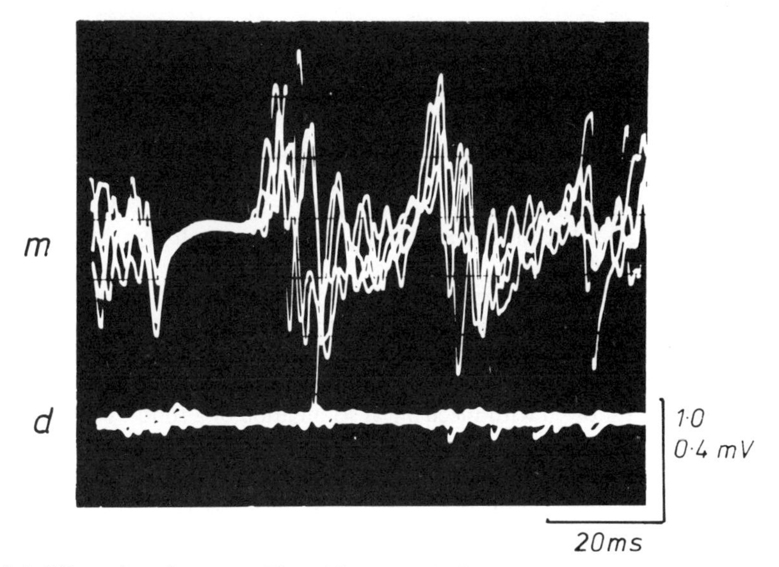

Fig. 17.5. Effect of tooth contact. The subject was asked to tap his teeth together and maintain them in a clenched position. This he repeated about once a second. The oscilloscope was triggered from a sensitive microphone taped to his cheek. The recording conditions were the same as for *Fig.* 17.1. *(From Matthews, B., 1975.)*

Discussion CHAIRMAN: D. J. ANDERSON

STOREY: I am surprised. Dr Yemm has shown that anxiety tends to raise the background activity in these muscles, but this does not seem to apply in him !
MATTHEWS: I suppose it reflects the confidence he had in the experimental technique !
SESSLE: One of the mechanisms we have postulated for the double silent period is that the two phases are due to two different groups of afferents. Another possibility is that the double silent period (in fact there are sometimes more than two), may be related to synchronization of motoneuron discharge, as you have suggested. We showed some evidence that the gamma-motoneurons tend to be synchronized with peripheral stimuli. The possibility of long loop 'reflexes' involving higher centres also cannot be discounted.
MØLLER: In *Fig.* 17.1B, what was the intensity of the on-going masseter activity with respect to maximal ? The response could be a sign of coactivation of alpha- and gamma-motoneurons ?
MATTHEWS: The contraction was submaximal; but the response does not prove the existence of coactivation.
MØLLER: Hagbarth has found that during very strong voluntary contractions, fusimotor innervation seems to stop. I wonder if you have made any observations with stronger degrees of contraction ?
MATTHEWS: No, we have not. I think to make any detailed investigation of this we would have to have some monitor of jaw movement. One would have to know that the stimulus one was applying was having a consistent effect. As you know, that is very difficult to do; but we have plans to use the kind of instrument which Dr Dibdin has described.

MØLLER: Regarding the inhibition you get during unloading. Can you offer an explanation for this in terms of defacilitation, as has been suggested (646).

MATTHEWS: The simple explanation would be that shortening of the elevator muscles causes the Ia drive to the motoneurons to switch off. Whether or not that is the correct and only explanation in this instance remains to be seen. At the same time as the muscles are shortening there is unloading of the periodontium and movement in the temporomandibular joint as well. I think it is fairly well established that the corresponding response in limb muscles is due to a decrease in spindle discharge.

SUMINO: You said there is an inhibitory period in the digastric muscle following electrical or mechanical stimulation. What are the latencies ?

MATTHEWS: About 12 msec.

BRATZLAVSKY: I was interested in the transient suppression of the digastric activity which you showed with mechanical stimulation of a tooth. I think it has also been reported by Beaudreau et al. (648). There can be the same response on painful stimulation of skin during active jaw opening.

MATTHEWS: Yes. Our experiment was done with the depressor muscles contracting isometrically.

SESSLE: If one records from cat digastric motoneurons that are excited by a peripheral stimulus and then introduces another peripheral stimulus one can actually see a prolonged inhibitory period.

MATTHEWS: The dominant response in the digastric muscle to peripheral stimulation in a decerebrate or lightly anaesthetized cat is activation.

SESSLE: But after the activation there is a period of inhibition, as we showed in our paper.

SCHEIKHOLESLAM: It may be that the first part of the silent period after a jaw jerk depends on inhibition from Renshaw cells. Have you any evidence on Renshaw cells in the trigeminal motor nucleus ?

MATTHEWS: In limb muscles, the corresponding silent period has been attributed variously to Renshaw inhibition, to inhibition from tendon organs and other mechanisms. Two things which we can add to that with respect to this response are that there are no collaterals in the trigeminal motonucleus (360) and, until Dr Lund's recent evidence at least, there is very little evidence for Golgi tendon organs in the mandibular elevator muscles.

Paper No. 18

The Connections between Muscle Afferents and the Motoneurons of the Muscles of Mastication

M. Bratzlavsky

The jaw-closing muscles (masseter and temporalis) in man and in different experimental animals are richly supplied with muscle spindles (42, 176, 299, 598). On the basis of anatomical and electrophysiological studies in the cat, it has been established that the primary afferents of jaw-closing muscle spindles run in the portio minor of the trigeminal nerve, join the trigeminal mesencephalic nucleus and project monosynaptically into the trigeminal motor nucleus. First Ramon y Cajal (469) noticed the predominant unipolar character of trigeminal mesencephalic nucleus cells. Johnston (294) showed evidence of their sensory function. This led several authors to compare the trigeminal mesencephalic nucleus cells with spinal ganglion cells and to point out their possible function as first order afferent neurons (13, 110, 569). Weinberg (620) reviewed the problem and suggested that the afferents in question have a proprioceptive origin and travel in the trigeminal motor root to reach the trigeminal mesencephalic nucleus. Corbin (120) fully confirmed this view by showing degeneration of fibres in the trigeminal mandibular branch after inflicting destructive lesions to the mesencephalic nucleus. These afferent fibres were subsequently proved to originate in jaw-closing muscle spindles, both on electrophysiological and on morphological grounds. Corbin and Harrison (122) evoked proprioceptive impulses in the trigeminal mesencephalic root upon stretching jaw-closing muscles. Szentagothai (556) observed degeneration of jaw-closing muscle spindles after destruction of the trigeminal mesencephalic nucleus. Furthermore, Szentagothai (556) confirmed Cajal's (469) description of a direct projection from axons of the unipolar mesencephalic neurons to the trigeminal motoneurons.

Evidence has been presented that this arc conveys the stretch reflex of jaw-closing muscles (the jaw jerk). In fact, this reflex is abolished by trigeminal mesencephalic root lesions (248) but is preserved after severing the trigeminal sensory root (376). Its monosynaptic nature has been established electrophysiologically by calculation of its central relay time (278). A similar anatomical arrangement has been suggested to occur in goats (118), dogs (317), birds (384), rats (31) and frogs (412). In man, the jaw jerk is not affected by section of the trigeminal sensory root (377). This suggests that the central connections of the reflex are much the same as those described in experimental animals. The human jaw jerk has a latency of approximately 7·5 msec (337). Its central relay time is suggestive of a monosynaptic pathway (377). Therefore, the jaw jerk can be used as a tool for testing the excitability of jaw-closing motoneurons. By this method, evidence has been obtained which suggests that the short and long latency pauses evoked in human jaw-closing muscle activity by peri- and intra-oral exteroceptive stimulation (79, 640, 643) are due to postsynaptic reflex inhibition of jaw-closing motoneurons (80). Similar conclusions have been reached concerning the reflex pause set up in human jaw-closing muscles by periodontal stimulation (81).

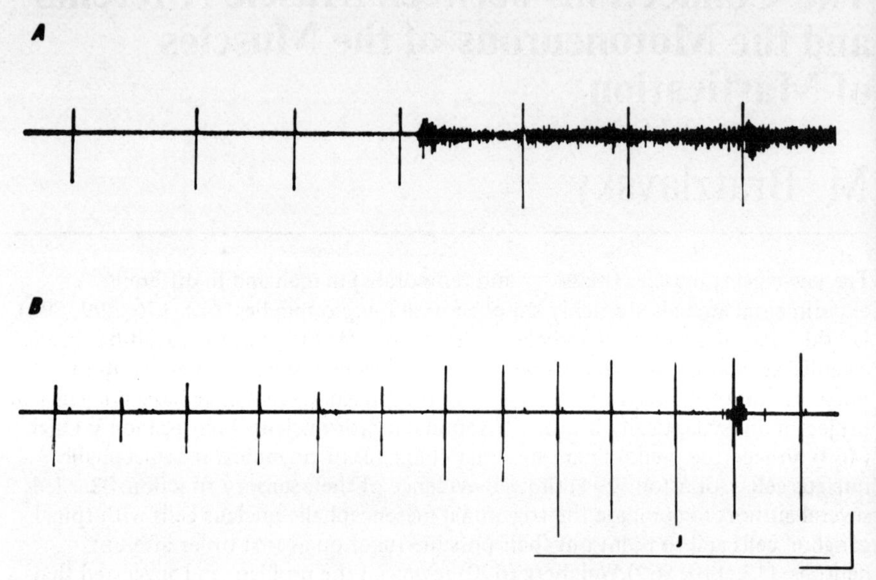

Fig. 18.1. Jaw jerk elicited in masseter muscle by downward taps of approximately 10 N (Newton) intensity, applied to the chin with an electrodynamic type force generator (mini-shaker Bruël & Kjaer, Type 4810). *A* Resting state followed by slight voluntary jaw closure; *B*. Resting state followed by Jendrassik manoeuvre (J). Calibration marks: 1 sec; 2 mV.

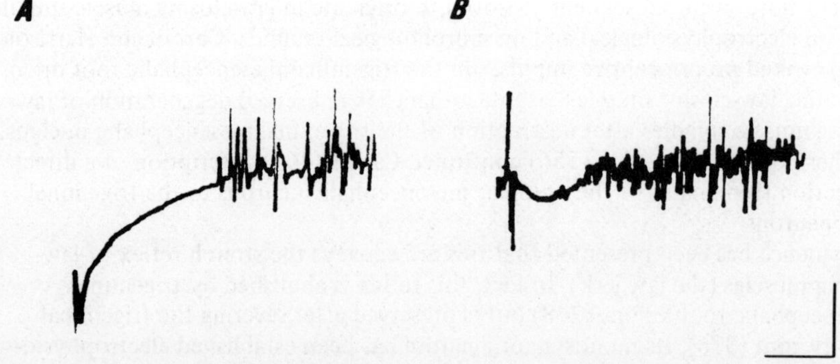

Fig. 18.2. Silent period in voluntary masseter activity following (*A*) electrical stimulation of masseter muscle, (*B*) jaw jerk. Calibration marks: 40 msec; 1 mV.

The human jaw jerk is potentiated during slight voluntary closure of the jaws and during the Jendrassik manoeuvre (*Fig.* 18.1). As in other skeletal muscles, the twitch contraction of jaw-closing muscles and the jaw jerk are followed by a silent period when occurring on a background of voluntary activity (*Fig.* 18.2).

Originally, it was suggested that this silent period resulted from a crossed autogenic inhibition arising from Golgi tendon organs (277). Later on, it appeared that this silent period could be explained on the sole basis of the pause in jaw-closing muscle spindle afferent discharges and that it was confined to the ipsilateral side (79). With the exception of Smith et al. (532), most researchers have been unable to elicit in cats clear contralateral effects at trigeminal mesencephalic or at trigeminal motor nucleus levels, upon stimulation of low threshold jaw muscle proprioceptors (122, 377, 427). A tonic vibration reflex is easily evoked in jaw-closing muscles of man by applying a downward vibratory stimulus (100–200 Hz) to the chin (*Fig.* 18.3*A*). Just as is the case in limb muscles (158, 180), the reflex is likely to arise mainly in spindle primary endings. Its development is progressive and it takes at least several seconds before the reflex is fully developed. In most instances, the jaw jerk exhibits a clear potentiation during the tonic vibration reflex (*Fig.* 18.3*B*). This finding strongly contrasts with observations made in limb muscles, where the phasic stretch reflexes are suppressed during the occurrence of the tonic vibration reflex (180, 234). Whereas the connections and the function of primary muscle spindle afferents of jaw elevators have been studied

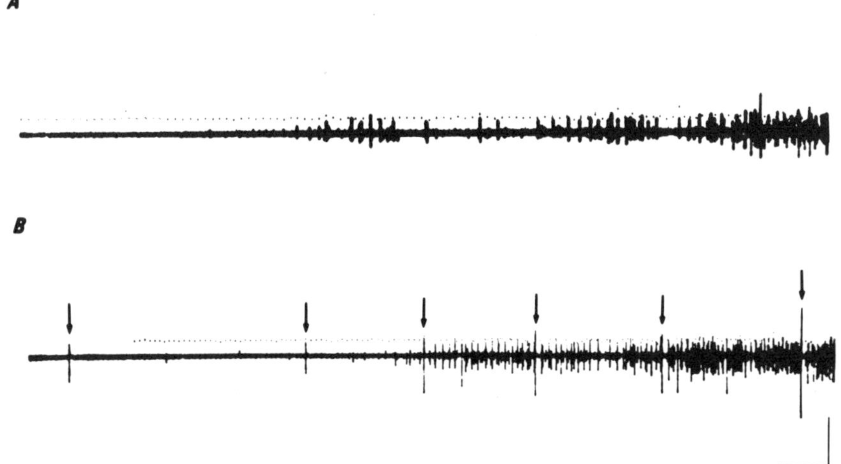

Fig. 18.3. (*A*) EMG activity evoked in masseter muscle by a vibratory stimulus applied to the chin (from onset of trace). An electromagnetic apparatus is used, with a vibrating surface of 1·5 cm diameter, beating at a frequency of 100 Hz with an amplitude of 1–2 mm; (*B*). Changes in amplitude of jaw jerk (↓) during tonic vibration reflex of masseter muscle. Calibration marks: 1 sec; 2 mV.

quite extensively, less data have been published concerning secondary spindle afferents of jaw-closing muscles. Nevertheless, the latter have been identified both by anatomical (299) and by electrophysiological (114) studies. Their central connections are still unknown.

Observations about Golgi tendon organs in jaw muscles are very scarce (291, 312, 556). In his degeneration studies, Szentagothai (556) showed that Golgi tendon organs in jaw elevators of cats are not affected by trigeminal mesencephalic lesions. Therefore, he suggested that their afferents travel centrally through the trigeminal sensory root. Although contested by some authors (529), several investigators confirmed Szentagothai's view through indirect electrophysiological observations (114, 289, 412). Nevertheless, until now, no tendon organ afferent impulses have been recorded from the trigeminal ganglion (44) and nothing is known about the possible central connections of such afferents. Moreover, with masseter nerve stimulation in cats, it has not been possible to elicit any IPSPs in jaw-closing motoneurons, which could be indicative of an input from Golgi tendon organ afferents (321).

In the cat, reflex movements of the tongue upon passive jaw opening have been reported (63, 508), which were attributed to the excitation of jaw-closing muscle spindles. Anatomical support of this assumption was provided by Szentagothai's (556) suggestion of a monosynaptic linkage of trigeminal mesencephalic neurons with hypoglossal motoneurons. However, subsequent investigations have been unable to confirm the assumption (411). Recent work in the cat (362) attributes to temporomandibular joint receptors the reflex influences of passive jaw opening on hypoglossal motoneurons. On the other hand, stimulation of the masseter nerve has been reported to induce postsynaptic inhibitory potentials in the cat's hypoglossal motoneurons (411), possibly mediated by Group II masseter afferents (551). The multisynaptic pathway which has been traced between masseter afferents and the hypoglossal nucleus (402) could be involved in this reflex. Furthermore, high threshold masseter afferents induce jaw opening in cats by postsynaptic excitation of jaw-opening motoneurons and reciprocal postsynaptic inhibition of jaw-closing motoneurons bilaterally (427). It has been suggested that some of the latter influences are conveyed by Group II masseter afferents of non-spindle origin (426). In man, no data are available concerning reflex influences of high threshold jaw-closing muscle afferents on hypoglossal and trigeminal motoneurons.

The digastric muscle seems to be devoid of muscle spindles in subprimates (42, 556) and in monkeys (146). In man, only a few muscle spindles have been described in the anterior belly of the digastric muscle (599). In other muscles which open the jaw, such as the lateral pterygoid, muscle spindles were found to be lacking in the monkey (146, 530) and present in man (185). Neither phasic nor tonic myotatic reflexes have been reported in jaw-opening muscles. The jaw-opening muscles can be regarded as having a flexor function, and this possibly accounts for the negative results. However, in cats the anterior digastric muscle could not be shown to have afferent connections with the mesencephalic tract (556). No data are available about any monosynaptic projections into jaw depressor motoneurons. Finally, until now no proprioceptive influences from jaw-closing muscles have been reported in digastric motoneurons and jaw-opening muscle afferents have not been reported to affect jaw-closing motoneurons (321).

In man and in Rhesus monkeys, the extrinsic and intrinsic tongue muscles are supplied with muscle spindles (72, 116, 604). In subprimates, the tongue muscles contain varieties of less organized sensory endings (345). Afferent impulses originating in lingual stretch receptors have been recorded in the hypoglossal nerve in Rhesus monkeys (73) and in cats (244, 409, 644). In man, several authors have shown that the tongue is not ataxic after bilateral anaesthesia of the lingual nerve and concluded that the hypoglossal nerve carries the tongue muscle spindle afferents (2, 618). In the Rhesus monkey these afferents leave the hypoglossal nerve via hypoglosso-cervical anastomoses and reach the medulla through C2 or C3 dorsal roots (74, 121). Thus, at least in a part of its course, the hypoglossal nerve is likely to function as an afferent pathway of lingual proprioception. However, the cell bodies of the primary afferents involved have not yet been identified. Furthermore, the existence of functional connections between tongue muscle afferents and hypoglossal motoneurons is uncertain. No anatomical data have been published concerning monosynaptic projections into hypoglossal motoneurons. Neither in cats (63), nor in man (82) has a phasic or a tonic stretch reflex been elicited in the tongue musculature. Nevertheless, some results have been reported suggesting a bilateral projection of Group Ia lingual spindle afferents into hypoglossal motoneurons in monkeys (74).

Discussion CHAIRMAN: B. MATTHEWS

FUJII: What was the amplitude of vibration you used on the jaw ?
BRATZLAVSKY: I used an electromagnetic apparatus vibrating at 100–200 Hz with an amplitude of 1–2 mm and with a vibrating surface of 1·5 cm.
CHAIRMAN: Did you monitor the movements of the jaw ?
BRATZLAVSKY: No, just the electromyographic activity.
CHAIRMAN: When you applied the standard tap to the chin, was that a standard tap in terms of movement ?
BRATZLAVSKY: No, in terms of force.
FUJII: I would like to present some experimental results of the tonic vibration reflex in jaw-closing muscles in man. In our investigation, we recorded electromyograms from the masseter and temporalis bilaterally, together with biting force and the vibratory stimuli. The latter were recorded as acceleration.

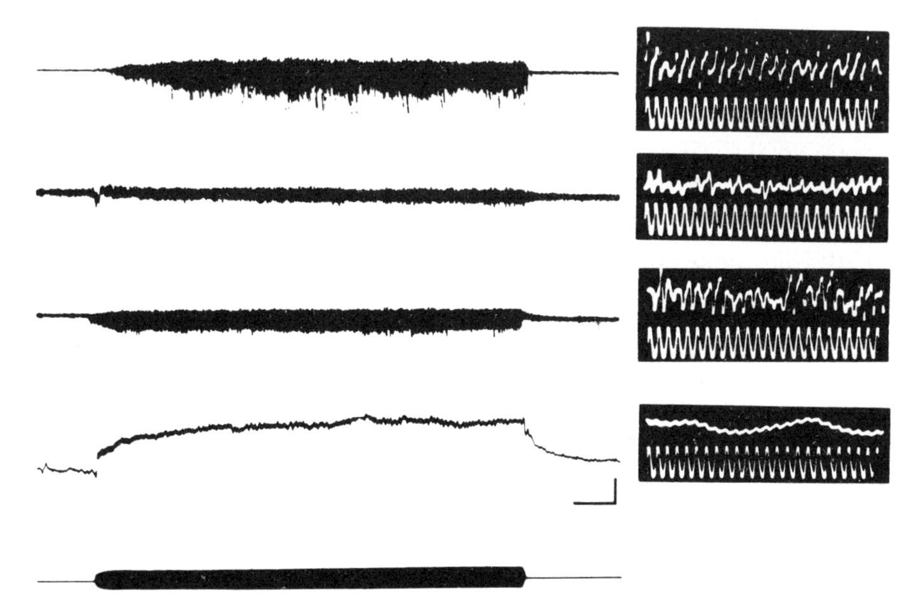

Fig. 18.4. The left-hand column of slower speed records, demonstrates changes in EMG and biting force before, during and after period of vibratory stimulation. The stimulus of 100 Hz and 21 dynes was applied unilaterally to the skin covering the mandibular angle at a position estimated to correspond with the insertion of the masseter muscle. Records from top to bottom: EMGs from the anterior portion of the temporalis on stimulated side, from the masseter on contralateral and on stimulated side, biting force and vibratory stimuli. In the right-hand column, high speed records, which were picked up from the corresponding left-hand record at random during the period of stimulation, are displayed as well as vibratory stimuli. During the experiment the subjects exerted a slight voluntary contraction upon the bite force recording apparatus. Calibration marks: horizontal: 2 sec; vertical: 400 μV in EMG or 2 kg in biting force.

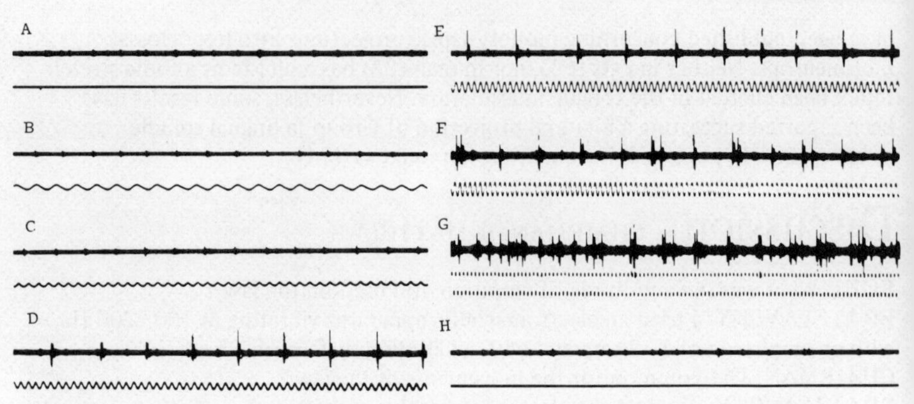

Fig. 18.5. Unitary EMG recorded from the masseter muscle under the same experimental conditions as those of *Fig.* 18.4: (A) before stimulation (B)–(G) during stimulation, (H) just after stimulation. Vibratory frequencies from (B)–(G): 31, 63, 84, 106, 147 and 170 Hz. Calibration mark: 200 μV.

Fig. 18.4 shows the bilateral effect of the vibratory stimuli. The intervals between each burst of electrical activity corresponded with an integral multiple of the interval between the stimuli. I believe that this is a clear confirmation of a tonic vibration reflex in jaw-closing muscles. *Fig.* 18.5 shows a unitary EMG, before, during and after application of vibratory stimuli. At low rates of stimulation the active units exhibited low amplitude action potentials. At higher rates of stimulation, other units with higher amplitude potentials were activated. It seems possible that the discharge of the latter is facilitated by the vibration (i.e. a tonic vibration reflex) and the lower amplitude units may be inhibited.

CHAIRMAN: What is the record under the EMG ?

FUJII: The output of an accelerometer attached to the vibrator, not to the jaw.

CHAIRMAN: So the movement of the jaw is not necessarily constant.

FUJII: I also wanted to say that under our experimental conditions the H reflex evoked by stimulation using surface electrodes above the mandibular notch was inhibited by vibratory stimulation.

THEXTON: You said that stimulation of low threshold afferents in the masseteric nerve would not produce effects in the digastric motoneurons. I wonder whether we might have got too much into the habit of thinking of muscle spindles projecting to motoneurons directly. It has been shown (660) that there are interneurons with muscle spindles projecting to them. I have reported (574) that stimulation of low threshold afferents in the masseter nerve produces enormous facilitation of temporalis motoneurons with a latency of about 25 msec, and a considerable depression in the excitability of digastric motoneurons with approximately the same latency.

BRATZLAVSKY: Recent work (672) on reciprocal innervation in limb muscles of man has shown that disynaptic Ia inhibitory pathway originating in flexor muscles is 'open' only during voluntary flexion movements. It is quite possible that the same thing happens in the masseter and digastric muscles, but there is no monosynaptic reflex for testing the excitability of digastric motoneurons.

KARLSSON: Would your data be compatible with the presence of relatively few spindles in the digastric muscle ?

BRATZLAVSKY: There have been no recordings from spindle afferents in the digastric muscle.

Paper No. 19

Electrotonic Coupling between Cells in the Mesencephalic Nucleus

C. F. L. Hinrichsen

The resolution of electron microscopy (14, 268, 270, 281) and micro-electrode recording techniques (31, 122, 267, 289) have done much to clarify many features of the central organization of the trigeminal mesencephalic nucleus. Ramon y Cajal (468, 469) studied this nucleus extensively in newborn material (*Fig.* 19.1) and noted synapses on the cells (*Fig.* 19.2) as well as dendrites extending from the soma. He considered that this nucleus supported his 'Law of Dynamic Polarization' and called it the 'Accessory Trigeminal Motor Nucleus'. Controversy arose when

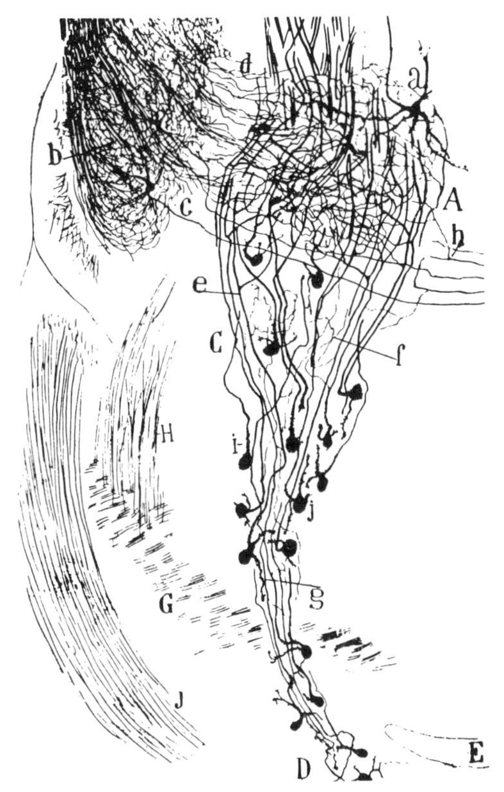

Fig. 19.1. Organization of the trigeminal mesencephalic and motor nuclei in mouse (Golgi method). (*From Ramon y Cajal, 1952.*)

the eminent neuro-anatomist Golgi (208) observed material from older animals, did not see synapses or dendrites on the cells and considered that this nucleus served to undermine Cajal's theory of basic neuronal function.

Prior to the work of Corbin and Harrison (122) which demonstrated the association of the mesencephalic nucleus with the jaw jerk reflex, attempts were made to deduce the function of the nucleus by comparing its cells (*Figs.* 19.15, 19.16) with those of spinal ganglia (110, 502, 507, 605). The nucleus was not thought to be homogeneous because of its dual embryological origin from neural crest and alar plate (235, 253, 330, 454) and its possible afferent innervation of extrinsic ocular muscles (175, 564, 606, 633).

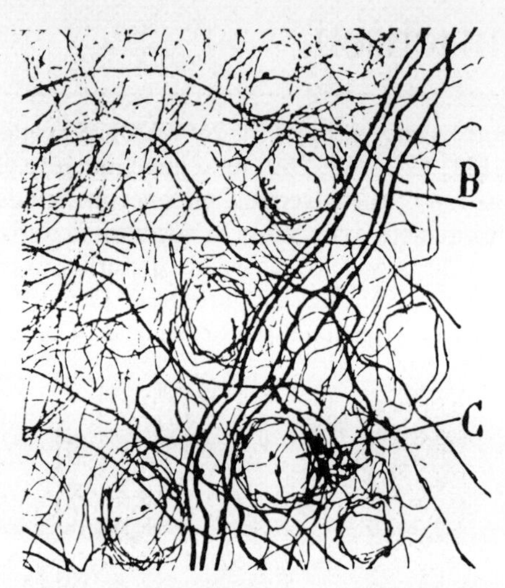

Fig. 19.2. Nerve plexus around cells of the inferior portion of the trigeminal mesencephalic nucleus. (*From Ramon y Cajal, 1952.*)

Cell clustering with soma–soma contacts in the mesencephalic nucleus (*Fig.* 19.3) has been observed in situ (23, 191, 620, 632) and in tissue culture (264) but its significance has not been established.

Recent investigations of the structural and functional organization of the trigeminal mesencephalic nucleus have therefore been directed towards three basic questions: (i) Are there synapses on these primary sensory neurons, a feature which does not appear in any other mammalian muscular afferent system, and, if so, what is their function ? (ii) Is the nucleus homogeneous or heterogeneous, and are there structural differences between cells ? (iii) What is the structural basis for cell clustering ? Is this a pathway for communication between cells and, if so, what is its role in the control of mastication ?

A survey of electron micrographs of the perimeter of mesencephalic nucleus cells in the mouse, rat, cat (268, 270) and hamster (14) has shown two major features which may work together in controlling the behaviour of the nucleus:

1. *Synapses on the Soma of some Cells (Figs. 19.4, 19.5 and 19.6)*

Sections show as many as five profiles of synapses per cell. This number appears to be small, but computation suggests that there may be up to 100 synapses on a cell, which could be quite significant. The cell surface is studded with many spinous

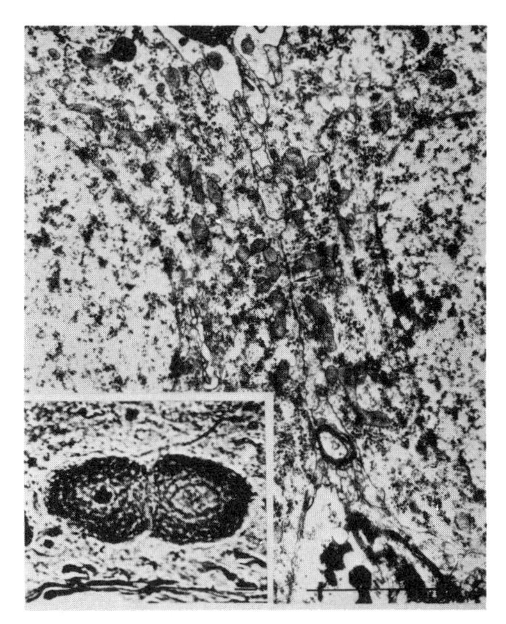

Fig. 19.3. Low-power electron micrograph showing cells in close apposition in the trigeminal mesencephalic nucleus in adult mouse (*inset:* light micrograph. Bar 5 μm).

processes (*Fig.* 19.7) but none of these supports synaptic contacts. A developmental study (14) indicates that their prime role is to increase the surface area of the cell for nutrient exchange.

Analysis of vesicles within the synapses (65, 342, 592) shows that small round

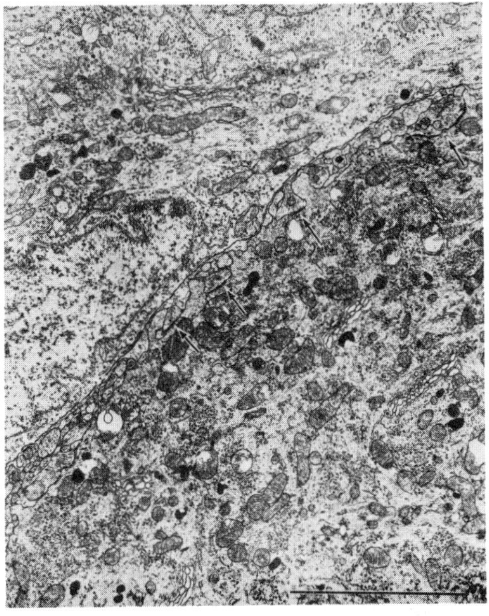

Fig. 19.4. Electron micrograph showing synapses on the soma of trigeminal mesencephalic cells. Bar 5 μm.

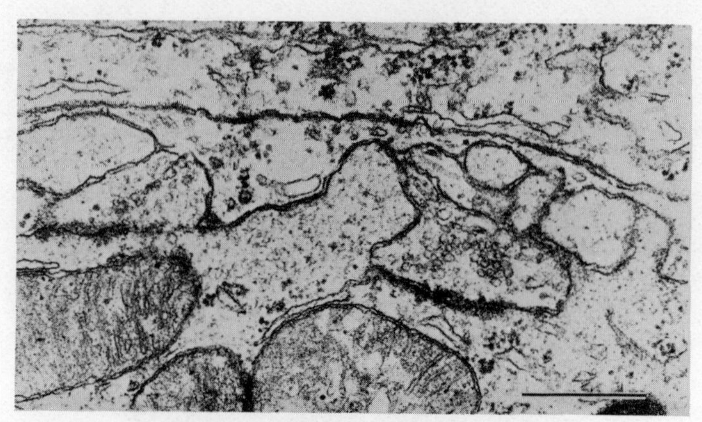

Fig. 19.5. Synapse on soma of mesencephalic cell. Bar 0·5 μm.

vesicles with clear cores, small elongated vesicles with clear cores and large vesicles with dense cores are present. Correlated studies in the cerebellum (153, 342) indicate that excitatory synapses contain round vesicles and inhibitory synapses, ovoid vesicles but it is not known whether this analysis can be extended to the mesencephalic nucleus. There is no correlation between a classification of synapses based on vesicle analysis and that based on synaptic gaps and pre- and postsynaptic densities (220).

The origin of the presynaptic neuron has not been established. I have found that small lesions in the rostral portion of the mesencephalic nucleus did not lead to a clear demonstration of degenerating terminals in the caudal part of the nucleus. This finding does not support Cajal's conclusion that the presynaptic neuron is in the rostral part of the nucleus.

Smith et al. (531, 532) have recorded bilateral activity in the mesencephalic nucleus and suggest that the nucleus could project contralaterally, either directly or through a synapse. Micro-electrode recordings have failed to show synaptic activity (31) and I have found that horseradish peroxidase injected into the masseter muscle or mesencephalic nucleus has not been taken up by the contralateral nucleus.

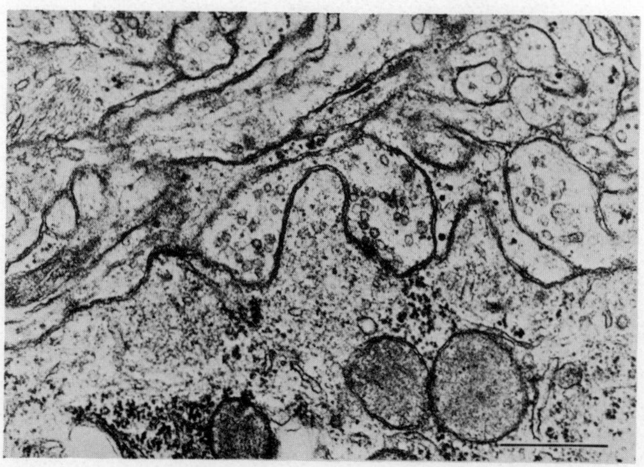

Fig. 19.6. Further example of synapses on soma of mesencephalic cell. Bar 0·5 μm.

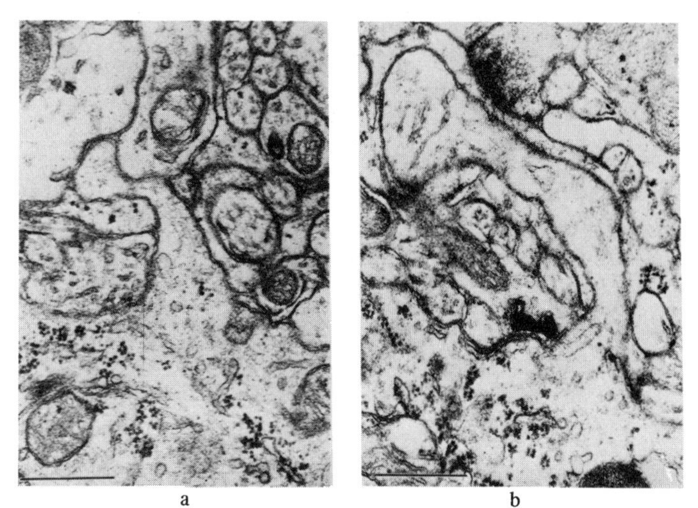

Fig. 19.7a, b. Spinous processes on the surface of trigeminal mesencephalic cells in mouse. Bar 0·5 μm. (*From Hinrichsen and Larramendi, 1970.*)

2. *Zones of Adhesion between Cells*

The membranes of cells associated in clusters show maculae similar in structure to a cross-section through a zonula adherens of epithelial cells (168) between their somas (*Fig.* 19.8) and between the soma and initial segment (*Fig.* 19.9). With the tissue processing methods used (306, 307), areas of the outer cell membrane leaflets were seen which appeared to merge to resemble zonulae occludentes (tight junctions) of epithelial cells. These comprised some of the junctions between clustering cells (*Fig.* 19.10). The mesencephalic nucleus has not yet been examined

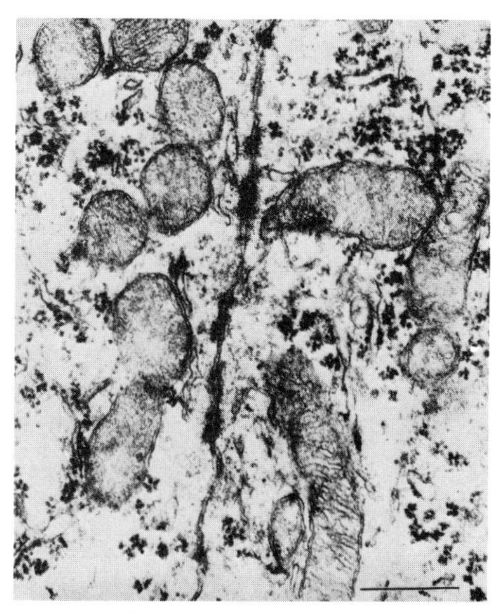

Fig. 19.8. Electron micrograph of the region of apposition between trigeminal mesencephalic cells showing maculae. Bar represents 1 μm.

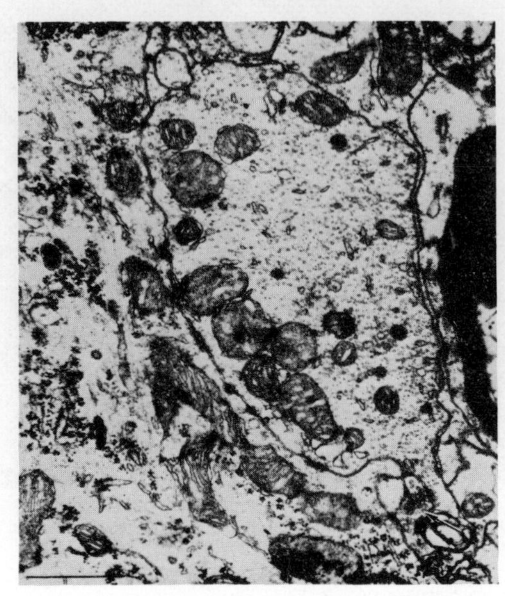

Fig. 19.9. Initial axonal segment cut in cross section where it contacts a trigeminal mesencephalic cell. Note close association of mitochondria with the zone of apposition which is studded with adhesion plaques. Bar represents 1 μm. (*From Hinrichsen and Larramendi, 1970.*)

with electron opaque tracers or freeze etching to establish whether these are tight junctions or gap junctions (87).

To test whether the close contacts between mesencephalic cells may act as electrically transmitting junctions, rats have been used in a study with micro-

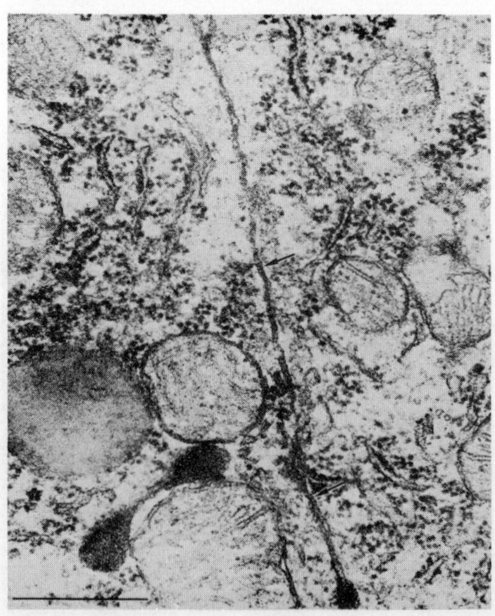

Fig. 19.10. Further example of the region of apposition between trigeminal mesencephalic cells showing a region of apparent membrane fusion. Bar represents 1 μm.

electrodes (31, 267). Electrotonic coupling would be established if, while recording from one cell, an action potential evoked by stimulation of the peripheral process of another cell could be recorded in the impaled cell. Also, the cell would be depolarized by the same procedure even if it had been made refractory by prior direct stimulation through the micro-electrode (*Fig.* 19.11). Intracellular recordings show that, in response to somatopetal stimulation, a small number (about 10 per cent) of cells show small, all-or-none, short latency depolarizations (*Fig.* 19.12). These are not seen in spinal ganglion cells (283, 284, 285, 495, 553). Baker and Llinás (31) have shown that the short latency potentials are not generated from the medullated axon (the M spike). Action potentials can be initiated from these short latency depolarizations. The latency for a particular cell is fixed and will depend on individual somatopetal conduction velocities and distances (*Fig.* 19.13). The latency for different cells was between 0 and 500 μsec.

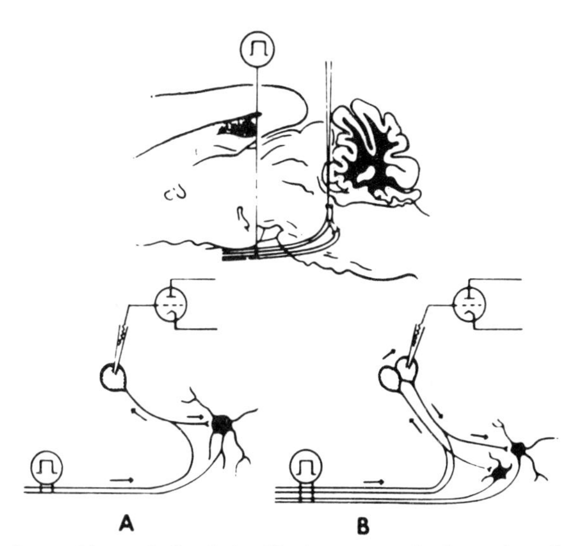

Fig. 19.11. Relative positions of stimulating bipolar concentric electrode and recording micro-electrode in rat brain stem. A. Intracellular recording from uncoupled mesencephalic cell. B. Intracellular recording from coupled mesencephalic cell. (*From Hinrichsen, 1970.*)

Direct activation of a coupled cell can be shown to block the somatopetal invasion of that cell but not the short latency depolarization arising from somatopetal invasion of a coupled neuron (*Fig.* 19.14). The electrotonic nature of the recorded short latency depolarizations has been differentiated from synaptic activity by their short latency, all-or-none nature and lack of alteration in magnitude with change in membrane potential.

To investigate whether coupled cells subserve more than one masticatory muscle I have made injections of horseradish peroxidase into the masseter muscle of newborn and adult rats. This technique relies on retrograde axonal transport of the protein to label the primary afferent and efferent neurons (95, 334, 344). These experiments show (*Figs.* 19.17, 19.18) that cells subserving masseter receptors occur singly in the mesencephalic nucleus, coupled together in clusters and coupled to unlabelled cells which subserve other peripheral sites. This work is supported by our recent studies in which the masseter nerve has been sectioned and primary afferent neurons were identified by retrograde changes in acid phosphatase content after 6 days survival (39, 40, 66, 309, 429, 534). Extracellular recordings (Linden, personal communication) from the mesencephalic nucleus have not shown units which respond to both muscle stretch and periodontal stimulation. It appears that coupled cells subserve only muscle receptors.

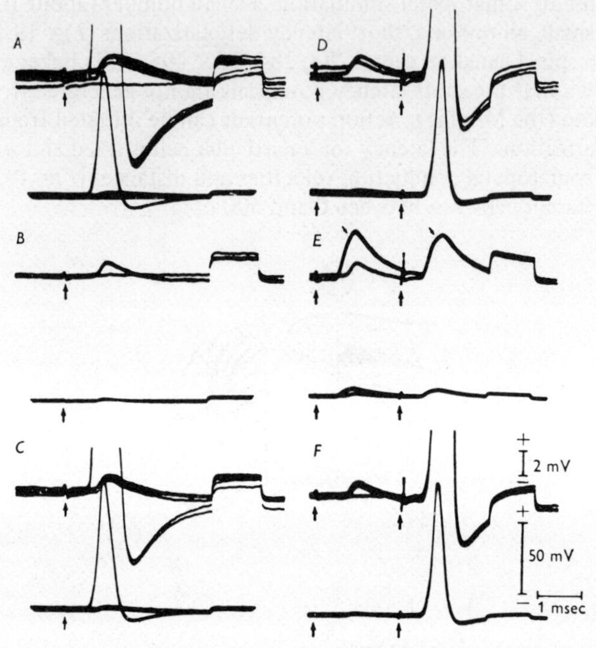

Fig. 19.12. Intracellular records from MSN neurons following threshold activation of trigeminal nerve (masseter branch) to show the short latency depolarizations (SLDs). The upper and lower traces in *A–F* were simultaneously taken with high gain RC-coupled and low gain DC-coupled amplifiers respectively. *A*, the stimulus (arrow) activates the neuron somatopetally. When the stimulus fails to generate a full action potential, a small all-or-none depolarization can be observed with a slightly longer latency. In *B* the stimulus is reduced in order to show the all-or-none SLD by itself. In *C* as in *A*, the stimulus is brought back to threshold level for somatopetal invasion of the cell. *D–F*, double trigeminal stimulation. The first stimulus is subthreshold and the second supra-threshold for the somatopetal invasion. In *D* and *F* the stimulus evokes a SLD in an all-or-none fashion. In *E* the cell is artificially hyperpolarized by 20 mV (the lower record ran off the screen, so the DC level had to be momentarily readjusted). Under this condition, the somapetal invasion of the soma and initial segment of the axon was blocked and only the action potential generated at the medullated axon (the M-spike) could be observed. Note that following the first stimulus in *E* a failure of the M-spike demonstrates the presence of a SLD, which can also be observed as a slight indentation at the peak of the M-spike (downward arrows). In *F* the membrane potential had returned to its resting value. (*From Baker and Llinás, 1971.*)

Functional Significance

One advantage of the monopolar ganglion cell over the bipolar ganglion cell is that the cell body does not lie in the sensory pathway and therefore is not a potential source of conduction block (283). It is difficult to ascribe any role to the ganglion soma cell in sensory conduction (284), although nodes on the monopolar axon do appear to favour conduction up to the cell body. It has been suggested that action potentials in ganglion cells are necessary for triggering metabolic activity to supply energy to its axon (283).

The presence of synapses on the soma of mesencephalic cells opens several possibilities not available to other proprioceptive systems. If the synapses are capable of initiating an impulse, they may either contribute to the central input and/or control the sensory input in a manner analogous to the efferent cochlear bundle in the auditory system (181, 473, 474, 528). In the jaw jerk reflex, sensory input could be modified from the cell body as well as at the presynaptic terminals.

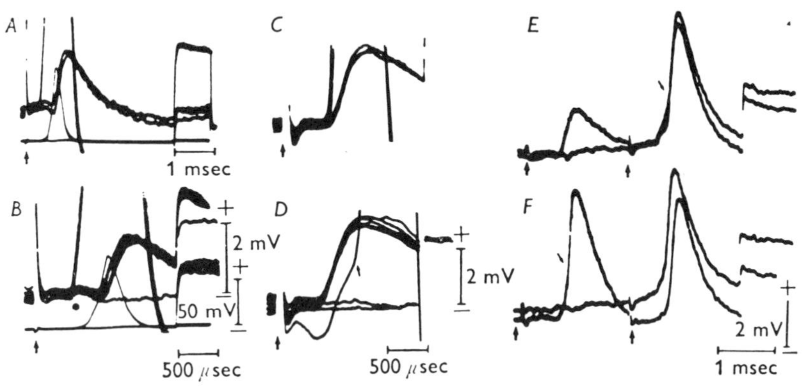

Fig. 19.13. Intracellular records from MSN neurons to show details of the SLDs. *A* and *B:* threshold separation of somatopetal invasion from SLD. Upper and lower traces, high and low gain respectively. *B* is recorded at twice the sweep-speed as *A* and shows the all-or-none character of the SLD. *C* and *D:* the SLD recorded from another neuron. The latency of the SLD was very close to that of the action potential (*C*). In *D* a subthreshold stimulus for somatopetal invasion and the all-or-none property of the SLD can be seen. In one case a full spike (arrow) is generated from this depolarization. *E* and *F*, SLD is shown to have two distinct components. In *E* the first stimulus is just threshold for a small potential, in *F* (first stimulus) a second all-or-none component is added to the first (downgoing arrow). In the second superimposed trace both failed. The second stimulus in both *E* and *F* is suprathreshold for the SLD but subthreshold for somatopetal invasion. (*From Baker and Llinás, 1971.*)

Although the junctions between mesencephalic cells have not been subjected to electron opaque tracers to determine whether they are in fact occluding (tight) or gap junctions, a growing body of evidence suggests that gap junctions rather than occluding junctions are the site of ionic and metabolic coupling. Recent studies using low molecular weight electron opaque tracers (46, 87, 147, 178, 293, 480), freeze fracture preparations (104, 188, 189, 209, 210) and iontophoretic introduction of dyes into coupled cells (49, 293) have done much to modify earlier concepts of the nature of cell membrane adhesion. Some epithelial receptor cells joined by occluding junctions are not electrotonically coupled and hence retain their spatial sensitivity. There appear to be no other examples of electrotonic coupling between cells where occluding junctions have been found to the exclusion of gap junctions. The function of gap junctions appears to be trophic. They are capable of permitting the transfer of substances up to the size of nucleotides but not nucleic acids or proteins (457) but this interdependence disappears in some developing systems.

In excitable tissues, varying degrees of coupling have been observed in electromotor, motor, pacemaker and relay neurons requiring close synchronization (47, 51, 52, 53, 54, 198, 328, 329, 333, 355, 430, 536). Coupling would only be functionally significant in the mesencephalic nucleus if impulses were propagated in an orthodromic or somatofugal direction. This could be possible only if the synapses on mesencephalic cells are functional or electrotonic junctions serve to recruit other cells and amplify the system.

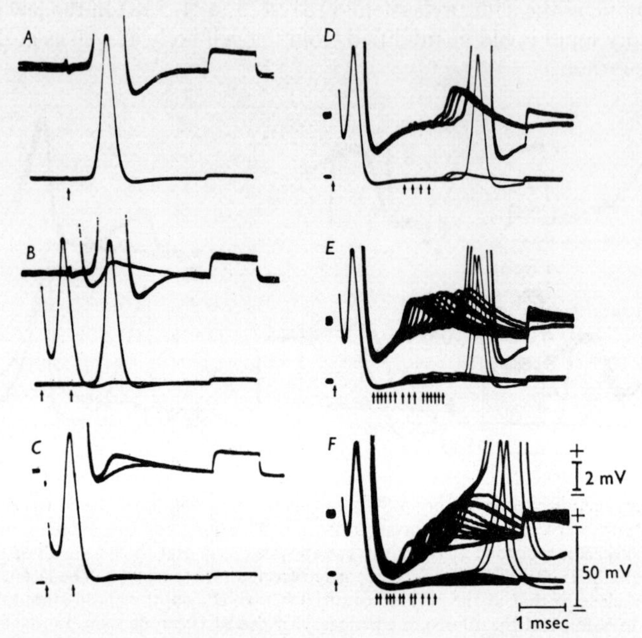

Fig. 19.14. Collision of somatopetal and somatofugal activation of MSN neurons. *A*, somatopetal invasion. In *B* a single direct activation of the impaled cell (first arrow) produces a block of the somatopetal invasion but does not block the SLD. The second sweep was taken without direct activation to show the latency of the SLD with respect to that of the somatopetal action potential. In *C* the all-or-none nature of the SLD is shown by reducing slightly the strength of the trigeminal stimulation. The somatopetal invasion was blocked by collision. *D–F*, examples of collision blockage of somatopetal invasion by somatofugal conduction following direct stimulation. Note that the SLD does not block or show refractoriness as the trigeminal stimuli are moved closer in time to the direct activation (arrows). (*From Baker and Llinás, 1971.*)

Wherever electrotonic coupling has been demonstrated between nerve cells, synapses are present on the cells involved (48). Since the synapses on mesencephalic cells may, from vesicle analysis, be both excitatory and inhibitory, they may serve to alter the degree of coupling between cells. This can be done not only by altering their degree of polarization but, as has been shown in one invertebrate, synaptic activity can cause uncoupling of cells (50, 537). The absence of satellite cells from mesencephalic cells may be an advantage as they tend to intrude on uncoupled junctions (24).

Discussion CHAIRMAN: B. MATTHEWS

GOLDBERG: Do you have any evidence using horseradish peroxidase on whether or not there is any coupling between the cell bodies of muscle and intra-oral afferents ?

HINRICHSEN: No, I do not. I think that has got to be done by recording and I understand that it is being done here in Bristol.
LINDEN: I have recorded from about 300 units that responded to mechanical stimulation of teeth and I have not seen any that also responded to opening of the mouth, in other words, that were coupled to spindle afferents.
TAYLOR: We have been interested in this problem, too. In trying to make some sense of this curious anatomical situation, we recently tried to determine first whether the periodontal cells are indeed first order, and secondly what is their distribution within the nucleus. We have looked at the first question by injecting horseradish peroxidase into the periodontal ligaments of upper and lower teeth and 24–48 hours later we found this well localized in the cells of the mesencephalic nucleus. Furthermore, the localization has confirmed our previous experiments in which we could not record from dental mechanoreceptor cells in the rostral part of the nucleus. These cells were entirely in the caudal half of the nucleus and their proportion of the total number of cells increased more or less linearly from the central part of the nucleus caudally. In the most caudal part, the dental cells constituted between 40–60 per cent of the whole. Also, we confirmed that where there are, side by side, both dental cells and muscle spindle cells, they do not seem to have any obvious interaction with each other. We have looked for evidence of synchronization between the firing of these cells, but so far we have not been able to find any.
CHAIRMAN: Would there be any effects which could be attributed to coupling if the cell was depolarized without producing an action potential ?
TAYLOR: Yes. In the rear part of the nucleus, the central processes of the muscle spindle afferents are very close to the motor nucleus (only a very few millimetres away) and it is just conceivable that the dental cells at that point can depolarize the spindle terminals and that the depolarization spreads the rather short distance into the terminals on the nearby motoneurons. In this way, they could exert a very fast form of presynaptic inhibition.
HINRICHSEN: I think the caudal part of the nucleus is the interesting part in some ways. This is where Cajal first showed the synapses and he subdivided the nucleus into rostral and caudal parts.
SESSLE: No one seems to have considered the possibility yet that some of these neurons in the mesencephalic nucleus may not in fact be primary neurons at all, but could be interneurons. It has recently been shown by Foster (653) and by Hubbard and Di Carlo (658), using degeneration and fluorescence techniques, that not all the neurons in the nucleus are in fact primary neurons. So there is a possibility that some of the neurons from which recordings were made were interneurons, perhaps linked synaptically or coupled electrotonically.
CHAIRMAN: Is it going to be possible to record from a cell one has labelled with peroxidase. Does peroxidase kill the cell ?
HINRICHSEN: No, it does not.
LUND: I believe Llinás has suggested that this electrotonic coupling is vestigial in mammals. He said that if you tap the teeth of crocodiles you get a very, very rapid jaw-closing reflex. Do you know anything about this ?
HINRICHSEN: I know a funny story about it. Llinás and his colleagues were working on a crocodile and when they had finished with the animal they chopped its head off. Then one of them turned around and the jolly thing, the head that is, bit him on the tail. So I guess it is monosynaptic.
MØLLER: I would like your comment on the interpretation Llinás put on the

electrotonic coupling. He suggested it served to provide simultaneous activation of a number of muscles in some activities, such as in sucking and swallowing, and that the normal synaptic pathways were used during other activities such as chewing.
HINRICHSEN: Coupling may be part of a generalized phenomenon or it may be concerned specifically with synchronization. It has some advantages over chemical transmission because both hyperpolarizing and depolarizing effects can be passed at the same junction.

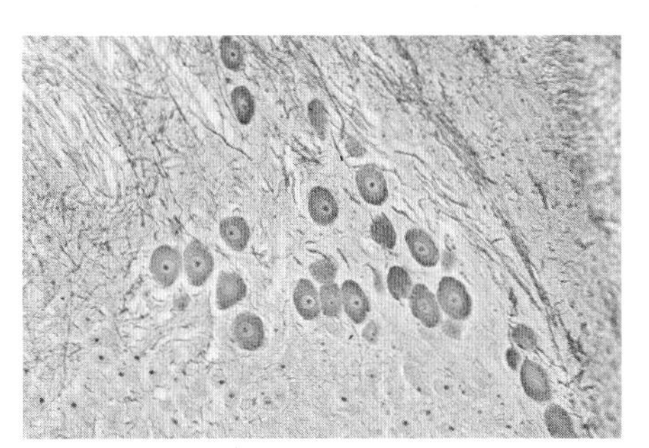

Fig. 19.15. Mesencephalic nucleus of adult rat. Ranson silver pyridine method. (× 83.)

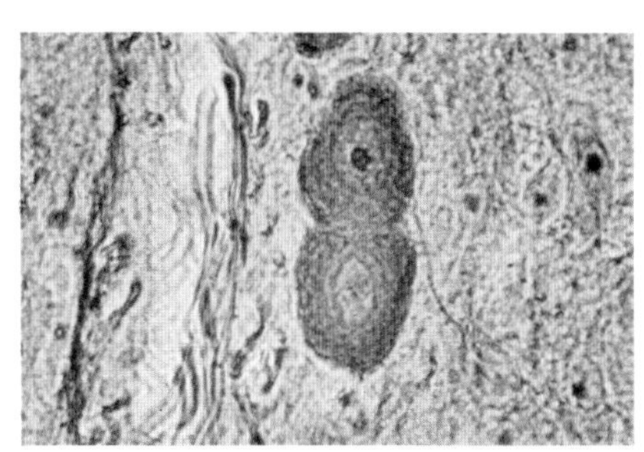

Fig. 19.16. Mesencephalic nucleus of adult rat. Ranson silver pyridine method. (× 633.)

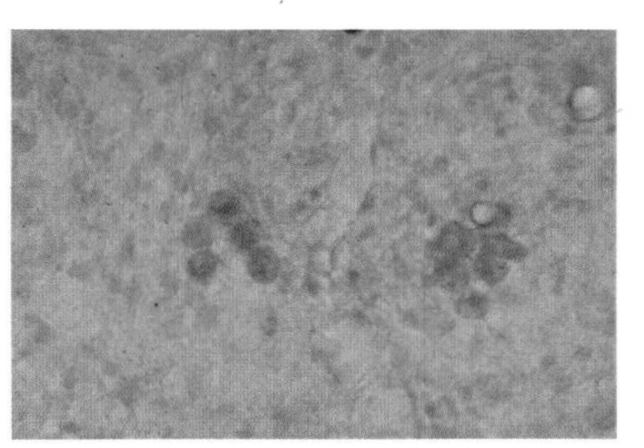

Fig. 19.17. Mesencephalic nucleus of 1-day-old rat after injection of horseradish peroxidase (HRP) into masseter (× 166.)

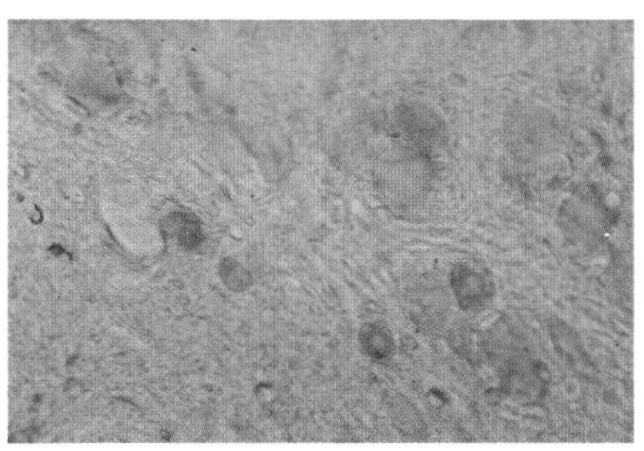

Fig. 19.18. Mesencephalic nucleus of adult rat after injection of HRP into masseter. Some coupled cells in clusters are labelled and some unlabelled. (× 166.)

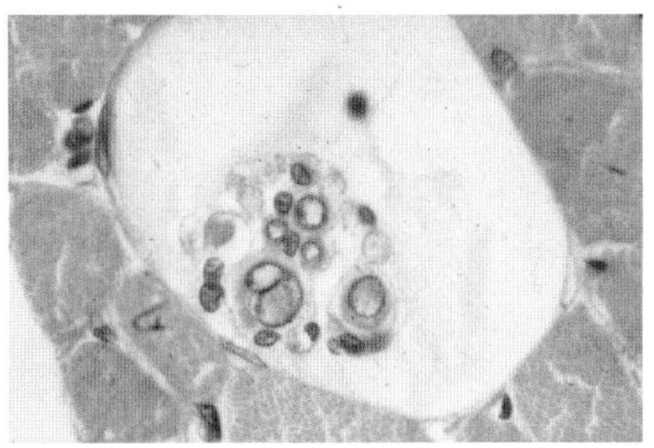

Fig. 19.19. Cross-section of a spindle in rat masseter. H & E. (× 333.)

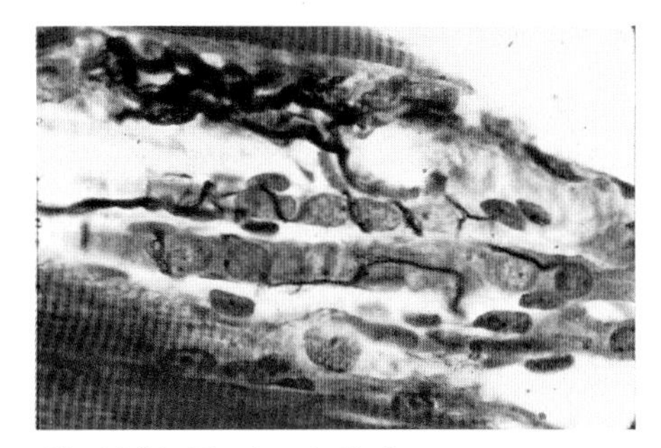

Fig. 19.20. Muscle spindle from rat masseter. Silver stain. (× 500.)

Paper No. 20

Changes in the Excitability of Elevator and Depressor Motoneurons produced by Stimulation of Intra-oral Nerves

L. J. Goldberg

INTRACELLULAR STUDIES

The best technique for determining excitability changes in elevator and depressor motoneurons is to impale a motoneuron with a micro-electrode, stimulate an intra-oral nerve and observe the results. There have, however, been relatively few such intracellular studies (205, 232, 233, 286, 321, 322, 427, 559).These experiments were performed on anaesthetized or decerebrate cats immobilized with a neuromuscular blocking agent and the results are based on recordings obtained almost exclusively from motoneurons innervating the masseter and digastric muscles.

Depressor Motoneurons

A digastric motoneuron in a cat gives a simple response to a single electrical shock delivered to an intra-oral nerve: it becomes excited. Stimulation of the lingual (233, 427) or inferior dental (233, 321) nerves evokes an excitatory postsynaptic potential (EPSP) which can lead to the generation of from one to four spikes in a digastric motoneuron. Gura et al. (233) reported that stimulation of the infra-orbital nerve, and Kidokoro et al. (321) that stimulation of the superior dental branch of the infra-orbital nerve, produced similar results. The latency to the onset of the EPSP evoked by ipsilateral stimulation of these nerves was approximately 2·0–4·0 msec with the total response lasting from 10 to 20 msec. The EPSP in digastric motoneurons could be evoked by stimulation of the lowest threshold fibres in either nerve (321, 427).These are fibres in the 10μ diameter range, with conduction velocities of approximately 40–60 msec, which innervate mechanoreceptors in the periodontal ligament and intra-oral mucosa (321, 428).

Elevator Motoneurons

Elevator motoneurons have much more complicated lives. Identical stimulation of the same low threshold afferents in the nerves discussed above results in a complex series of responses in motoneurons of the masseter muscle in the cat. The responses can be separated into two categories, inhibitory and excitatory, and the inhibitory response can itself be differentiated into two distinct types.

1. *The Inhibitory Response*

Beginning approximately 2·0–3·0 msec after a single shock stimulus is delivered to the ipsilateral inferior dental, superior dental, lingual or infra-orbital nerves the

inferior dental nerve after the administration of strychnine resulted in a depolarization of the membrane beginning approximately 2·5 msec after the stimulus; this depolarization sometimes resulted in spike generation. Takata and Kawamura (559) reported that stimulation of the maxillary nerve resulted in a depolarization of masseteric motoneurons which preceded any observable hyperpolarization, with the depolarization beginning at a latency of approximately 2·5 msec. They stated that this depolarization was not present after stimulation of the lingual or inferior dental nerve; however, in their *Fig.* 5, II A, B and C, they illustrate the response of a masseteric motoneuron to inferior dental nerve stimulation. This figure clearly shows a small initial depolarization to be present prior to membrane hyperpolarization; and the latency of the depolarization is approximately 2·5 msec. In recent experiments in our laboratory, on cats anaesthetized with α-chloralose, we observed an initial depolarizing response to inferior dental nerve stimulation when recording intracellularly in masseteric motoneurons (*Fig.* 20.2B, C). In one case, the inferior dental nerve stimulus resulted in the initiation of a spike after a latency of 2 msec, as illustrated in *Fig.* 20.2A.

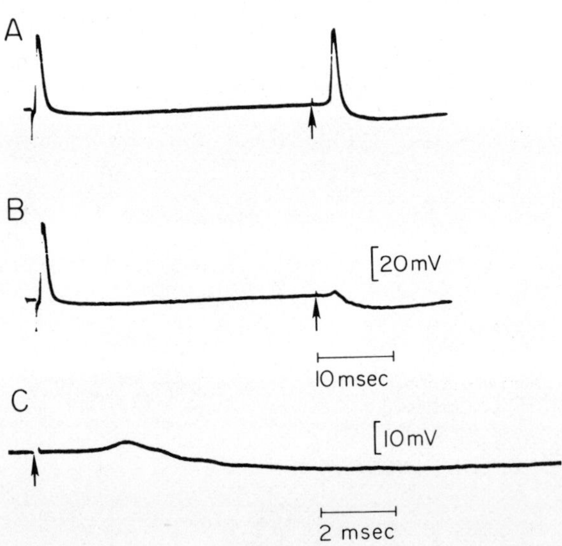

Fig. 20.2. Short latency depolarization and spike induction evoked in a masseteric motoneuron by inferior dental nerve stimulation. A, the first spike is evoked by antidromic stimulation of the masseteric nerve and the second by stimulation (at arrow) of the ipsilateral inferior dental nerve. In B, the inferior dental nerve stimulation fails to produce a spike but a depolarizing potential is seen to precede the hyperpolarization. C, the response to inferior dental stimulation of the same masseteric motoneuron shown in A and B but at a faster sweep speed and at a higher gain. Time base and voltage calibrations in A are the same for B.

Membrane depolarization following intra-oral afferent stimulation is not observed in all masseteric motoneurons. However, it is clear that in many cases, either preceding the early phase of hyperpolarization or between the early and late phase of inhibition, depolarization can be observed, and may in fact result in spike generation.

STUDIES ON ORAL REFLEXES

Information concerning changes in motoneuron excitability can be obtained by recording from motor nerves in the periphery. The monosynaptic excitatory connection between muscle primary afferent fibres (whose cell bodies are located in the mesencephalic nucleus of V) and motoneurons of jaw-elevator muscles (again primarily masseteric motoneurons) has been used as an indirect test of motoneuron excitability. Studies concerning the effects of intra-oral nerve stimulation on this monosynaptic reflex (the masseteric reflex) have generally supported the results of the intracellular recording studies which have been described above.

Of particular interest are the results supporting the view that stimulation of low threshold afferents in intra-oral nerves evokes a short latency excitatory effect which is masked by the simultaneous occurrence of the early inhibitory hyperpolarizing influence (201, 205, 550). Electromyographic studies in both animals (179, 202, 571, 576) and man (199, 518) have further demonstrated that excitatory effects can be evoked in elevator muscles by stimulation of intra-oral receptors.

Stimulation of intra-oral afferent nerves also inhibits the monosynaptic masseteric reflex in a diphasic manner which corresponds to the early and late phases of hyperpolarization observed in masseteric motoneurons (201, 205, 321, 550). The question has been raised as to whether some component of the late phase of inhibition was produced by a presynaptic inhibitory mechanism via depolarization of the primary afferent terminals mediating the monosynaptic excitatory effect (321, 550). Evidence has been presented supporting this possibility from experiments which demonstrated that picrotoxin reduced the effectiveness of the late inhibitory phase without apparently affecting the early phase (201). Further evidence from studies in our laboratory on decerebrate cats indicating that presynaptic inhibition of the monosynaptic reflex results from intra-oral nerve stimulation is illustrated in *Fig.* 20.3. The methods involved in these experiments are the same as those described in a previous paper for testing the excitability of the central terminals of afferents in nerves supplying the intra-oral mucosa (201). In the present case, a bipolar concentric electrode was placed in the brain stem at the rostral border of the trigeminal motor nucleus. A single shock, at the intensity indicated by the arrow in *Fig.* 20.3B, delivered through the electrode evoked the compound action potential illustrated in *Fig.* 20.3D, which was recorded from the cut central end of the masseteric nerve in the periphery. The potential had a similar latency to the antidromic volley in Group I afferent fibres which could be recorded from the masseteric nerve after stimulation of the trigeminal mesencephalic nucleus (*Fig.* 20.3C, first peak). Stimulation through the electrode at higher intensities can evoke the monosynaptic reflex which is then observed as a second compound action potential similar in shape and latency to the monosynaptic reflex shown in *Fig.* 20.3C (second peak), evoked by mesencephalic nucleus stimulation. When a conditioning stimulus was delivered to the ipsilateral or contralateral lingual nerve at 2X threshold (*Fig.* 20.3E, c), and the test stimulus delivered through the electrode in the motor nucleus, the compound action potential recorded in the masseteric nerve was increased with a time course illustrated in *Fig.* 20.3A. The amplitude of the potential began to increase at a CS–TS interval of approximately 10–15 msec, reached a maximum at an interval of approximately 25–45 msec, and then gradually decreased and reached the control level after approximately 120 msec. The results indicate that the conditioning stimulus increases the excitability of

Recent studies have attempted to confront some of these problems by using more physiological intra-oral stimuli while observing jaw movements and electrical activity in jaw muscles, rather than by looking at excitability changes in motoneurons of jaw muscles in paralysed preparations (367, 370, 575, 576).

One of the limitations of the experiments described in this paper is the almost exclusive focus on masseteric and digastric motoneurons in the cat. The cat has no appreciable lateral jaw movements and its patterns of masticatory movements are obviously quite different from those of man. Therefore, these studies have not, up to the present time, provided us with information concerning the effects of intra-oral nerve stimulation on the excitability of motoneurons of the pterygoid muscles. This is, obviously, information which is very important in our attempt to understand the role of feedback from intra-oral receptors on co-ordinated jaw movements in man. Nevertheless, the experiments have provided us with the means of beginning to understand what mechanisms the nervous system has at its command for the control and co-ordination of jaw movements. Without an understanding of these elements we cannot know how jaw movements are controlled in the process of normal mastication and we will not be able to determine what elements have been disrupted when we are confronted with neurologically-based malfunctions of the masticatory system.

Discussion CHAIRMAN: B. MATTHEWS

MØLLER: Dr Öwall and I (441) have some evidence which fits your hypothesis very well. We have measured tactile sensibility during conscious biting using metal foil and found the threshold to be 20 μm. We then investigated the threshold during mastication by putting steel balls into different foods and instructing the subjects to chew and tell us when they felt the balls. The threshold was reached with steel balls 0·6–0·9 mm in diameter. This was a 30–60 fold increase in threshold. Detection occurred at a significantly later stroke than the first make of tooth contact in the intercuspal position. I think we are justified in concluding that the results were due to a true rise in threshold and not a measure of the approximation of the teeth during chewing. The rise in threshold was most likely due to the cortical inhibition you proposed, but I would like to change your interpretation a little bit. I suggest that the inhibition is a means of adjusting (reducing) the gain in the afferent pathways to fit with the forces exerted during the function in question (swimming of your fish, chewing in our subjects).

GOLDBERG: If you bite your tongue or your fork or a piece of grit while eating, there is complete elevator shut off. The stimulus goes through the presynaptic inhibition and produces a reflex effect which stops the central programme.

SUMINO: As I mentioned earlier, peripheral inputs from the alveolar and lingual nerves inhibited the masseteric motoneurons and excited the digastric motoneurons disynaptically through the interneurons in the supratrigeminal nucleus and in the spinal nucleus respectively. On the other hand, Dr Sauerland and his co-workers (668) have reported that transection of the brain stem just caudal to the trigeminal motor nucleus could abolish the cortically induced depression of the masseteric monosynaptic reflex and recently Dr Nakamura and his associates have observed the same effect on the medial bulbar reticular neurons which were activated by cortical stimulation but received only weak excitatory inputs from the lingual nerve. Furthermore, they also showed (666) that localized stimulation of the medial bulbar reticular formation exerted monosynaptic reciprocal effects on the trigeminal motoneurons bilaterally; inhibitory on the masseteric and excitatory on the digastric motoneurons. Therefore, during mastication, the peripheral input and the central input would be programmed in different ways.

LUND: Your comments about inhibition are very important because, if you look at the afferent information arising during mastication, it seems to be out of phase with the on-going movements, as Taylor in this Symposium and Goodwin and Luschei (212) and others have shown. The greatest firing of the spindle afferents from the jaw-closing muscles occurs during opening, and the greatest firing from the periodontal receptors which activate jaw-opening muscles will occur during closing. Now there must be some mechanism which is stopping these very powerful inputs from activating their target muscles. If there is strong inhibition along the pathway, a lot of afferent information could continue to come in, but be turned off before it gets to the motoneurons; and the information could go to the cortex or elsewhere. As examples, the spindle input could be used to control jaw opening (e.g. Lund and Lamarre (661)) and the input from the teeth could be used to control the force of closure (369). Conversely, a small amount of input from say the muscle spindles during closure (they often continue to fire during closure) on to already active elevator motoneurons could have a great effect.

The absence of evidence for DIG gamma motoneurons, or of central effects of mesencephalic stimulation that might be attributed to DIG muscle spindles (also *see below*), is consistent with previous findings (114, 117, 321). An input from the trigeminal mesencephalic nucleus was established for 40 of the 43 elevator muscle alpha motoneurons tested and for 16 of 18 gamma motoneurons tested (*Figs.* 21.1, 21.2). Spinal gamma motoneurons appear to lack at least an autogenic muscle spindle input (152, 390). The mean latency of the response of the gamma motoneurons to mesencephalic stimulation was significantly longer than that of the alpha motoneurons which had latencies compatible with a monosynaptic pathway. The gamma motoneuron latencies indicate a polysynaptic input from the trigeminal mesencephalic nucleus, perhaps from the periodontal afferents in the nucleus (114) since the gammas were also excited by tooth tapping (*see below*).

The gamma motoneurons frequently showed a tonic discharge (*Fig.* 21.1) that could not be suppressed by closing the jaw. Sometimes tonic discharge was recorded from elevator muscle alpha motoneurons, but this could be completely suppressed by closing the jaw to tooth contact. This should not be taken as evidence

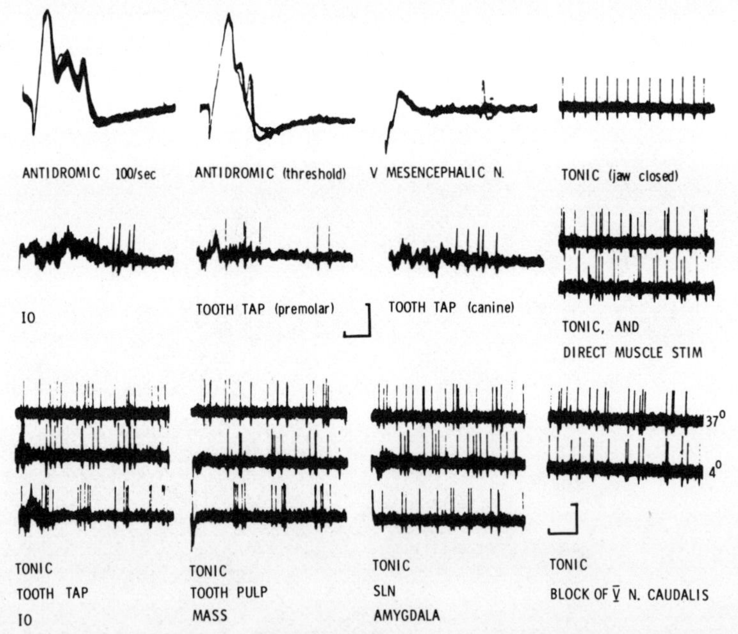

Fig. 21.1. Characteristics of a masseter gamma motoneuron. The threshold for antidromic activation was relatively high, as evidenced by the large antidromic field potential present (*top left*). The second record shows the motoneuron firing twice to five successive masseter nerve stimuli of threshold intensity and delivered at 1 per sec. The motoneuron had no input from the trigeminal mesencephalic nucleus (5 traces superimposed), was tonically active, and could be activated by infra-orbital nerve stimulation (IO) and tooth tap. The tonic firing of the unit was not affected by direct muscle stimulation but, as shown in the lower half of the figure, it could be suppressed by stimulation of a number of sites, including the masseter (MASS) and superior laryngeal nerves (SLN). Note that some of the single stimuli tended to cause synchronization of the tonic discharge; this feature was also seen in some other gamma motoneurons. Blocking synaptic transmission in nucleus caudalis by cooling (516) had no appreciable effect on the tonic discharge (cf *Fig.* 21.3). Voltage calibrations: 0·2 mV, time calibrations: 1 msec (antidromic and mesencephalic records), 5 msec (IO and tooth tap records), and 40 msec (tonic records).

per se that mandibular rest position is determined actively. The spontaneous discharge of the alpha motoneurons probably resulted from passive stretching of the jaw-closing muscles in the animal which had been paralysed with a muscle relaxant which allowed the jaw to hang open beyond the rest position. Unlike the elevator muscle alpha motoneurons, the gamma motoneurons may be truly tonic, like gamma motoneurons at spinal levels (390).

The tonic discharges of both alpha and gamma motoneurons could be increased by small amounts (2–3 mm) of jaw opening. This movement apparently activated muscle spindle afferents, and perhaps even temporomandibular joint afferents, and thus increased excitatory input to the motoneurons. Pressure applied manually with a wooden probe over part of the appropriate muscle could also increase the tonic discharge of the alpha motoneurons (312). A direct electrical stimulus which caused the muscle to contract was found to decrease the tonic activity of alpha motoneurons. Presumably this was due to unloading of the muscle sufficient to decrease the stretch that excited its spindles, a view supported by the similar response of spontaneously active elevator muscle afferents to direct stimulation of the muscle.

Tooth tap also excited alpha and gamma motoneurons of elevator muscles (*Figs.* 21.1, 21.2), and DIG motoneurons (*Fig.* 21.3). This stimulus activated 25 of 39 MASS alphas tested (mean latency 6·1 msec, s.d. 1·40), 8 of 12 TEMP alphas (7·6, s.d. 2·15), 8 of 10 MASS gammas (6·7, s.d. 1·68), 6 of 6 TEMP gammas (8·8, s.d. 1·35), and 13 of 27 DIG motoneurons (8·4, s.d. 2·28). Taking into account the rise-time of the mechanical stimulus, some of these latencies, especially of the MASS alphas, suggest a monosynaptic pathway. The intracellular studies of Kidokoro et al. (321) indicated a disynaptic path for excitation of MASS motoneurons by intra-oral afferents, and most of our latencies would be compatible with this finding. However, the possibility of a monosynaptic pathway is supported by findings in man (199, 518). Our findings also provide evidence of peripherally induced coactivation of alpha and gamma elevator motoneurons.

The short-latency responses to tooth tap in both jaw-closing and DIG motoneurons could be depressed or abolished by local anaesthetic infiltration around the stimulated tooth (*Figs.* 21.2, 21.3). This finding indicates that receptors in, or more likely around, the tooth are involved. Local anaesthesia could also reversibly abolish the inhibitory effect of tooth tap (*see below*), and this supports recent findings in man (518).

Stimuli applied to IO, SLN and IX could occasionally activate jaw-closing alpha and gamma motoneurons (*Fig.* 21.1) but not as often as with tooth tap. In MASS reflex experiments we have also found these, and palatal (cf. Thexton (575)), stimuli to have excitatory effects. The significance of such reflex activation of jaw elevators invites speculation, particularly in terms of factors affecting jaw motility. The results of other studies (179, 200, 370, 575) suggest that stimulus strength, direction, duration and force, as well as jaw position, could all be factors influencing the jaw muscle response and the type of movement produced.

In contrast to their apparently weak excitatory effects on the elevator motoneurons, stimuli applied to IO, MASS and TEMP (high threshold), SLN, IX, and tooth pulp had marked excitatory effects on single DIG motoneurons (*Fig.* 21.3), in agreement with previous observations on single motoneurons (321, 427) and reflexes (19, 202, 240, 506, 550, 575). The mean latency of the effects on single

usually about 50–60 msec in duration, but inhibitory effects lasting as long as 200 msec were not uncommon. These prolonged inhibitory effects may have a presynaptic inhibitory component (201, 517) although postsynaptic inhibition is involved, at least in the early part of the inhibitory period (321, 427, 550).

Occasionally a facilitatory effect, or return towards the control level, was noticeable in conditioning sequences at 20–40 msec following the delivery of the conditioning stimulus. This has also been noted previously on MASS activity (201, 321, 427, 550). It may reflect the normally suppressed effect of a second type of excitatory afferent input, perhaps of a nociceptive or startling nature, in contrast to the early excitation that can be produced with muscle spindle activation and light tooth taps (*see above*). We suspect that the

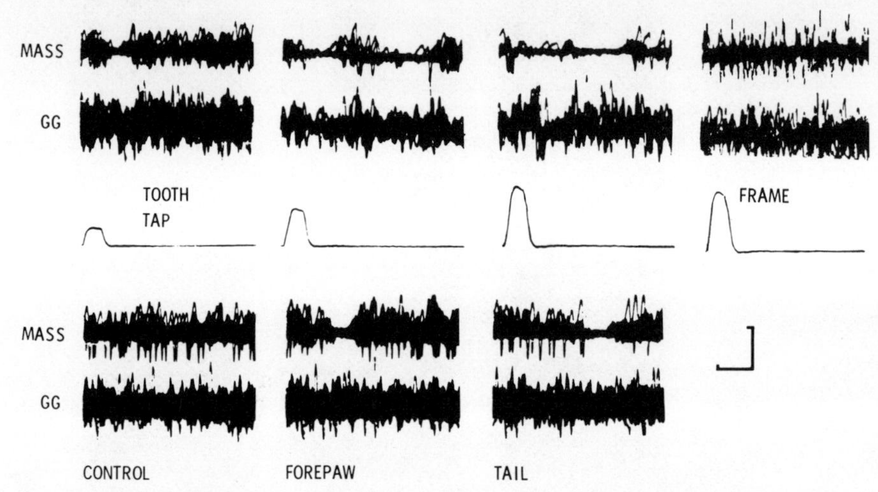

Fig. 21.4. Electromyographic recordings of masseter (MASS) and genioglossus (GG) muscles of a rhesus monkey anaesthetized with phencyclidine hydrochloride (Sernylan, Bio-Ceutic Laboratories, Mo.). The effects of 10 successive tooth taps of different intensities delivered (1 per sec) to the ipsilateral maxillary canine on MASS and GG activity induced by maintained jaw opening are illustrated in the top half of the figure. Note the increase in duration of the inhibitory period in MASS, and perhaps in GG, as the intensity is increased from 3 to 8 to 18 g. Tapping the stereotaxic frame in which the animal's head was placed induced no apparent inhibition (upper right record). The lower half of the figure shows the control or background level of MASS and GG activity (again during maintained opening of the jaw) and the effects of 30 successive bipolar stimuli (single pulses, 5 mA, 0·5 msec) delivered (1 per sec) to the skin of the ipsilateral forepaw and of the tip of the monkey's tail. Only MASS showed obvious inhibition. Voltage calibration: 0·1 mV; time calibration: 20 msec.

prolonged inhibitory influence also reflects effects produced by a dual afferent input although long loop effects, involving higher centres, might also be involved. Yu et al. (643) indicated that a late period of inhibition in the MASS EMG can be produced by a noxious orofacial stimulus in man. We have also found, in preliminary experiments in which EMGs were recorded from the masseter of cats and monkeys that, if the force of the tooth tap is greatly increased, a later inhibitory period can be produced in addition to the early phase (*Fig.* 21.4). This requires further study and may have possible clinical significance, not only in normal masticatory movements, but also in conditions such as the temporomandibular joint pain dysfunction syndrome.

Inhibition of MASS alpha (26 of 42) and gamma (12 of 14) motoneurons was particularly noticeable with tooth tap stimulation, although other stimuli were also effective. DIG nerve stimulation caused inhibition in 4 of 11 MASS alpha motoneurons tested, but only at high stimulus intensities. The lack of low threshold effects is not surprising in view of the lack of spindles in DIG, but serves to emphasize that Ia reciprocal inhibition, present at spinal levels (390), is lacking in the trigeminal system. But, as shown above, reciprocal effects can occur as a result of stimuli (tooth tap, tooth pulp, SLN, etc.) that activate afferents of Group II diameter and less (427). Stimulation of SLN, for example, produced inhibition in 6 of 26 MASS alphas tested, and even forepaw stimuli had inhibitory effects in 2 of 9 alpha motoneurons of elevator muscles and 1 of 3 gamma motoneurons tested (*Figs.* 21.1, 21.2).

The danger of over-emphasizing the importance of periodontal receptors in the control of masticatory movements, which was stressed recently (643) is apparent from these findings. Indeed, in our preliminary EMG studies of MASS activity in cat and monkey, we found quite reproducible and powerful inhibitory effects following electrical stimulation of the big toe or tip of the tail (*Fig.* 21.4)! But, conversely, we do not wish to over-emphasize the significance of peripheral influences on jaw movements at the expense of central effects (e.g. *Figs.* 21.2, 21.3).

Discussion CHAIRMAN: D. J. ANDERSON

WATT: In human experiments we have produced inhibition of isometric masseter contraction by tapping the vertebral column even down as far as the lumbar vertebrae. We consistently produced silent periods by tapping over the cervical vertebrae.
SESSLE: We have gone all the way down to the tip of the tail! Moreover, an inhibitory effect with tooth tapping can also be shown in other muscles. We have seen some evidence earlier today that it can occur in digastric, and we have evidence that it can also occur in some tongue and pharyngeal muscles.
SUMINO: Did you find any gamma motoneurons in the digastric or glossopharyngeal nerves ?
SESSLE: We have not looked at the glossopharyngeal. There was no evidence of gamma motoneurons supplying the digastric.
TAYLOR: Do you think there is any possibility of distinguishing static and dynamic fusimotoneurons in this situation ?
SESSLE: I think it may be possible, using vibratory stimuli or suxamethonium.
TAYLOR: I was thinking of the fusimotoneurons.
SESSLE: Yes, it may be possible, but this is very difficult even at spinal levels (390).
TAYLOR: I was wondering if they could be distinguished on the basis of conduction velocities.
SESSLE: No, they do not seem to fall into two distinct populations, but again, the evidence at spinal cord level is not very good in terms of conduction velocity as a criterion.
BULLER: But you showed one or two conduction velocities above what are classically considered to be gamma. Also, could I support Professor Taylor in using the term 'fusimotor' rather than gamma.

SESSLE: You are suggesting the use of the term 'fusimotor' because of its functional implications rather than more classic nomenclature ?
BULLER: Erlanger and Gasser based their classification on conduction velocities and some of the fibres you showed had conduction velocities outside the range associated with gamma fibres. The point you were trying to make was that they are fusimotor in function. I think it is wiser to differentiate.
GOLDBERG: Previously, you have shown trigeminal mucosal afferent depolarization and reduction of transmission following superior laryngeal nerve stimulation. You also showed some digastric inhibition which you attributed to presynaptic mechanisms. Have you ever seen any of this kind of depolarization in relation to any functional situation involving the superior laryngeal nerve in swallowing?
SESSLE: You are apparently thinking along similar lines to those developed in your presentation in which you suggested that during mastication there is concomitant inhibition of orofacial sensory input into the brain. I have not as yet tested to see if there is presynaptic depolarization of the laryngeal input actually *during swallowing*, but I have previously suggested (515) for swallowing, as you now have for mastication, that there is suppression, perhaps largely presynaptically derived, of sensory inflow into the brain stem which might otherwise upset the ongoing neural synergistic events of swallowing. Apart from my previous (514, 517) observations of presynaptic depolarization and inhibition of trigeminal and solitary tract neurons, and the similar time course of these effects and the duration of a swallow, this view receives support from recent observations in my laboratory that the elicitation of trigeminal and laryngeal reflexes may be depressed during swallowing.
GOLDBERG: Is this in a decerebrate or anaesthetized cat ?
SESSLE: Both. I have also observed these effects in anaesthetized monkeys.
HEADLEY: I note that most of the pauses in firing that you have seen following these various inputs follow on excitation. I am unhappy about the use of the word inhibition as applied to these pauses in firing since one can get a similar pause, and particularly the subsequent synchronization of firing, simply as a result of after-hyperpolarization. This is a property of the motoneurons themselves rather than being due to any synaptic input. Without doing any intracellular studies it is virtually impossible to distinguish between these two forms of hyperpolarization.
SESSLE: The inhibition frequently occurs without preceding excitation. Moreover, many people have recorded intracellularly from masseter and temporalis alpha motoneurons and have shown that a true IPSP is produced by some of these stimuli (205, 312, 321, 427). So I have no doubt in my mind that, at least in the alpha motoneurons, what we are seeing is true inhibition. However, some of the masseter gamma motoneurons showed synchronization in their tonic discharge as a result of peripheral stimuli and perhaps, as you suggest, this may not always have been associated with true inhibition, although again, preceding excitation was not necessary for its production (*see Fig.* 21.1).
FUNAKOSHI: I also have recorded facilitatory responses from peripheral gamma efferents in the masseter nerve produced by mechanical stimulation of upper incisors in rat.
SESSLE: How did you know they were gammas ?
FUNAKOSHI: I stimulated the trigeminal motor nucleus and measured the latency of the evoked potentials recorded from the peripheral nerve.
SESSLE: We have shown that there is quite a considerable overlap between the

conduction velocities of alpha and gamma motoneurons in the trigeminal system. I do not think you can necessarily assume that you are dealing with a gamma motoneuron on the basis of conduction velocity or antidromic latency alone. Nevertheless, tooth tap can certainly excite both elevator alpha and gamma motoneurons; this is evidence of peripherally induced coactivation.

MATTHEWS: What is the distance and time involved in your latency measurements? What is the error in your estimates of conduction velocity ?

SESSLE: The distance from the masseter or temporalis is approximately 40 mm and for the digastric nerve approximately 45–50 mm.

MATTHEWS: Are you measuring time to an accuracy of 10 μsec ?

SESSLE: Approximately, certainly within 50 μsec.

Paper No. 22

Mechanisms of Responses of Masseteric Motoneurons to Intra-oral Stimulation in the Cat

R. Sumino

In 1917, Sherrington (524) demonstrated beautifully relaxation of the jaw-closing muscles during the jaw-opening reflex. Mechanical and electrical stimulation to gums, teeth or hard plate evoked a jaw-opening reflex followed by rebound closure. After the jaw-opening muscles had been detached from the mandible and the mandible split in the midline, the jaw remained closed as a partial phenomenon of decerebrate rigidity. In this condition, stimulation caused the ipsilateral half of the mandible to drop immediately. He suggested that a central inhibitory mechanism was involved. Many investigators have observed reduction of activity of the jaw-closing muscles with the jaw-opening reflex, initiated by electrical and mechanical stimulation of intra-oral structures in animals (316) and human subjects (79, 242, 272, 639, 640). Recently, however, activation of the jaw-closing muscles prior to the suppression phase has been reported (179, 199, 243, 576).

In the present paper, I will describe the responses of masseteric motoneurons to intra-oral stimulation, and the central pathways involved in these responses.

METHODS

Experiments were performed in cats anaesthetized with pentobarbital sodium (35 mg per kg). The inferior alveolar nerve was stimulated to evoke the jaw-opening reflex. The changes in excitability of the masseteric motoneurons were examined by intracellular recording from the motoneurons with glass micropipettes filled with 2M K citrate or 3M K chloride solution, as well as by recording monosynaptic reflex discharges (201, 205, 278, 335, 424, 425).

The details of surgical, stimulating and recording procedures have been described elsewhere (321, 322, 550).

RESULTS AND DISCUSSION

Responses of masseteric motoneurons to stimulation of the inferior alveolar nerve

The amplitude of the monosynaptic masseteric reflex discharge was suppressed by supramaximal conditioning stimuli (c.s.) applied to the inferior alveolar nerve (*Fig.* 22.1A). *Fig.* 22.1B, C illustrate the time course of the excitability change in the masseteric motoneuron following inferior alveolar nerve stimulation. Two phases of depression, early and late, were induced. The early depression started at about 3 msec and was absolute between 4·5 and 35 msec. The late, weaker depression began at approximately 45 msec, reached a maximum at 60–70 msec and lasted for more than 100 msec. A temporary recovery was often seen in the transition between the two phases. Sometimes, a small facilitation preceded the early depression.

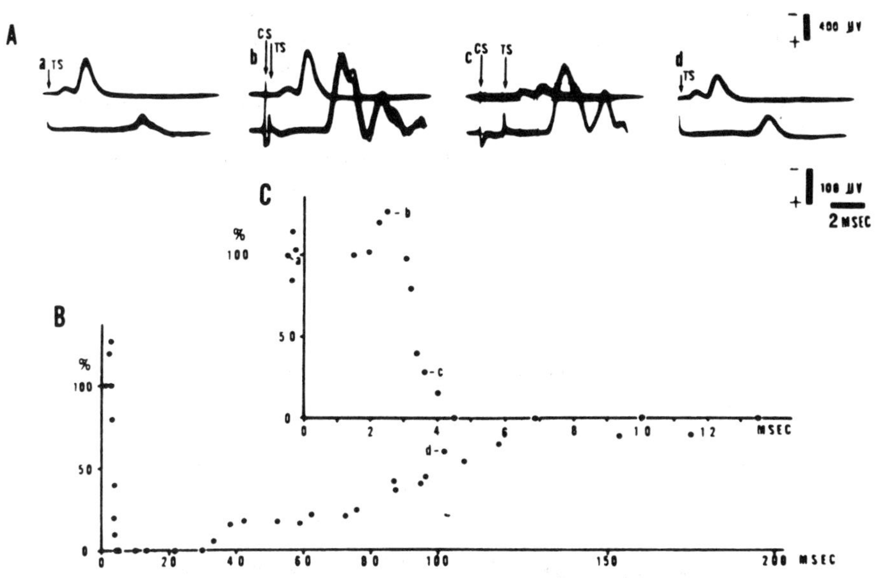

Fig. 22.1. Monosynaptic reflex potential recorded from the masseteric nerve by stimulation of the ipsilateral mesencephalic tract nucleus and the time course of the change in amplitude of the monosynaptic reflex potential induced by preceding stimulation of the ipsilateral inferior alveolar nerve. A, a in the upper trace shows a control masseteric monosynaptic reflex (stimulation of the mesencephalic tract nucleus (TS): P. 3·0, R. 2·0, V. –1·5, 9·2 V, 0·01 msec). The lower trace represents reflex potentials recorded from the digastric nerve. b to d show modification of the amplitude of the masseteric monosynaptic reflex induced by preceding stimulation of the ipsilateral alveolar nerve (CS) (6·0 V, 0·01 msec). The corresponding points in graphs B and C are indicated. B, amplitude of the monosynaptic reflex (expressed as percent of the control) plotted against the intervals between the conditioning stimulus to the inferior alveolar nerve (CS) and the peak of the monosynaptic reflex. C, The early part of B on an enlarged time scale. (*From Sumino, 1971.*)

Intracellular potentials recorded from masseteric motoneurons following inferior alveolar nerve stimulation are shown in *Fig.* 22.2 in which records A and B show the time course of the hyperpolarization of two motoneurons. The hyperpolarization has two phases, early and late, which corresponded approximately in time with the two phases of depression of the masseteric monosynaptic reflex shown in *Fig.* 22.1. Transition from the early phase to the late phase occurred at about 18 msec. The magnitude of the depolarization at this transition point varied from cell to cell, and sometimes there was sufficient depolarization to generate a spike (*Fig.* 22.2B).

The latency of the early hyperpolarization was measured by subtracting the extracellular field potential from the intracellular potential (*Fig.* 22.2D). It was 2·5 msec. In 42 cells, the latencies were 2·40 ± 0·21 msec. The estimated conduction time along the inferior alveolar nerve was about 0·8 msec in our experiments and therefore the central delay for the early hyperpolarization was 1·6 msec. This value coincides exactly with the central delay for the digastric

EPSP induced by inferior alveolar nerve stimulation (321). It therefore appears that a disynaptic linkage is responsible for the early hyperpolarization.

Recently, several investigators (179, 199, 243, 576) have reported that an early excitatory masseteric reflex precedes the inhibitory period evoked by stimulation of periodontal and gingival receptors in subjects maintaining a steady bite. As described above, in cats, a conditioning stimulus applied to the inferior alveolar nerve sometimes (4 out of 10) caused slight facilitation of the monosynaptic masseteric reflex immediately before the early depression (*Fig.* 22.1). However, a clear depolarization preceding the early IPSP could not be detected in intracellular records when the extracellular field potential was subtracted from it. A possible reason might be the lowered excitability of the masseteric motoneurons during nembutal anaesthesia.

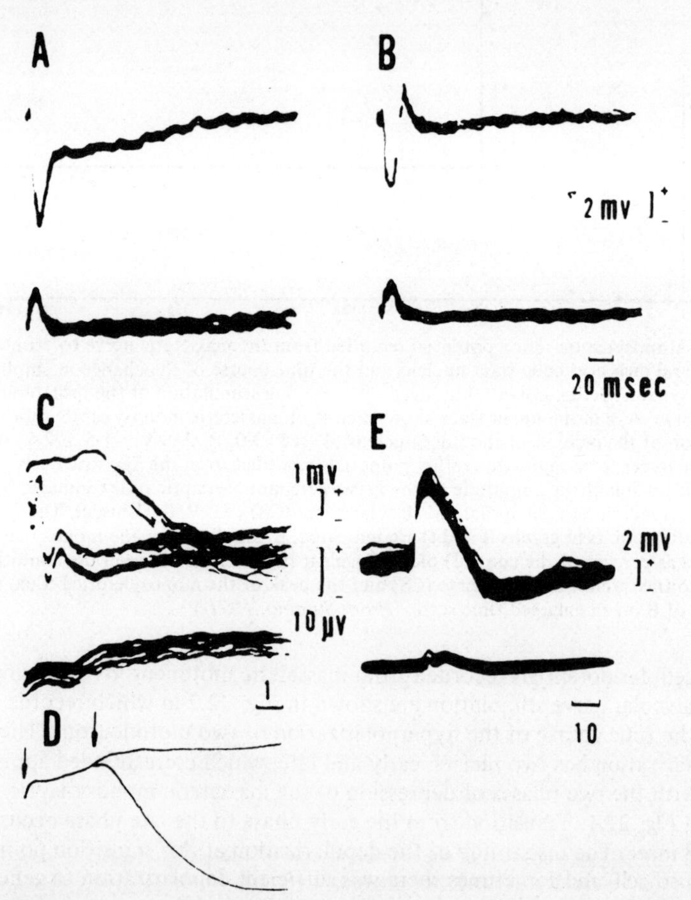

Fig. 22.2. IPSPs of masseter motoneurons, induced by stimulation of the inferior alveolar nerve. A and B are examples of IPSPs evoked from two different masseter motoneurons at supramaximal stimulus intensities. The extracellular field potentials are shown in the lower traces, C. Upper trace, initial portions of the IPSP of a different cell at a faster sweep speed; middle trace, the pontine surface potential; lower trace, the extracellular field potential. D, tracings from intra- and extracellular potentials in C. Arrows indicate timing of the shock and beginning of the IPSP. E, The reversed IPSP recorded with a KCl electrode. The lower trace is the corresponding extracellular potential. (*From Kidokoro et al., 1968.*)

The nature of the two phases of hyperpolarization

When recording with a K chloride electrode, the early hyperpolarization was reversed into a depolarizing potential within several minutes of penetration of the cell, but the late hyperpolarization did not reverse (*Fig.* 22.2E). The change in membrane conductance of masseteric motoneurons during the two phases of hyperpolarization was studied by superimposing either the monosynaptic EPSP induced by mesencephalic stimulation or rectangular current pulses of short duration (*Fig.* 22.3). It was shown that the conductance of the membrane increased during both phases of hyperpolarization but more during the early than the late phase. Therefore, both phases are IPSPs, though their characteristics are different. Hannam et al. (242, 243) and Yemm (639, 640) have demonstrated, in man, two phases of suppression in averaged EMGs recorded from jaw-closing muscles. These were induced by tooth contact or by electrical stimulation of oral mucous membrane. The time course of the two phases of suppression corresponded approximately with the two phases of hyperpolarization in our experiments.

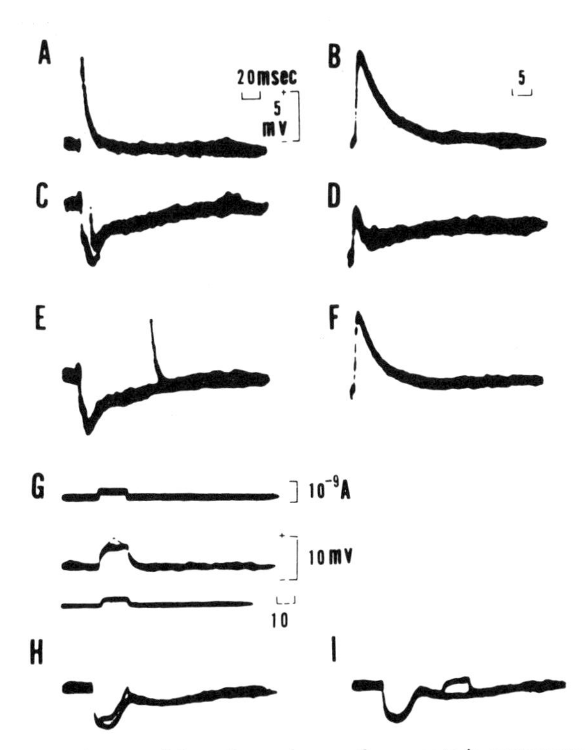

Fig. 22.3. Conductance changes of the cell membrane of a masseteric motoneuron during the IPSP induced by inferior alveolar nerve stimulation. A to F, Changes in the monosynaptic EPSP configuration during IPSPs. In A and B the monosynaptic EPSPs were evoked with a shock to the mesencephalic tract. They were superimposed upon IPSPs in C and E. D and F. the EPSPs of C and E with a faster sweep speed. G to I, the changes in an intracellularly applied pulse during IPSPs. G, Upper trace: current pulse alone, 15 msec duration; middle trace, resultant membrane potential change; lower trace, extracellular potential change. H and I, superimposed records of the IPSPs induced by inferior alveolar nerve stimulation and records in which the intracellular current pulse has been added during the early (H) or late (I) phase of depolarization. (*From Kidokoro et al., 1968.*)

Relation between IPSPs and stimulus intensity

IPSPs were recorded from masseter motoneurons during stimulation of the inferior alveolar nerve at different intensities together with incoming nerve volleys and the extracellular field potentials. The amplitude of the late IPSP and the nerve potential reached their maximum with a stimulus intensity about 2·5 X threshold (T), whereas the early IPSP increased even up to 7 or 8 T (*Fig.* 22.4). This finding suggests that, in addition to low threshold afferents, high threshold fibres in the inferior alveolar nerve are involved in the production of the early inhibition of the jaw-closing muscles. Indeed, about half of the fibres in the inferior alveolar nerve are less than 6μ in diameter (77). Thus, the component of the early inhibition due to these high threshold fibres might form the basis for the jaw-opening reflex to function as an avoidance reflex (272, 316, 382). However, the lowest threshold fibres of the inferior alveolar nerve could also initiate a jaw-opening reflex.

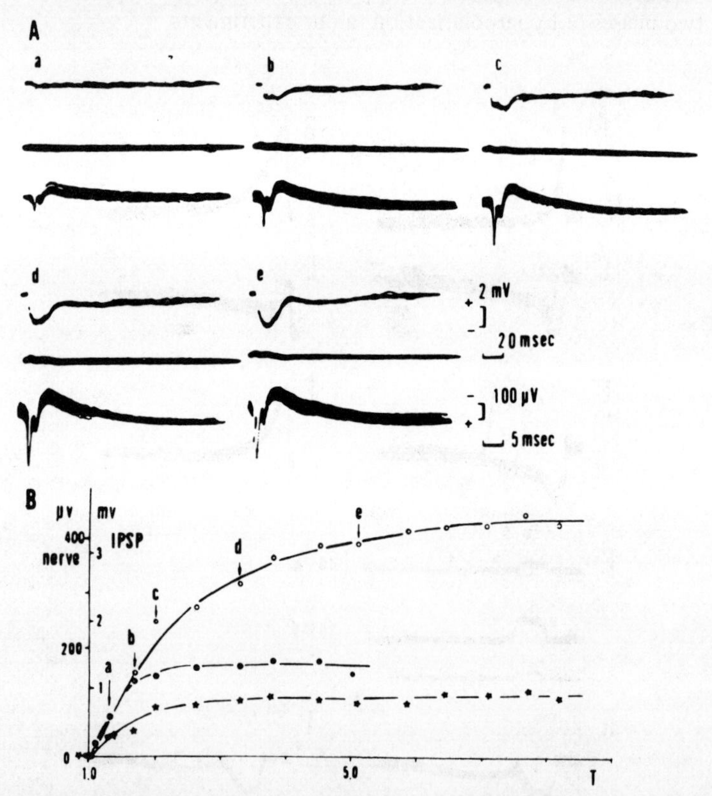

Fig. 22.4. Relationship between stimulus intensity to the inferior alveolar nerve and evoked IPSPs from a masseter motoneuron. Records in A a–e show the superimposed synaptic potentials (upper traces), extracellular field potentials (middle) and the nerve potentials (lower), evoked by shocks to the inferior alveolar nerve at intensities of 1·3 T (a), 1·7 T (b), 2·0 T (c), 3·3 T (d) and 5·1 T (e). Amplitude and time calibrations are shown at the right of e, 2 mV and 20 msec for the upper and middle, and 100 μV and 5 msec for the lower. In B, peak amplitudes of the early phase of the IPSP (○), late phase of IPSP (★) and nerve potentials (●) were plotted against inferior alveolar nerve shock intensity in multiples of the nerve threshold. Scales for the synaptic potentials are shown on the right side of the axis and for the nerve on the left side. Small letters above the curve (a–e) with arrows correspond to responses shown in A. (*From Kidokoro, 1968.*)

According to Stewart (544) and Pfaffman (453) the low threshold afferent fibres of the inferior alveolar nerve respond to mechanical stimulation of the periodontal membrane and lower lip skin, and have conduction velocities of 24—60 m per sec. We estimated the conduction velocities of the lowest threshold fibres in the inferior alveolar nerve to be 56·2 ± 2·6 m per sec. It appears therefore that thick fibres originating from periodontal receptors constitute an important afferent limb of the jaw-opening reflex.

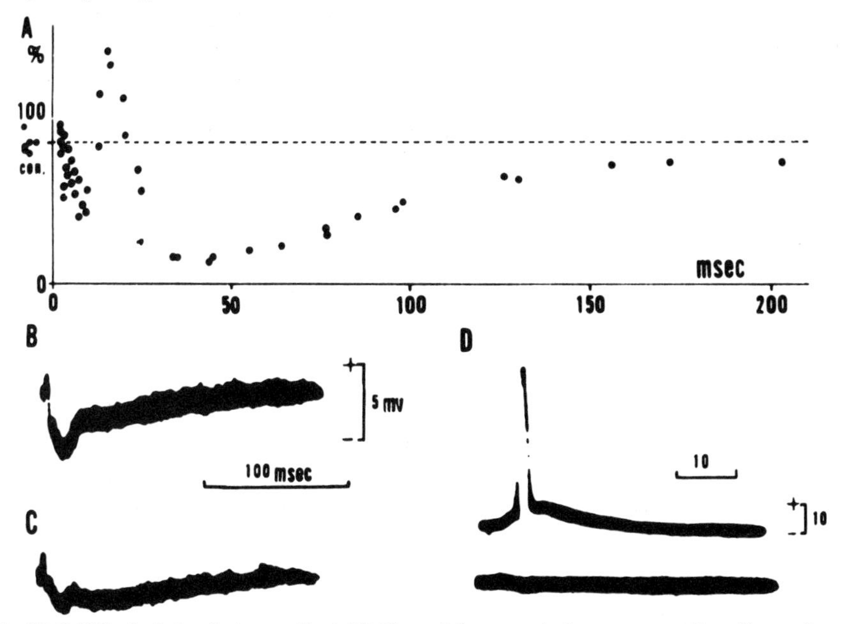

Fig. 22.5. Effect of strychnine on the inhibition of the masseteric monosynaptic reflex and IPSPs of masseter motoneurons induced by inferior alveolar stimulation. The masseteric monosynaptic reflex in the masseter nerve induced by stimulating the mesencephalic tract was conditioned with a supramaximal inferior alveolar shock after strychnine sulphate, 0·1 mg per kg, given intravenously. Data in A were recorded after strychnine. The IPSP in B was recorded in a masseter motoneuron before strychnine and trace C after the drug. D upper trace illustrates a response obtained after strychnine in which an inferior alveolar nerve shock induced a depolarization with a spike (upper trace). Lower trace: the extracellular potential. (*From Kidokoro et al., 1968.*)

Nature of transition between the two phases of inhibition

Fig. 22.5 illustrates the effect of intravenous administration of strychnine (0·1—0·4 mg per kg) on the two phases of suppression of the monosynaptic reflex and the IPSPs induced by ipsilateral inferior alveolar nerve stimulation. Following strychnine, facilitatory effects on the monosynaptic reflex became evident during the transition between the early and late phases of depression. Finally, the early depression was reversed to a complete facilitation with almost the same latency as the early depression (*Fig.* 22.6A, C). On the other hand, the late depression was often slightly enhanced (*Fig.* 22.6A, C). Strychnine exhibited the same effects on the inferior alveolar induced IPSPs of masseter motoneurons (*Fig.* 22.5B, C). Also, after strychnine, inferior alveolar nerve stimulation occasionally evoked depolarizing potentials starting at 2·54 ± 0·27 msec (n = 11), which sometimes resulted in spike generation (*Fig.* 22.5D). The resting potentials of motoneurons were not altered by strychnine.

Bradley et al. (75) observed that, following intravenous strychnine, the direct inhibitory action of Group Ia afferents on spinal motoneurons of the cat was reduced, and a secondary facilitation with a little longer latency appeared. They postulated that strychnine blocked the action of the inhibitory transmitters by occupying their receptor sites and revealed the masked EPSPs from Group Ib afferents. Curtis (131, 132) confirmed Bradley's proposal and showed that strychnine selectively blocked glycine transmission in the central nervous system of mammals. On the other hand, Pollen and Ajmone-Marsan (458) and Stefanis and Jasper (538) have inferred that strychnine has a direct effect on the subsynaptic membrane, permitting the passage of large ions, including presumably sodium, so that IPSPs are reversed to become depolarizing potentials of similar latency and duration. Either of these two mechanisms might explain the reversal of the inferior alveolar induced IPSPs to depolarizing potentials following strychnine administration.

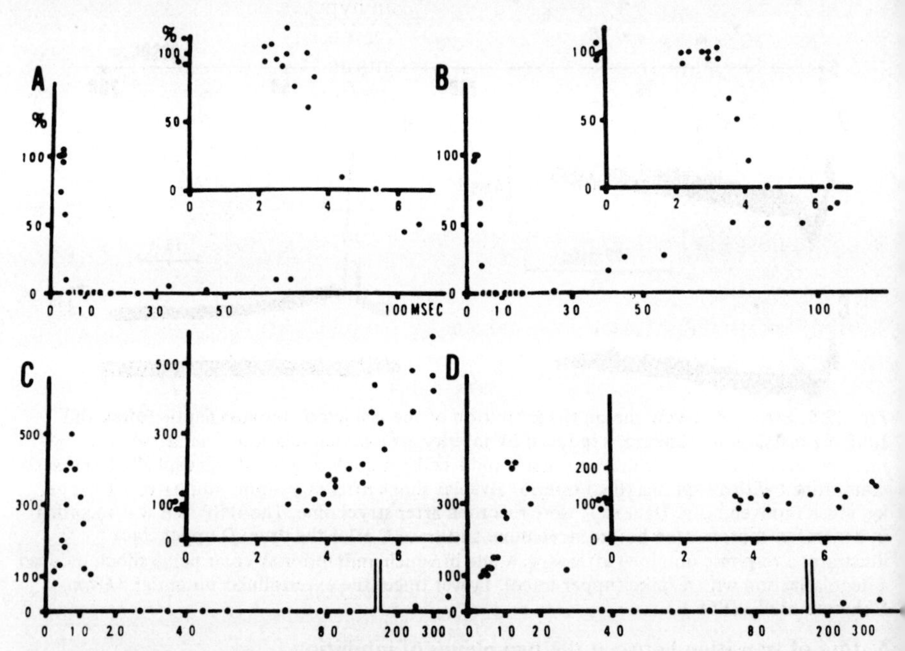

Fig. 22.6. Effects of intravenous administration of strychnine (0·4 mg per kg) on the time course of the two phases of depression of the masseteric monosynaptic reflex. A, time course of the two phases of depression induced by stimulation of the ipsilateral inferior alveolar nerve before administration of strychnine, B, time course of the two phases of depression induced by stimulation of the contralateral inferior alveolar nerve before administration of strychnine. C, ipsilateral nerve stimulation after administration of strychnine. D, contralateral nerve stimulation after administration of strychnine.

Figures on the upper right show the early parts of A to D on an enlarged time scale. (*From Sumino, 1971.*)

As shown in *Fig.* 22.6, although the latency of both the early depression before strychnine and the facilitation afterwards produced by ipsilateral inferior alveolar nerve stimulation were similar, the depression prior to strychnine produced by contralateral stimulation had a latency which was 4 msec less than the corresponding facilitation after strychnine. The latter appears to correspond with the latency of digastric reflex discharges induced by stimulation of the contralateral inferior alveolar nerve. Therefore, the secondary facilitation was not due only to reversal of the early phase of depression. It would be more reasonable to assume that strychnine blocked the early IPSP of masseteric motoneurons induced by stimulation of the inferior alveolar nerve and that a latent EPSP from the same afferent was revealed. The transition between the two phases of IPSPs seems to be part of the latent EPSP. Since the latency of the depolarization after strychnine (2·54 ± 0·27 msec) was similar to that of the early phase of IPSPs before strychnine (2·40 ± 0·21 msec) with ipsilateral stimulation (*Fig.* 22.5), it could be concluded that, in masseteric motoneurons, EPSPs and IPSPs coexist which are activated almost simultaneously by afferent impulses from intra-oral structures. Eccles and Lundberg (155) and Wilson and Kato (629) have reported alternative inhibitory and excitatory pathways from flexor reflex afferents to ipsilateral extensor spinal motoneurons in the cat.

Considering these results, we could propose an hypothesis that at first motoneurons of both jaw-opening and jaw-closing muscles might be excited disynaptically by low threshold intra-oral afferents and that, later, more powerful disynaptic inhibitory pathways from the same afferents to the motoneurons of jaw-closing muscles became active to suppress them during the jaw-opening reflex. Even in the normal condition with a high level of background activity, the early disynaptic excitation can be detected, and this is a possible explanation for the early excitation of the masseter muscles preceding the inhibitory period in the periodontomasseteric reflex (179, 199, 243, 576).

Identification of the interneurons responsible for the early phase of inhibition

In a previous paper (322), we have concluded that inhibitory interneurons responsible for the early phase of IPSP were localized in the region dorsomedial to the trigeminal motor nucleus, which presumably corresponds to the supratrigeminal nucleus. This conclusion was based on the following findings:

1. Within this region, unit spikes were recorded which exhibited the following properties:
   a. Extracellularly recorded spikes were mainly negative-going potentials and intracellular recordings sometimes exhibited small EPSPs, indicating that recordings were derived from somata, not from axons.
   b. Spikes fired repetitively with a frequency of about 700 per sec in response to a single shock and the intervals between spikes (approx. 1·3 msec) corresponded with the interval between the steps in the initial falling phase of the early IPSP.
   c. The latency of the first of a train of spikes was about 0·7 msec later than the arrival of the afferent volley and 0·5 msec earlier than latency of the early IPSP.
   d. Spikes were activated by low threshold afferents in the inferior alveolar nerve and there was close correlation between the number of spikes and the amplitude of the early IPSPs.

   These properties are similar to those of neurons classified as inhibitory interneurons in the central nervous system of mammals (151).

2. Localization of these units was established by electrophoretic injection of dyes. The identified inhibitory interneurons were well localized in the region

dorsomedial to the trigeminal motor nucleus (*Fig.* 22.7), which Lorenté de Nó (358) called the supratrigeminal nucleus and in which Jerge (290) recorded units activated by pressure stimulation of intra-oral structures (teeth, gingiva, palate and tongue).

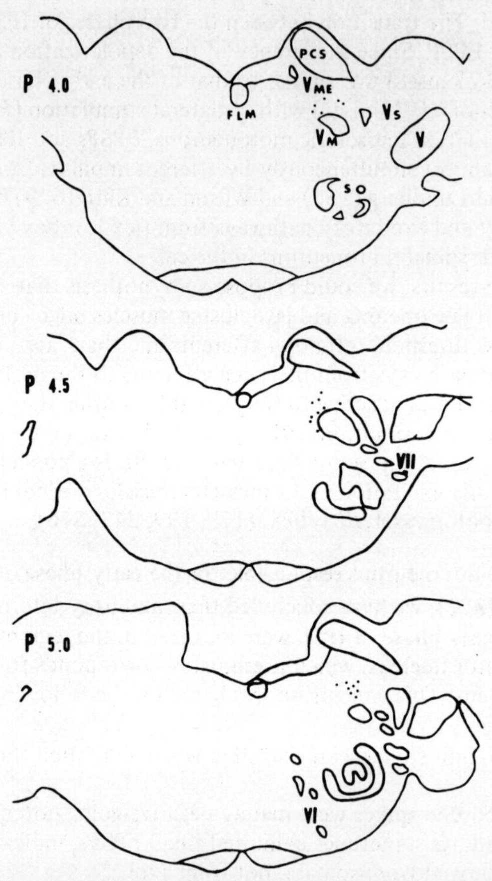

Fig. 22.7. Distribution of points from which records of repetitively active units were derived. Three transverse planes are shown corresponding to the stereotaxic levels indicated to the left. PCS, superior cerebellar peduncle; FLM, medial longitudinal fasciculus; Vm, trigeminal motor nucleus; Vs, trigeminal nerve; So, superior olive; VII, facial nerve; and VI, abducens nerve. (*From Kidokoro et al., 1968.*)

3. Stimulation of the same region produced monosynaptic IPSPs in masseteric motoneurons and the effective area for initiation of the monosynaptic IPSP was in this region.

Goldberg and Nakamura (205) have recorded repetitively firing units dorsal to the trigeminal motor nucleus which were activated by low threshold stimulation of the lingual nerve and they postulated that these were the inhibitory interneurons for the early IPSP in masseteric motoneurons.

Central pathways for the late phase of inhibition

Transection of the spinal tract and nucleus at the level of P 7·5 almost completely abolished the late phase of depression of the monosynaptic masseteric reflex but did not impair the early depression (*Fig.* 22.8). Transection at the level of the obex (P 15) did not produce any significant change in either the early or late depressions (*Fig.* 22.9). Goldberg and Nakamura (205) have reported a similar dissociation of the two phases of depression induced by lingual nerve stimulation following transection of the brain stem at a level 2 mm caudal to the trigeminal motor nucleus. Consequently, it may be assumed that the spinal tract, the spinal tract nucleus and the adjacent reticular formation at the level between P 7·5 and P 15 are necessary for production of the late phase of depression.

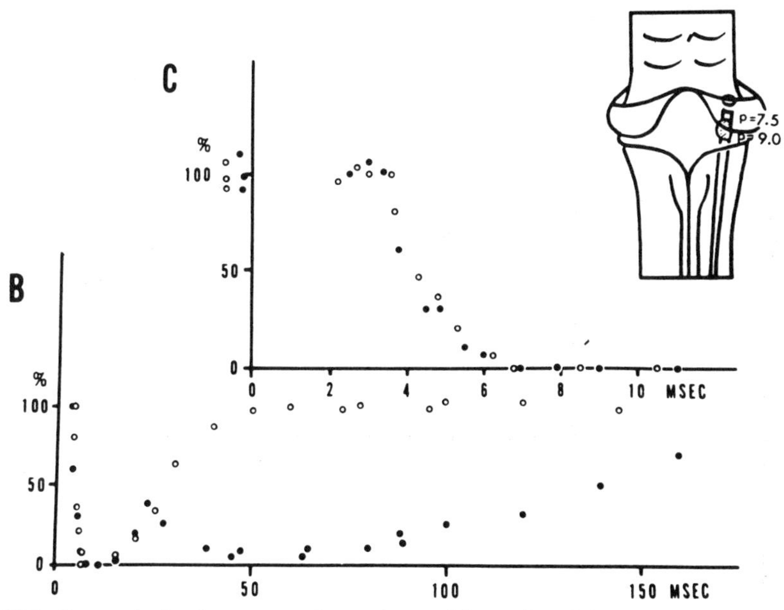

Fig. 22.8. Changes in the time course of two phases of depression of the masseteric monosynaptic reflex by the destruction illustrated in the right upper figure. Closed circles (•) show the time course before destruction; open circles (○) after destruction. (*From Sumino, 1971.*)

The long-lasting, late depression of the monosynaptic masseteric reflex could be evoked not only by stimulation of the ipsi- and contralateral trigeminal nerves (the inferior and superior alveolar nerve, the lingual nerve and the infra-orbital nerve) but also the hypoglossal nerve (425) the superficial radial nerve (335) and even by stimulation of the orbital gyrus of the cortex (424).

Llinás (354) reported that strychnine almost completely blocked the direct inhibition from Group Ia muscle afferents on extensor alpha motoneurons but did not appreciably affect the reticular inhibition on the same motoneurons. It can be concluded that the bulbar reticular formation, activated via nucleus oralis and interpolaris, is involved in the production of the late phase of depression in view of (1) its long-lasting action, (2) the increase in conductance of the cell membranes, (3) its resistance to strychnine and intracellular Cl^- diffusion and (4) the multiple input sources. Presynaptic inhibition may also play some role (154, 201, 505).

Central pathways for the latent disynaptic excitation

Transection of the spinal tract and nucleus at the boundary between the main sensory nucleus and the rostral pole of the spinal tract nucleus, P 6·5–7·0, almost eliminated the secondary facilitation unmasked by strychnine (*Fig.* 22.10). Further, after transection at the level of P 7·0 as shown in the inset of *Fig.* 22.11, stimulation of the spinal tract or the spinal nucleus caudal to the level of section produced the early excitation with a latency of about 3·5 msec, which was followed by the late long-lasting depression (★). This time course closely resembled that of the facilitation following administration of strychnine (*Figs.* 22.6, 22.10). Accordingly, we can conclude that excitatory interneurons in the nucleus oralis, the nucleus interpolaris or the adjacent reticular formation may project directly, at most disynaptically, to the ipsilateral motoneurons of the jaw-closing muscles.

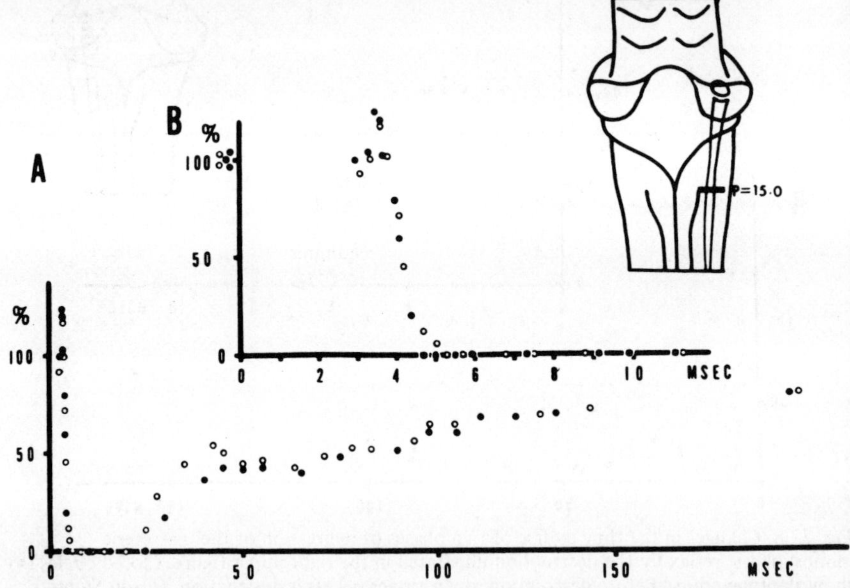

Fig. 22.9. Changes in time course of two phases of depression of the masseteric monosynaptic reflex by transection of the spinal tract and the spinal tract nucleus at the level of the obex illustrated in the right upper figure. Closed circles (●) before destruction; open circles (○), after destruction. B: the early part of A on an enlarged time scale. (*From Sumino, 1971.*)

On the other hand, several investigators (114, 122, 289) have reported that some of the mesencephalic tract neurons can be activated by pressure on teeth or the hard plate. Recently, Funakoshi and Amano (179) have shown that, in the rat, responses to mechanical stimulation of teeth with a latency of 4–8 msec were eliminated by transection of the brain stem between the trigeminal motor nucleus and the mesencephalic tract nucleus. At the same time, slow responses with latencies of 10–30 msec remained. These observations suggest a possible fast excitatory pathway from periodontal receptors to the motoneurons of jaw closing muscles via the mesencephalic tract.

Finally, *Fig.* 22.12 illustrates diagrammatically the central pathways involved in the jaw-opening reflex and the associated tongue movement induced by

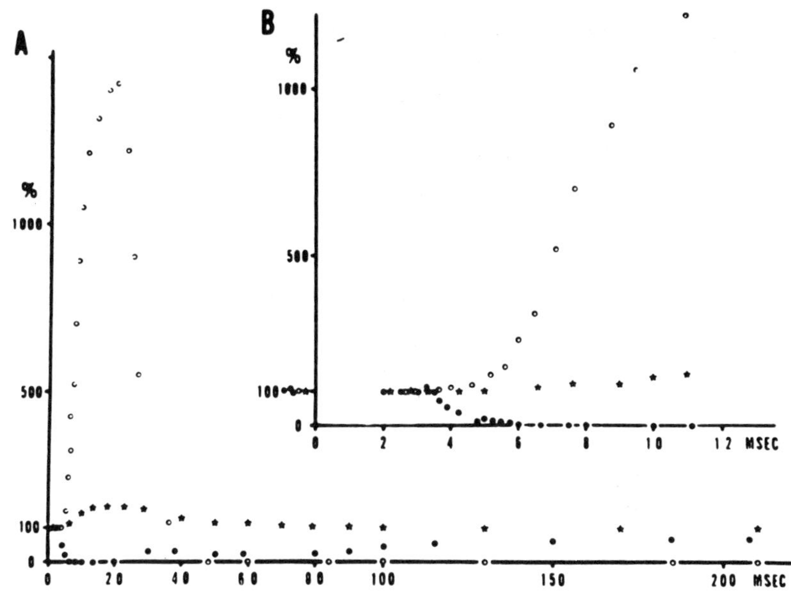

Fig. 22.10. Abolition of the secondary facilitation by transection of the spinal tract and the spinal tract nucleus at the level of P 6·5. Closed circles (•) show the time course of two phases of depression of the masseteric monosynaptic reflex. Open circles (○) show the two phases (early facilitation and late depression) after strychnine administration. Stars (★) show the abolition of the facilitation by the transection. (*From Sumino, 1971.*)

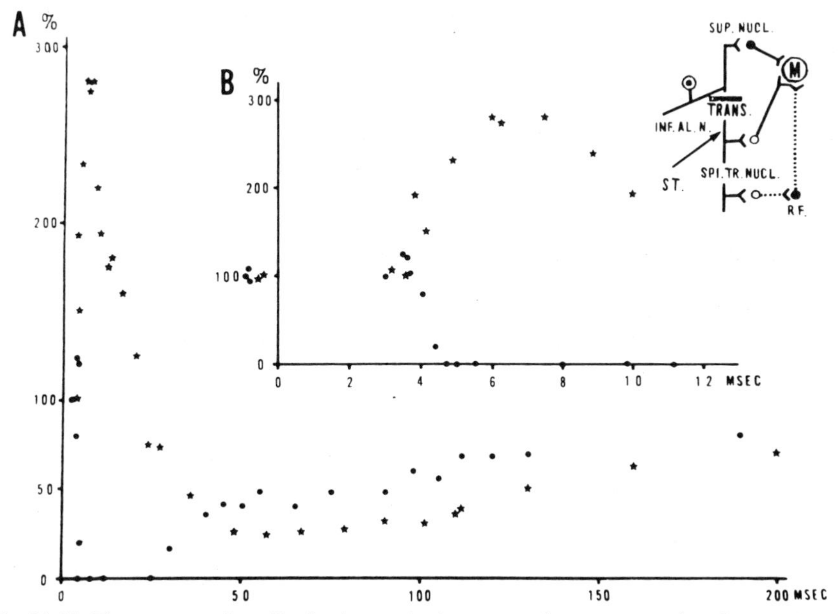

Fig. 22.11. Time course of amplitude changes in the masseteric monosynaptic reflex produced by stimulation of the spinal tract nucleus (P. 9·0, R. 4·0, V. – 6·0) after section of the spinal tract at the level of P 7·0. (upper right). Closed circles (•) show the time course of the two phases of depression of the masseteric monosynaptic reflex evoked by stimulation of the ipsilateral inferior alveolar nerve before the transection. Stars (★) represent the results after the transection. (*From Sumino, 1971.*)

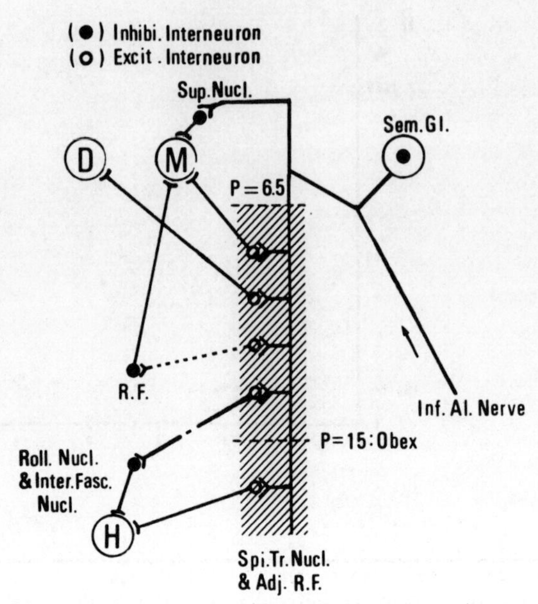

Fig. 22.12. Diagram of the central neural pathways involved in the jaw-opening reflex and the associated tongue movement. D, M and H indicate the digastric, the masseteric and the hypoglossal motoneurons respectively.

stimulation of the inferior alveolar nerve. Details of the trigemino-digastric and the trigemino-hypoglossal reflexes have been described elsewhere. (550, 551).

Acknowledgements
The author expresses an appreciation to Professor Y. Nakamura and Dr S. Nozaki for their criticism and assistance with the manuscript.

Discussion CHAIRMAN: B. MATTHEWS

SESSLE: Did you suggest that, when you stimulated intra-oral nerves, the inhibition was more powerful than the excitation of masseter alpha motoneurons and that the inhibition can overcome the excitation ?
SUMINO: Yes. In our experiments, the disynaptic inhibitory pathway was more powerful than the disynaptic excitatory pathway.
SESSLE: In our studies we sometimes found, when we were tapping the teeth, that the excitatory effects were just as powerful as the inhibitory effects. I showed some examples in which we activated masseter and temporalis alpha motoneurons and also gamma motoneurons by peripheral stimuli. So presumably the peripheral or central state of the animal might play a part in determining whether excitation or inhibition is manifested.
CHAIRMAN: Are there any clinicians with any comments on the suggestion that there is excitatory feed-back on elevator motoneurons from intra-oral mechanoreceptors ?
THEXTON: One can get excitatory responses with pressure applied to the palate. You said that the lowest threshold afferents in the inferior dental nerve are

responsible for the jaw-opening reflex. You were using, I presume, an anaesthetized animal and a single shock stimulus.
SUMINO: We used cats anaesthetized with pentobarbital sodium in a dose of 35 mg per kg.
THEXTON: At least in the infra-orbital nerve in decerebrate cats, 2 or 3 very weak, brief stimuli evoke the excitatory response in temporalis without there being any response in digastric. However, this does depend on the fibre diameter spectrum of the particular branch you are stimulating.
GOLDBERG: The evidence for strychnine blocking and reversing the inhibition, and for the inhibition being postsynaptic is, at least in the trigeminal system, circumstantial. Because it is at least a disynaptic pathway, there is always a possibility that the drug affects the inhibitory pathway selectively and does not affect the excitatory pathway. Although the evidence from the spinal cord suggests that it has postsynaptic effects, that is not conclusive.
SUMINO: There are two possible pathways, one going to the mesencephalic tract and the other going to the spinal tract. After strychnine facilitation, I cut the spinal tract and nucleus at about P 7·5 and the unmasked facilitation was almost but not completely blocked. Furthermore, I cut above this level to block the input to supratrigeminal nucleus, because this has inhibitory neurons. After this, stimulating the caudal part produced changes in the excitability of the masseter motoneurons with a similar time course. This was without strychnine.

Paper No. 23

The Role of Jaw Elevator Muscle Spindles

A. Taylor

In considering the spindles of the masticatory muscles we are here primarily interested in their role in the special motor control tasks of mastication and regulation of the posture of the jaw. However, work in this area has come to have a special place in the study of spindle function because it is only here that a technique has become available for recording their afferent discharge during entirely normal, unrestricted movements. It happens that the cell bodies of the first order afferents from the jaw-closing muscle spindles are located in the mid-brain in the so-called mesencephalic nucleus of the fifth nerve (MeNV). This curious anatomical situation was recognized by Cajal (469) and recordings were made of spindle type activity from the nucleus by Corbin and Harrison (122). Jerge (289) confirmed the presence of cells responsive to jaw opening and to force applied to the teeth. At the same time a view became popular that cell bodies of eye muscle proprioceptors were also present (172), though the evidence for this was never strong. The physiological constitution of the nucleus in the cat was studied in some detail by Cody et al. (114) and it was concluded that it contained the first order cell bodies of spindle afferents, both primary and secondary, from masseter, temporalis and pterygoid (but no evidence could be found for the presence of tendon organ cells nor for eye muscle proprioceptors). The presence of tooth mechanoreceptor cells was confirmed, but it appeared that they were restricted to the caudal part of the nucleus (113). This provided a good foundation for extracellular unit recordings to be made in chronic preparations, the feasibility of which had been suggested in previous work on lightly anaesthetized animals (135, 136, 568). The first published recordings in unanaesthetized animals were obtained by Matsunami and Kubota (387) from the monkey with rigidly restrained head and successful experiments in completely free conscious cats have been reported by Cody and Taylor (115, 566). Now the subject has been supported by important extensions of the experiments with restrained monkeys by examining normal movement patterns (373), the effects of lesions in the tract of the MeNV (211) and the firing patterns of spindle afferents (212). The purpose of this paper is to review the recent observations and to say what we now can of the role of the spindles in jaw movement control.

THE SPINDLES AND THEIR REFLEX CONNECTIONS

Estimates of spindle numbers in man (652) show them to be plentiful in masseter and temporalis with some 155 and 210, respectively. As with the majority of other muscles, this is very approximately 10 per cent of the total number of myelinated nerve fibres in the muscle nerve. The small bulk of the jaw muscles relative to the larger limb muscles places them high in the order of richness of endowment with spindles, excepting lateral pterygoid, which has none (299). There seems to be nothing unusual about the structure of the jaw muscle spindles (299) and functionally there is reasonable evidence for the existence of roughly equal numbers of primary and secondary afferents, judging by their dynamic response to stretch when their intrafusal fibres are excited with succinyl choline (114). Because of the technical difficulties of dissecting the muscle nerves and of providing perfect mechanical recording conditions, there is less detailed documentation of the properties of these spindles than those in the hind-limb muscles of the cat, in which so much of the basic work has been done (*see* Matthews (390)). Nevertheless, it appears quite safe to apply general findings for spindle properties in this particular instance, always recognizing that it would be highly desirable to check them whenever possible.

The existence of phasic and tonic stretch reflexes in the jaw elevators is well known (524), a monosynaptic reflex involving spindle afferents is accepted (556) and the afferent fibres are recognized as entering the brain stem in the special branch of the fifth nerve known as its motor root (376). There seems every reason to suppose that the segmental autogenetic stretch reflex is as important here as anywhere and the shortness of the reflex pathway and the high speed of the muscles (567) indeed, make this particular stretch reflex one of those potentially best suited for negative feedback control in a reasonable bandwidth with minimal problems of instability. At first sight it might seem a simple matter to invoke the stretch reflex as the basis for a length servo drive, via the input to the alpha motoneurons (for reviews *see* e.g. Taylor (565) and Stein (540)). In this scheme, a central programme activated by a variety of possible generalized inputs (142) would feed a pattern of impulses to the motoneuron pools representing a temporal pattern of intended jaw displacement. This would then be executed more or less faithfully depending on the degree of loading provided by the food and the gain around the servo loop. There may indeed be much to be said for this scheme, but we recognize that there is room for far more sophistication of control than this via the spindles (leaving aside other receptors), since the central properties of the reflex can be altered (62, 112) and semi-independent control of incremental sensitivity of spindles to displacement and velocity may be exerted continuously via the fusimotor system (390).

DOES THE STRETCH REFLEX DO ANYTHING FOR NORMAL MASTICATORY MOVEMENTS ?

Wisely recognizing the difficulties of interpreting the significance of spindle activity, Goodwin and Luschei (211) started by asking this simple question. They made unilateral and bilateral lesions in the tract of the MeNV so as to block all spindle input to the motoneurons. That they had achieved this was demonstrated by the loss of the phasic jaw stretch reflex on one or both sides. The effects of the lesion were a transient clumsiness in manipulating food and a preference for chewing on the opposite side. Certainly there was no evidence that the spindle input was necessary for the operation of a normal masticatory movement programme and whatever the spindle input may have been doing normally, it could evidently be effectively replaced by other sensory modalities. The alternative possibility that it does nothing at all is scarcely credible in view of the elaborateness of the provision of spindles emerging from the evolutionary process.

WHAT INFORMATION DO THE SPINDLES NORMALLY PROVIDE ?

It might well be concluded that the sophistication of the nervous system, with its multiple parallel mechanisms for any given task, has defeated the simplistic assault of trying to see what one mechanism normally contributes by cutting it out. The alternative approach is to see what information is normally provided by the spindles and then to deduce what it must be doing from a knowledge of the anatomy and physiology of their connections. Clearly this knowledge is incomplete, but we can start by recording from the spindles in normal movements.

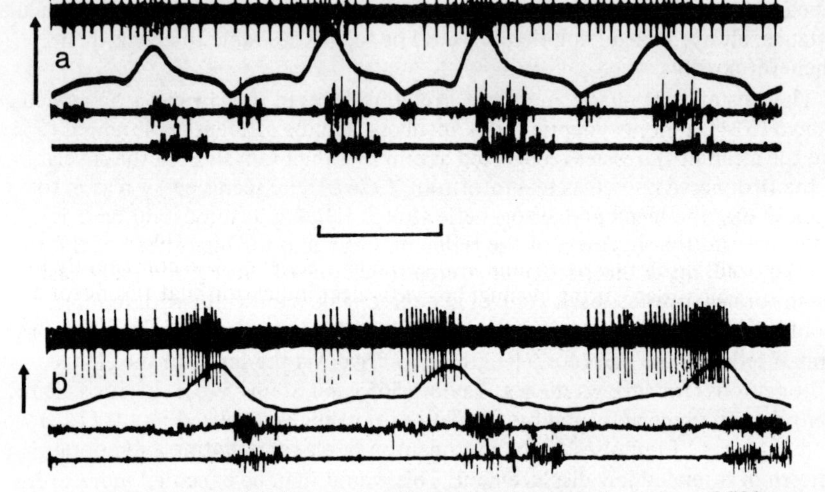

Fig. 23.1. Responses of spindle afferents recorded from mesencephalic nucleus of fifth nerve in an unrestrained cat during eating. a, 'Low frequency' unit from masseter. b, 'High frequency' unit from masseter. In each case the traces from above downwards are spindle afferent, jaw movement (arrow: 25° opening), masseter and temporalis EMG. Time: 200 msec.

In unanaesthetized cats, Taylor and Cody (566) were able to divide spindle units into two groups according to their maximum instantaneous frequencies (MIF) during eating movements. The high frequency group (MIF 220–400 per sec) were highly phasic whereas the low frequency group (MIF 120–180 per sec) were essentially tonic in behaviour. They were tentatively identified as belonging to primary and secondary afferents respectively on the basis of an extension of the observations in the anaesthetized cat (114).Generally the 'primaries' were either silenced or greatly reduced in frequency during active muscle shortening and speeded greatly during stretch (*Fig.* 23.1a). The 'secondaries' were much less strongly modulated and commonly continued to fire at reducing frequency during active shortening (*Fig.* 23.1b). We have since been able to show that the behaviour of the 'secondaries' is essentially that of length transducers with incremental sensitivity adjusted, apparently by static fusimotion, to be 2 to 3 times greater during small amplitude movements of lapping than during the much larger chewing movements. In addition, the relationship of length to firing frequency is the same during closing as during opening. By contrast, the 'primaries' are generally purely velocity sensitive with very little dependence on actual length. It has not so far been possible to recognize systematic changes in the velocity sensitivity between eating and lapping. The general conclusions are that, during normal eating and lapping movements in the cat, fusimotion proceeds so as to make secondaries and primaries into tolerably linear length and velocity transducers respectively. There is no evidence in this case to support the idea (394) that the movements might be 'driven' via the fusimotor route to any substantial degree, as this would require increased firing in jaw-closing spindles during active closure. Indeed, the original idea related particularly to the reflex effects of primaries, which here seem to be completely silenced during the closing movements of mastication. Neither is there any real support for α–γ coactivation (*see* Granit (219)) as a useful interpretation of the data because there appears to be less static fusimotor drive during the strong contractions of eating than during the weaker movements of lapping. Rather, one is impressed by the apparent flexibility of the relationship between muscle activity and fusimotion. Matsunami and Kubota (387) had also found a very variable relationship between jaw motion and spindle firing, but understanding of their results is limited by inadequate movement recording and characterization of units.

The recent work of Goodwin and Luschei (212) has taken into account the added complexities of movement of the jaw in monkeys with a system of recording vertical and lateral displacement in animals trained to accept rigid head stabilization. The general finding was again that spindle firing frequency was consistently higher during opening than during closing, but with clear indication that fusimotor activity during closing was usually present. In their general attitude Goodwin and Luschei favoured existence of α–γ coactivation and argued against 'γ leading'.

THE POSSIBILITIES FOR THE SPINDLES IN MOTOR CONTROL

In trying to decide the role of the muscle spindles in a particular task we must recognize that the control problem to be faced varies from one part of the body to another, so that generalizations are risky. It is well established for example that the extra-ocular muscles contain spindles (or equivalent receptors) yet they have no reflex effect at all on the extra-ocular motoneurons. The postural muscles, on the other hand show well marked tonic and phasic stretch reflexes. Respiratory muscle spindles exert some autogenetic excitatory effects (513) and in the case of the small muscles of the hand, though an H reflex can sometimes be evoked, it is not normally possible to demonstrate a stretch reflex at the segmental level (386). The masticatory muscles appear to lie somewhere in the middle of this array. The existence of a tonic stretch reflex is beyond doubt, and this must function as elsewhere to regulate the posture of the jaw. However, such a function can scarcely justify such an abundant provision of receptors with facilities for regulation of their static and dynamic sensitivities. The lack of reflex connection for the extra-ocular muscle spindles is theoretically justified since no postural action is required and the load to be moved is constant and predictable in its behaviour. In the case of jaw elevators, the load is not constant so there could be definite advantage in having some degree of servo action if a centrally determined programme is to be executed. This does not mean (contrary to the view often expressed) that the drive to initiate movement should be via the fusimotor route. It only requires that the spindles should not cease to function during active shortening of the muscle and that there should be some appreciable gain in the reflex arc. The observations on the cat reported here show that the presumed primaries, by silencing during closure, are essentially unfitted for this role much of the time. This is disturbing, because it is the primaries which have a long standing claim to be the receptors for the monosynaptic stretch reflex.

However, help has arrived in the form of the recent finding by Kirkwood and Sears (323) that spindle secondaries can provide important monosynaptic excitatory input. This brings attention back to our demonstration that the presumed secondaries of the jaw elevators can fire throughout the movement cycle. The reduced frequency in shortening may thus merely signify the operation of negative feedback in the length servo driven via input to the α motoneurons. On this basis, the changing incremental sensitivity to stretch in eating relative to lapping must signify changing loop gain as well as a linearization of spindle response over the appropriate range of movement.

The tentative conclusion is that the potential significance of the spindle secondary afferents should be emphasized first in the tonic regulation of length as a potential function and secondly as adjustable linear length feedback transducers in the regulation of movement. Clearly we still need subtle means of estimating reflex gain throughout movements and to challenge the control with small, realistic disturbances of loading while monitoring spindle discharge. Finally, it should be pointed out that whether or not the spindles of the jaw elevators are used in the way suggested, they must surely also be used in other mechanisms at a higher level in the central nervous system.

This work was supported by the Wellcome Trust and the Medical Research Council. It is a pleasure to acknowledge the collaborations of Dr F. W. J. Cody and Dr Linda M. Harrison.

Discussion CHAIRMAN: B. MATTHEWS

BULLER: Presumably the short latency of this reflex makes it difficult to see changes in the dynamic index. Have you tried changing the load on the system ? This is where you would expect it to act as a stabilizer.

TAYLOR: That is true. We have not tried loading the system because we have been nervous of trying to set up experiments which would be difficult to interpret. If we were suddenly to load the jaw muscles during movement it is very likely that the animal would suddenly change the programme. In experiments in which finger movements are loaded (394), it is quite obvious that the subject does not just go on applying servo correction blindly, he gives up increasing the force after quite a short while. The system is sophisticated enough that, as soon as there is a slight deviation from normality (presumably outside some predetermined level which is acceptable), the programme is changed. We feel we need some means of applying a rather subtle change in load, the results of which we could interpret but which would not disturb the cat.

GREENFIELD: Is there any point in looking at the digastrics in this experiment ?

TAYLOR: There is no evidence for any significant number of spindles in the digastric. We have never seen any evidence of spindle afferents of jaw-opening muscles in the mesencephalic nucleus. They all appear to be in the jaw-closing muscles. So we have not had any opportunity of recording from them.

KARLSSON: Is it very important that you know whether you are dealing with primary or secondary afferents ?

TAYLOR: We were concerned in the first place in this work with trying to record from primaries because, at that stage, primaries were believed to be the most important, if not the only spindle afferents involved in the stretch reflex in this situation. They were the only ones which were well documented as contributing to the stretch reflex, but now we are quite happy that the secondaries are very likely to exert a quite powerful monosynaptic excitatory input.

BRATZLAVSKY: In the work in which recordings have been made from Ia afferents in humans (654), units were found which were silenced during voluntary activation of the muscle and other units were found which fired more rapidly, the latter ones fitting with the coactivation theory. I wonder if all your units were silenced during jaw closing, and whether you have a significant sample of Ia afferents.

TAYLOR: I have never thought that the idea of coactivation was a particularly illuminating one. It assumes that there is some special advantage in providing a fusimotor system to keep the muscle spindles firing during active shortening. The amphibian muscle spindles manage perfectly well with branches of alpha axons to innervate their intrafusal muscle fibres. The result is that, during contraction of a muscle, the spindles are automatically biased to keep them firing. There seems little advantage in the mammalian system of a dual fusimotor arrangement independent of the alpha motoneurons, if all it is used for is to coactivate the intrafusal muscle fibres. So one must look for something more subtle than that. We know that the fusimotors can more or less independently alter the incremental sensitivity of the spindle to length and to velocity and we have some evidence for it in operation in jaw muscles. I would suggest that, in the mammal, coactivation is not an important part of the mechanism. It is probably more important that the fusimotors regulate the transducing characteristics of the spindles.

LUND: What percentage of the units actually go silent and how many continue to fire ?

TAYLOR: This work was based on about 29 good recordings. Approximately half were believed to be primaries and half secondaries. In all cases the spindles reduced their frequency during shortening in eating and drinking.

LUND: How many actually went silent ?

TAYLOR: The majority of those we labelled as primaries went silent, at least for an appreciable part of the cycle. Of course it is difficult to say what length of pause should be called silence.

Paper No. 24

Evidence for a Central Neural Pattern Generator Regulating the Chewing Cycle

J. P. Lund

For many years, most investigators of masticatory control studied the jaw-opening and jaw-jerk reflexes first described by Sherrington (524) on the assumption that these reflexes are the basic units of mastication. Sherrington showed that mechanical and electrical stimuli applied to the gum bordering the upper and lower teeth or the front part of the hard palate evoked jaw opening in the decerebrate cat. This jaw-opening reflex, which can also arise from periodontal receptors (240) was then sharply followed by active closure, the jaw-jerk reflex, initiated by stretch of the spindles within the elevator muscles (376). Sherrington pointed out that biting into food tends to initiate the jaw-opening reflex, which in turn activates and alternates with the jaw-jerk reflex 'so long as there is something biteable between the jaws' (124, 524). This suggestion prompted Rioch (486) to formulate a theory explaining

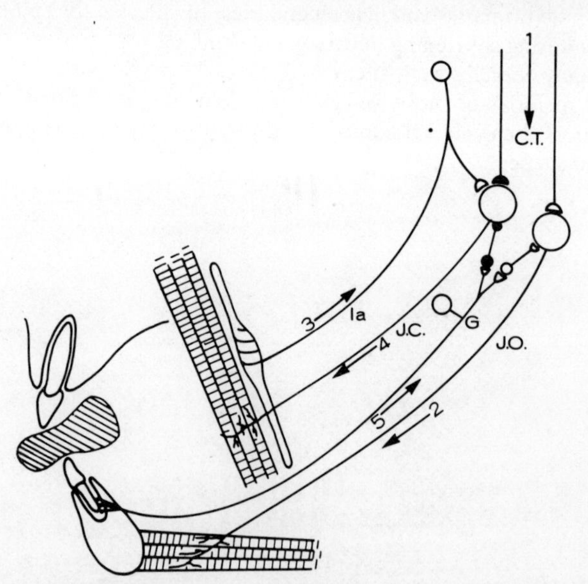

Fig. 24.1. Diagram of the chain reflex theory of masticatory control as elaborated by Rioch (1934). C.T., corticobulbar tracts; J.O., jaw-opening motoneurons; J.C., jaw-closing motoneurons; G, periodontal or mucosal afferents; Ia, muscle spindle afferents from jaw-closing muscles; white neurons–excitatory; black neurons–inhibitory. The sequence and direction of hypothetical events are given. No attempt has been made to depict accurately the number and importance of neurons or synapses.

how electrical stimulation of the masticatory area of the motor cortex and by inference voluntary control, could produce rhythmical mastication. She suggested that the masticatory area of the cortex acts directly upon the trigeminal motor nucleus to excite the jaw-opening motoneurons and inhibit the jaw closers (*Fig.* 24.1). Jaw opening would thus occur, stretching the jaw-closing muscles; this would in turn excite the jaw-jerk reflex. As soon as the jaw was raised, opening would occur again and the cycle could then be repeated as long as the jaw-opening motoneurons were tonically activated from above. Food between the teeth would additionally reinforce the jaw-opening phase. This theory of peripheral patterning of mastication was generally accepted (105, 291, 310, 311).

The basic control system for mastication does, in fact, appear to lie below the mesencephalon since animals decerebrated above the pons are capable of chewing. The additional removal of the cerebellum in a previously decerebrated rabbit does not prevent rhythmical mastication from occurring (*Fig.* 24.2). Bazett and Penfield (43) maintained cats for up to 3 weeks after decerebration. These animals not only swallowed but chewed their food or stomach tube if it was inserted into the back of the mouth. Chewing can also be provoked in the decerebrate cat and rabbit by rubbing the mucosa of the buccal commissures (83) or by electrical stimulation of the corticobulbar tracts during their passage through the pons (142, 367, 381). The intra-oral and corticobulbar stimuli facilitate one another (367).

However, are these rhythmic alternating movements patterned basically by alternating reciprocal reflexes, or by a central motor programme which directs the sequence of activation of specific motoneurons without the obligate participation of returning information from peripheral sensory receptors? Such motor programmes

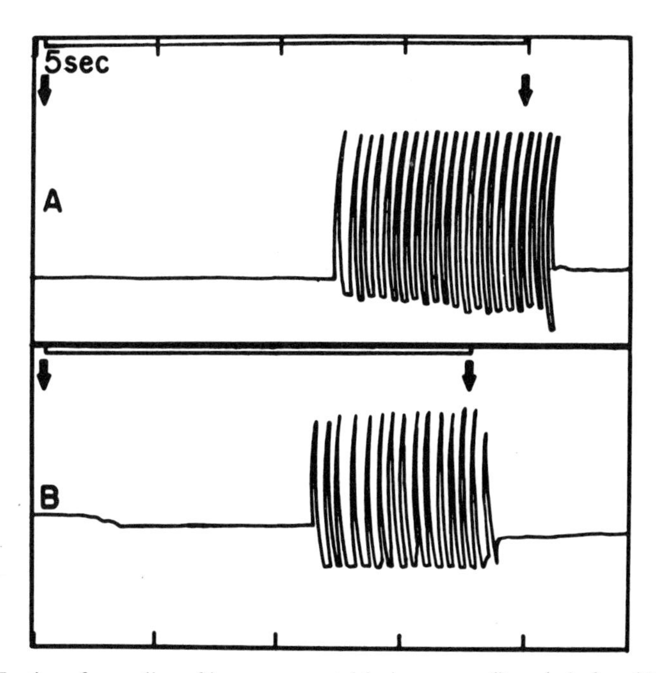

Fig. 24.2. Tracing of recording of jaw movement (closing: upward) made before (A) and after (B) removal of the cerebellum in a decerebrate rabbit when stimulating adjacent to the corticobulbar tracts (6V, 1 msec, 10 Hz).

are known to control many repetitive sequences of movement, particularly in invertebrates (*see* review by DeLong (140)). Sumi (549) and Dellow and I (142, 368) have attempted to provide an answer.

During repetitive electrical stimulation of the motor cortex or descending corticobulbar pathways of rabbits, bursts of activity can be recorded from the cut mylohyoid and masseteric motor nerves which are in phase with jaw-opening and jaw-closing respectively (142). If Rioch's peripheral control theory is correct, then paralysis of the animal should be followed by tonic activation of digastric motoneurons and a lack of activity in masseteric motoneurons, since the masseteric monosynaptic reflex cannot occur without muscle stretch. However, as can be seen in *Fig.* 24.3A, throughout the period of stimulation alternating bursts of discharge continue to be recorded from these two nerves after the injection of the neuromuscular blocking agent, gallamine triethiodide.

Rhythmic bursts of activity in phase with mastication can also be recorded from the hypoglossal nerve or motor nucleus which persist after paralysis (*Fig.* 24.3B, C). Sumi (549) has shown that the pattern of discharge of single hypoglossal motoneurons is essentially unchanged by paralysis. The majority of these units discharged in phase with jaw closing, during which tongue retraction occurs, while others fired during jaw opening and tongue protrusion.

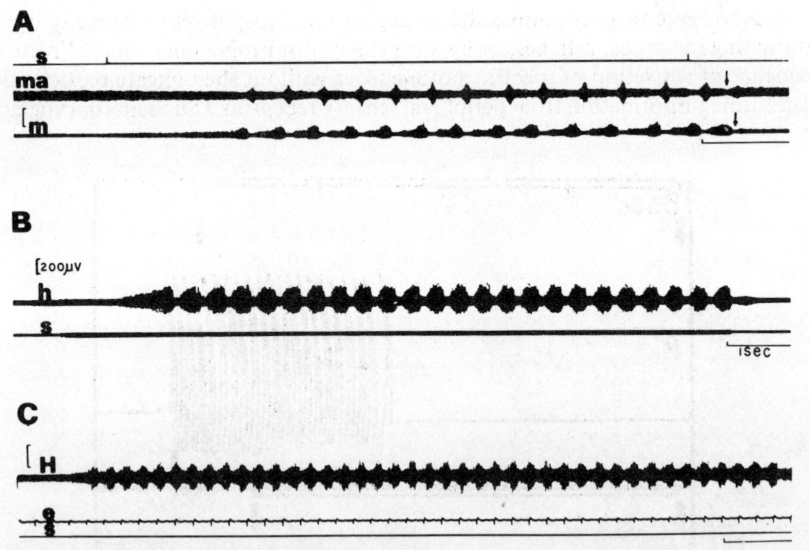

Fig. 24.3. Neurograms showing multi-unit activity recorded from three paralysed animals while stimulating (A) cortical white matter, 30 V, (B) pedunculus cerebri, 9·9 V, (C) internal capsule 12 V. The frequency of stimulation was 40 Hz and of pulse duration 1 msec. Neurograms: Ma, masseteric; M, mylohyoid; h, hypoglossal; H, hypoglossal nucleus; s, stimulus; e, ecg. The arrows indicate that neurogram activity is depressed below resting levels when the antagonist is active. (*From Dellow and Lund, 1971.*)

In a number of animals, the branchial and cervical nerves and the spinal cord were cut to test the possibility that afferent inputs from unblocked intrafusal muscle fibres, peripheral respiratory or vascular mechanoreceptors were responsible for the rhythmical output. These procedures were without effect on the rhythmic activity recorded from the hypoglossal nucleus.

The above results prove that neurons within the brain stem can pattern the alternating activity in jaw-opening and jaw-closing muscles and in tongue protractors and retractors without feedback from the periphery. However, to what extent are the rate and regularity of the repetitive bursts of discharge dependant on the intrinsic properties of this central mechanism and independent of the input parameters ? Increasing the stimulus frequency and strength does increase the frequency of its output to the masticatory muscles but a limiting output frequency exists (*Fig.* 24.4).

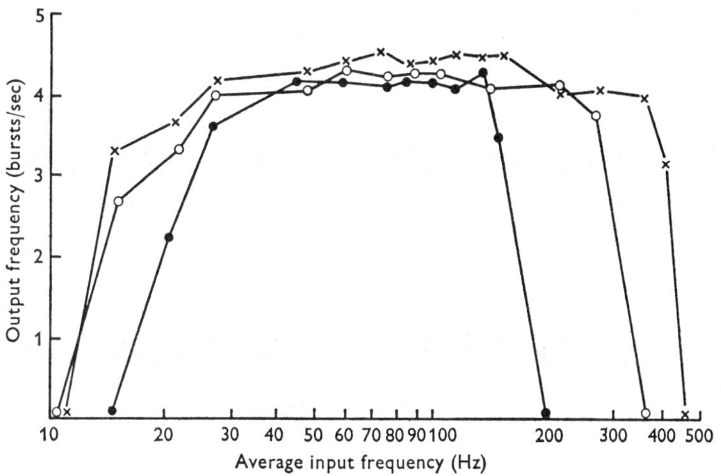

Fig. 24.4. Semilog. plot of the relationship between the average stimulation frequency using random trains of monophasic pulses and the rate of occurrence of bursts of neural activity. The stimulus was delivered to the internal capsule. The three stimulus voltages used were 1·5 T, 6·75 V (•); 2T, 9V (○); and 3T, 13·5 V (x). (*From Dellow and Lund, 1971.*)

However, unlike certain other known central pattern generators (275, 627), the addition of proprioceptive feedback during unobstructed movements does not increase the output frequency produced by a given input (*Fig.* 24.5).

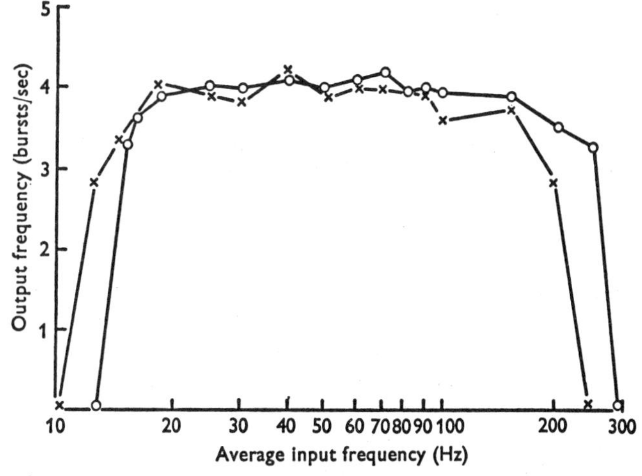

Fig. 24.5. Semilog. plot of the relationship between the average stimulation frequency and the output bursts before (○) and after (x) paralysis with gallamine. The subcortical white matter was stimulated with random trains of monophasic pulses at a voltage 1·5 times threshold (10·5 V). (*From Dellow and Lund, 1971.*)

It is possible that the constant frequency of the bursts of motoneuron activity is a consequence of regularity in the input. A train of shocks given at a constant rate could induce a regular output pattern. Alternatively, the oscillator may be capable of giving a rhythmic output even from an unpatterned input. It was for this reason that the repetitive shocks were given in random trains to evoke masticatory activity. After paralysis, bursts were still recorded from the hypoglossal nerve or nucleus and these occurred at apparently regular intervals (*Fig.* 24.6). The signals were also integrated and gated to allow autocorrelations to be performed (*Fig.* 24.7). These show clear, well separated peaks, indicating a regular process.

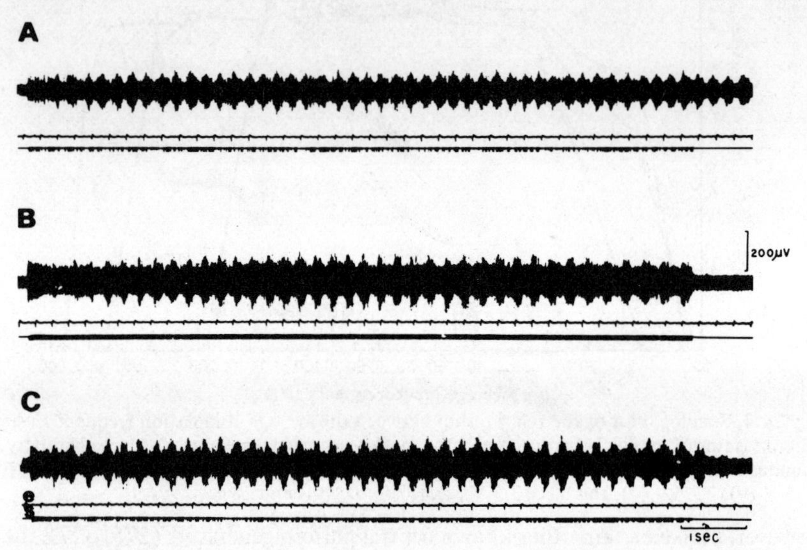

Fig. 24.6. Recordings from the hypoglossal motor nucleus made during monopolar, monophasic stimulation of the internal capsule (A) 3T (24V), average frequency 152 Hz; (B) 2T (16V), average frequency 285 Hz; (C) 1·5 T (12V), average frequency 221 Hz. e, ecg; s, stimulus. All three records were taken from the same animal. (*From Dellow and Lund, 1971.*)

Before we could accept that the regular masticatory pattern was generated basically within the central nervous system, it was necessary to eliminate the possibility that timing information could arise from vascular pulsations within the brain or from respiratory movements (622). Turning off the respirator did not abolish the activity and the cross-correlation histograms relating the appearance of a hypoglossal burst to the cardiac cycle showed the type of sample to sample variations that one would expect from independent pacemakers (450).

Blowing up a small balloon in a rabbit's mouth also caused rhythmical bursts of activity in the masticatory system which persisted after paralysis (*Fig.* 24.8). Summation occurred between central and peripheral excitatory stimuli and inhibitory peripheral stimuli interrupted the cycle; these interactions also persisted after paralysis (368). Thus, even the oral stimuli which provoke mastication do not do so by the alternate activation of reflexes (366).

It can thus be concluded that the chewing cycle is controlled basically by a pattern generator within the brain stem. The site of the neurons making up this structure is not presently known. Ramon y Cajal (469) suggested that the reticular formation medial to the trigeminal complex was a likely site, while Kuypers (339) proposed that it may lie within the spinal trigeminal nucleus. Both these regions receive cortical and trigeminal afferent projections and themselves project to the necessary cranial motor nuclei (221, 359, 403, 469, 545, 595, 596, 631, 632).

A simple model for the central masticatory controller can be proposed in which both the short latency trigeminal and hypoglossal responses to stimulation of nerves or cortex, and the long latency response of rhythmical mastication, are mediated through common interneurons. Jaw-opening motoneurons discharge about 5 msec after single shocks to the cortical masticatory area (105, 106) and about 3 msec after stimulation of trigeminal sensory nerves, with concurrent inhibition of jaw-closing motoneurons (205, 321, 550). Porter (460) and Sumi (549) identified single

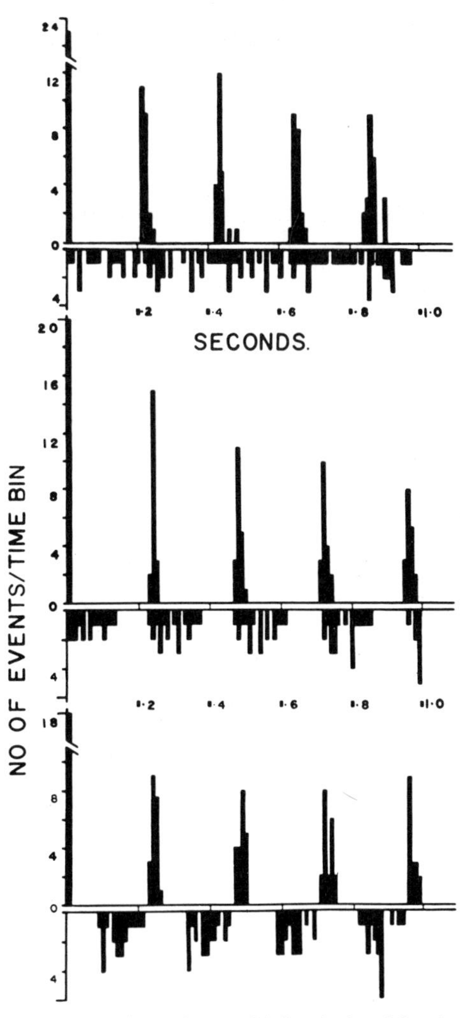

Fig. 24.7. Analyses of neurograms shown in *Fig.* 24.6, calculated for the 5 sec of stimulation in which the neural pattern was most regular. The cross correlation histograms between heart beats and the following bursts of neural activity are shown beneath the autocorrection histograms of the integrated neurograms. Both analyses were continued for 1 sec following the initial event. Bin width 20 msec. (*From Dellow and Lund, 1971.*)

hypoglossal neurons which discharged 5–15 msec after each cortical shock; these cells are connected to both corticobulbar and lingual nerve inputs by interneurons in or near the spinal trigeminal nucleus. As previously shown, both corticobulbar and oral stimulation will activate rhythmical mastication but bursts of

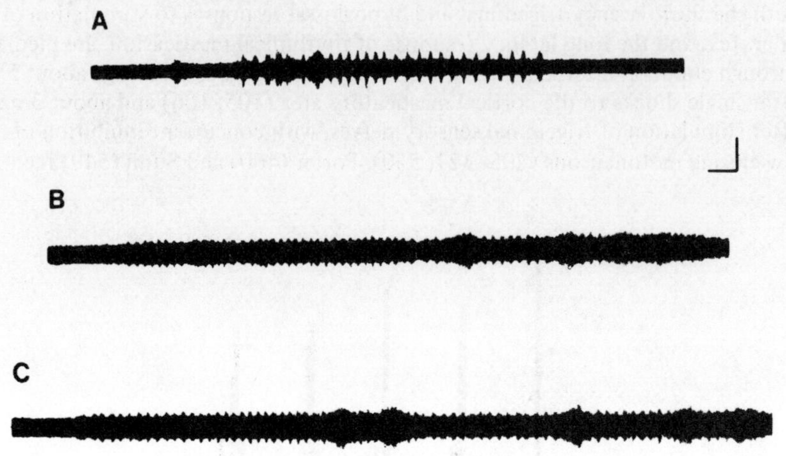

Fig. 24.8. Three recordings from the hypoglossal nucleus of two paralysed rabbits. B and C were taken from the same animal. The rhythmical masticatory discharges, which were interrupted by longer lasting swallowing activity, occurred in response to inflation of a balloon in the mouth. Calibrations: 1 sec; 200 μV.

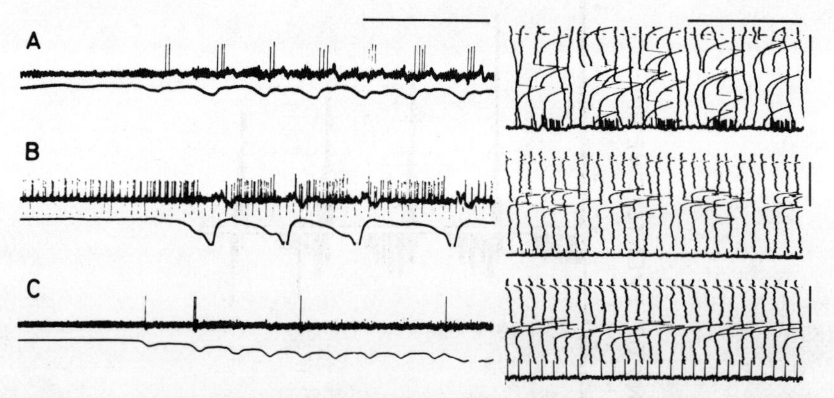

Fig. 24.9. A–C. Hypoglossal nerve fibres displaying different patterns of activity to cortical stimulation. Records from two examples of each of three types of unit (A, B and C). On the left, the discharge of the unit is shown with a record of jaw position (opening: upward deflection). On the right, the discharge of the unit is shown on a slow time-base running horizontally and on a fast time-base, running from above downwards. Cortical stimuli were applied throughout the period of recording and each of the vertically swept traces on the right was triggered just before a stimulus was applied. A,Rhythmically grouped discharge in phase with jaw opening (left). The firing of a unit of this type was not synchronized to the stimulus (right). B, Units of the type which discharged both after each stimulus and in phase with rhythmic chewing (jaw closing in the example shown on the left). Within the burst of impulses which occurred at a particular phase of jaw movement, the greatest probability of firing was after a stimulus (right). C, *Left:* a unit which discharged sporadically, neither locked to the rhythm of chewing nor to each stimulus. C, *Right:* a unit which followed each stimulus regularly, without locking to the rhythm of chewing. Time bars: 1 sec (horizontal) and 10 msec (vertical) respectively. (*From Sumi, 1970.*)

neuronal activity occurred during repetitive CNS stimulation only after a latent period of more than 80 msec (365). One could suggest that during the relatively long period that precedes the first burst of masticatory activity the responses to each single stimulus are gradually changed into a rhythmic output, as neurons of the pattern generator become coupled together.

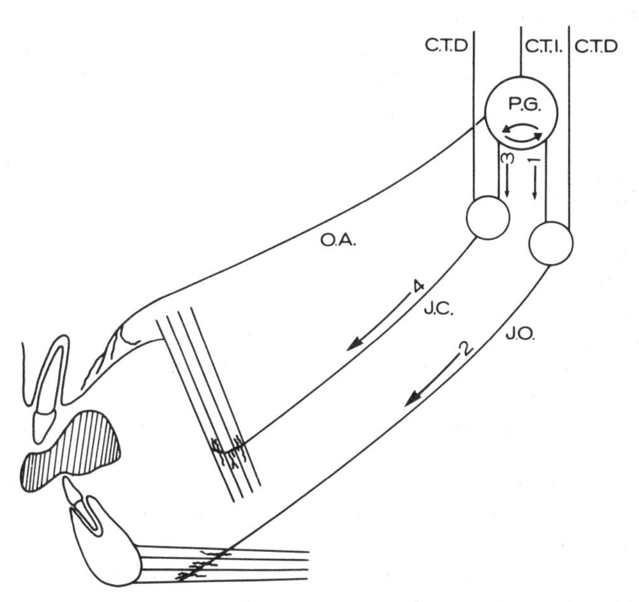

Fig. 24.10. Diagram of a masticatory control system based on a central neural pattern generator (P.G.). C.T.I., indirect corticobulbar input to bulbar motoneurons via P.G.; C.T.D., direct corticobulbar input (monosynaptic or via interneurons); J.O., jaw-opening motoneurons; J.C., jaw-closing motoneurons; O.A., oral afferents capable of activating P.G. (*See* Lund, 1976). Inputs to motoneurons are both excitatory and inhibitory. It is probable that the input from P.G. is inhibitory during the activation of the antagonistic muscle group (e.g. arrow in *Fig.* 24.3). Both the pacemaker and motoneurons are probably susceptible to peripheral inputs.

There is evidence, nevertheless, that the masticatory pattern is not formed by the interneurons in the fast corticobulbar pathway. During repetitive stimulation, most hypoglossal motoneurons continued to fire after each stimulus pulse and also in bursts in time with each chewing cycle, while another group of hypoglossal motoneurons discharged repetitively only in time with mastication Sumi (549) (*Fig.* 24.9). There are two alternative explanations for these observations. Either the hypoglossal motoneurons are themselves capable of elaborating the masticatory pattern; or there are two corticohypoglossal inputs, one responsible for the discrete, short-latency discharge and another which activates the interneurons which generate the masticatory pattern (*Fig.* 24.10). Neither the hypoglossal nor the trigeminal motoneurons appear to have the properties necessary for pattern generation. Strong excitatory or inhibitory coupling is needed to account for the regular appearance of synchronized bursts in many neurons (626). The motoneurons probably do not have the collaterals required for this activity (359, 360, 469) and they are not synaptically activated during antidromic stimulation of adjacent cells (321, 356, 410, 459, 551).

Thus, the evidence indicates that the basic pattern of rhythmical mastication is generated amongst a pool of interneurons at the bulbar level. Activation of this structure would occur from higher centres of the brain and during stimulation of the mouth by a food bolus.

Acknowledgements

J.P.L. is a Scholar of the Medical Research Council of Canada. Certain figures were reproduced by courtesy of the *Journal of Physiology, Experimental Neurology* and Dr Tadaki Sumi. I wish to thank Messieurs E. Rupnik, D. Cyr, R. Péloquin and Mlle C. Laurier for their help in preparing this manuscript.

Discussion CHAIRMAN: B. MATTHEWS

KARLSSON: Do you envisage that there are two brain stem pattern generators, one on each side ?
LUND: I would think there are probably two centres intimately linked together across the midline.
MEYER: Did you do cross correlations with the EKG and the arterial pulse ?
LUND: No, just the EKG.
MEYER: Could the discharge have been in phase with the blood volume pulse, let us say, in the muscle?
LUND: We got wide fluctuations (90–180°) in phase relationships. The phase shifts were too large for blood volume pulses, which would be in phase with the EKG, to be involved.
BOSSMAN: Over what period did you do the correlations ? Although two signals may be totally uncorrelated, you can show some correlation if the sample period is too short.
LUND: The sample time for those correlations was 20 sec. The total observation period is limited because, if you continue to stimulate, the activity eventually dies away.
BOSSMAN: I think that 20 sec is too small for signals of such low frequency.
LUND: I agree, but we are incapable of driving the system for too long with electrical stimulation because of polarization of the electrodes and because the masticatory activity eventually dies away with repetitive stimulation. It was for this reason that we carried out a number of cross-correlations and attempted to show variations in phase relationship from one series to another.

Paper No. 25

To what Extent is Mastication Programmed and Independent of Peripheral Feedback?

A. J. Thexton

There is very little firm information on the relative importance of central programming and peripheral feedback in the production of masticatory movements and what information is available tends to be inconsistent. However, this inconsistency may be more apparent than real in that it may be due to an underlying assumption that the relative importance of the two influences is fixed and immutable; as will become evident from the examples given, there is little justification for such an assumption. It will also be evident that the experimental approach used can bias the conclusions reached. The disparate results obtained in physiological experimentation, in normal biting and in normal chewing are therefore considered separately; the final section of the paper suggests a reconciliation of some of the data.

EFFECT OF PERIPHERAL FEEDBACK IN EXPERIMENTAL SYSTEMS

It has been shown that, in the rabbit, some form of brain stem 'rhythm generator' exists which is capable of producing cyclic jaw movements (142). Whether the activity is elicited by central electrical stimulation or peripheral mechanical stimuli (367), it is independent of subsequent peripheral feedback, i.e. the centrally generated activity can be recorded during motor paralysis. A very similar response, which can also be elicited in the absence of sensory feedback from the periphery, has been found to occur when steady pressure is applied to the hard palate of the

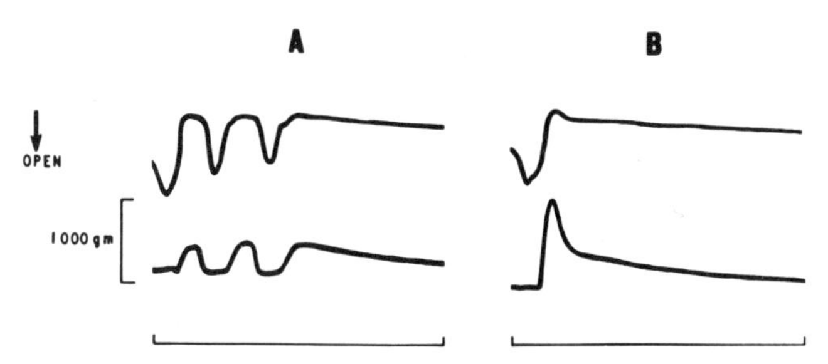

Fig. 25.1. The 'masticatory' response elicited in a decerebrate cat. Record A demonstrates the control response. Record B was obtained after heavy infiltration of the upper and lower incisor regions with local anaesthetic. The top traces show jaw movement. The bottom traces show the force of closure as registered by a bite gauge placed between the upper and lower incisors. Note that in B the loss of sensory feedback from the anaesthetized force bearing areas does not prevent initiation of the response but does prevent its continuation.

decerebrate cat (573, 575). This response is manifest as cyclical (4–5Hz) movements of the tongue with rhythmic digastric EMG activity and rhythmic changes in the central excitability of the temporalis reflex arcs (570, 572). In some of these decerebrate preparations a second response, consisting of cyclic jaw movements (43, 572) repeating at 4–5Hz, can also be elicited by, for example, a series of repetitive very brief applications of force to molar teeth and gingivae. These 'masticatory' responses continue for a variable period after removal of the eliciting stimulus. The responses can also be elicited after a bite gauge is placed between the incisor teeth to take the entire force of closure (*Fig.* 25.1). However, when the incisal areas are heavily infiltrated with local anaesthetic, the continuation of the response is reduced or abolished although the initial opening and closing movement is not affected (*Fig.* 25.1). In the case of the 'masticatory' response it would appear that the rhythmic activity is highly dependent upon sensory feedback, unlike the rhythmic activity occurring in the palatally elicited response; furthermore the patterns of digastric EMG activity differ in the two rhythmic activities (*Fig.* 25.2). Parallel routes of activation of a system for producing rhythmic oral function do not necessarily therefore give the same results, particularly where one route involves phasic sensory feedback generated by jaw closure.

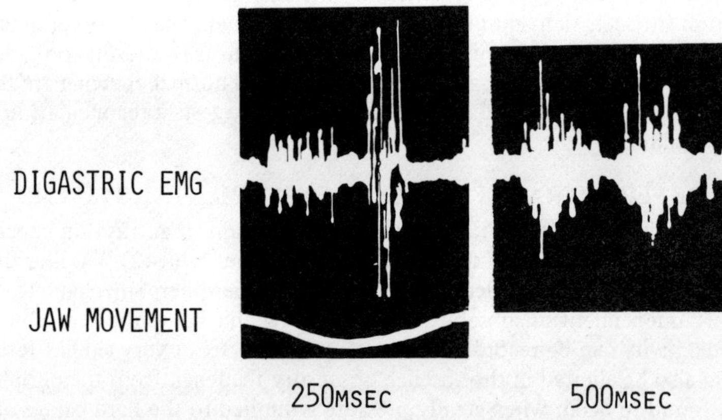

Fig. 25.2. The patterns of digastric EMG activity in two rhythmic oral responses that can be obtained in the decerebrate cat. The left record shows a single cycle of digastric activity in the 'masticatory' response, with the associated profile of jaw movement (opening downwards). The right record shows two cycles of digastric activity occurring in the rhythmic response, which is elicited by steady pressure on palatal mucosa.

According to the classic concept, the sensory feedback arising from the teeth when they meet resistance during jaw closure has the effect of producing reflex jaw opening which is automatically followed by 'rebound' jaw closure, as a central phenomenon (524). However, in whatever way 'rebound' jaw closure is elicited (83, 524, 575, 576), the period of the open and close movement seems to be about 80–100 msec. This is far from the 200–400 msec period of natural function, or of the 'masticatory response'. Furthermore, the 'rebound' phenomenon can be adequately accounted for on the basis of a change in motoneuron excitability (576) caused by muscle spindle stretch or by an increased fusimotor drive of the type described by Lewis (350). One cannot therefore argue, from the existence of 'rebound' phenomena, that a brief stimulus capable of activating simple reflex pathways is also one capable of bringing about activation of a centre for generating normal rhythmic jaw activity. Evidence that sensory input can cause activation of neuronal circuits showing longer term and slower oscillations can however be obtained from conditioning–testing (C–T) curves constructed to show the change in the central excitability of the temporalis and digastric excitatory reflex arcs following a sensory input (550, 576). Such curves can be produced simultaneously in both arcs, without interrupting other afferent pathways, by processing the EMG activity evoked by test stimuli following a single (conditioning) sensory input (542). The C–T curves shown in *Fig.* 25.3 demonstrate opposing swings of

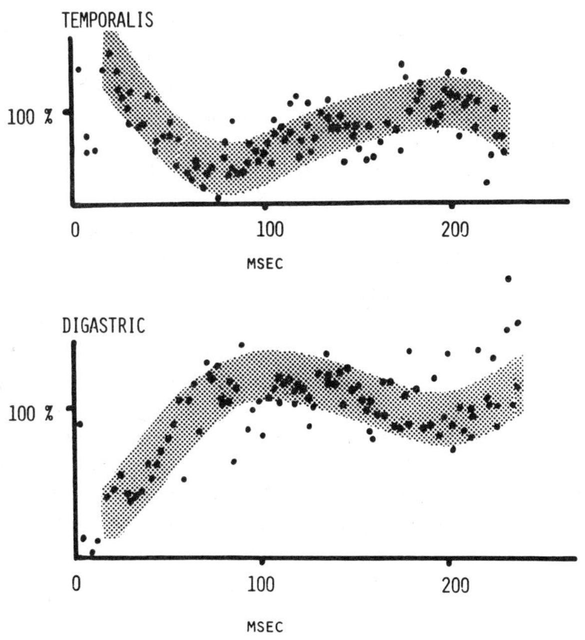

Fig. 25.3. Changes in excitability of the temporalis and digastric reflex arcs following a single electrical stimulus. These conditioning testing curves were obtained simultaneously in the decerebrate rat by processing the EMG activity. Conditioning and test stimuli were both applied to labial fibres of the infra-orbital nerve. The general tendency, indicated by shading, is for reciprocal swings of excitability to occur in the two reflex arcs with a period of about 200 msec, although other periods of oscillation may be superimposed.

excitability in the two opposing reflex arcs in the rat; the periodicity, which appears to be about 200 msec, at least corresponds to the rhythm of normal function in that animal (Weijs, personal communication). Such swings of excitability continue for at least 400 msec after the conditioning stimulus. This suggests that a single stimulus can in some measure 'switch in' or 'drive' a pre-programmed oscillation. The single brief sensory inputs used do not, however, have the ability to cause any externally manifest oscillation of the jaw. One might speculate whether the brief sensory input produced either by contact with food when it enters the mouth, or by contact with a bite guage in the 'masticatory response', is substantially the same in effect. If the sensory input is similar in effect then, although insufficient to cause continuing rhythmic jaw movements, such inputs could facilitate existing weak activity in the central oscillator; they would therefore help to generate rhythmic movements which in turn would produce sensory inputs having exactly the same facilitatory effect.

PERIPHERAL CONTROL IN BITING

There is little information on what actually does happen when an animal closes its teeth on something. It has been reported that the attacking bite of the opossum is associated with a brief burst of EMG activity in the elevator muscles which is largely over before the jaw has fully closed (578). It is difficult to see what peripheral control could be exerted over this activity since, whatever reflex responses are generated by oral contact with the object, they can only have an effect after an appreciable delay. During this time, forces produced e.g. by the kinetic energy of the jaw, can cause damage to oral structures. The digastric muscle, because of its small size, is unlikely to be very effective in decelerating or 'off loading' the jaw and its effect could in any case not be significant for 16 msec (reflex latency, say, 8 msec plus a time to peak muscle contraction of at least 8 msec). A similar argument applies to the inhibition of the elevator muscles; the motoneuron firing occurring at the instant of contact (or, due to reflex latency, up to 8 msec after contact) still has to be expressed as force exerted by the muscle even if all subsequent motoneuron firing is reflexly suppressed.

The profiles of force generated in attacking bites are broadly of two types, simple ones lasting about 80 msec and compound ones of much greater duration. The simple bites reach a relatively reproducible peak force of about 4 kg in approximately 30 msec, whereas the compound bites reach much higher values but take a correspondingly longer time to do so. This division corresponds to two types of snap seen using electromyography. In one there is a short burst of elevator muscle activity which is largely over by the time the jaws meet (*Fig.* 25.4) and in the other there are repeating bursts of activity continuing after jaw closure. It is clear, therefore, that in a biting attack, the animal does not immediately produce a large motoneuron discharge, capable of inducing maximum jaw-closing force, and then adjust it with negative feedback, but appears to 'try out' some pre-programmed value which is safely within the structural limits of the jaw apparatus. Increase of this force then appears to occur provided there has been no negative feedback from an 'overload' state.

PERIPHERAL CONTROL IN CHEWING

During the early stages of jaw closing in each masticatory cycle, the EMG activity in the elevator muscles is of relatively low amplitude. As the teeth meet the food and resistance is offered to further closure, the velocity of jaw closing falls. At this stage EMG activity shows a sudden increase in amplitude (*Fig.* 25.5). However, as the food is softened by mastication, the large amplitude component of the EMG disappears. It can be argued that the large amplitude component is the reflex result of an oral sensory input (580). This would imply that the low amplitude activity represents mainly the fundamental or pre-programmed component.

If one considers the type of stimuli that might commonly be present in mastication, they can be broadly divided into two groups on the basis of the food consistency. First, jaw closure on lumps of hard food will tend to produce localized pressures on oral structures. These pressures will rise rapidly with jaw closure due to lack of deformation of the food. Conversely, jaw closure on soft food will readily cause its deformation so that pressures will be lower, less localized and will rise more

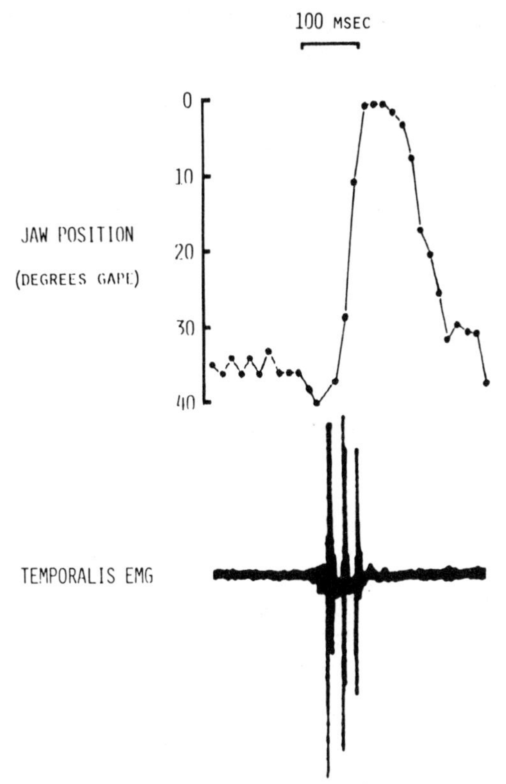

Fig. 25.4. Temporalis EMG and the profile of jaw movement (obtained from cineradiographs) occurring in an opossum during an attacking bite. Note that the muscle activity is over by the time the jaws meet.

slowly with jaw closure. It has been shown experimentally that rapidly changing localized pressure on oral mucosa, such as might be produced in chewing hard food, will elicit transient reflex jaw closing and at higher pressures, transient reflex jaw opening (575). On the other hand, slowly rising or steady pressure on large areas of mucosa, such as might be produced by chewing very soft food, inhibits elevator muscle activity and elicits pharyngeally directed waves of contraction in the tongue (575). The former group, the transient reflexes of opening and closing, represent the responses to stimuli with a minimum of A δ fibre activity (576). Whereas largely A α fibre inputs are capable of facilitatory interactions and could perhaps summate with the central rhythmic activity in mastication, small diameter fibre afferents cause prolonged inhibition of elevator motoneurons (576) and would therefore oppose rhythmic jaw closure. It is suggested therefore that the central effects of stimuli capable of eliciting a jaw-closing reflex might summate with the centrally programmed activity generating closure. Since the stimuli in question would themselves be generated by jaw closure on food, there is good reason to postulate that a positive feedback mechanism operates. One might suggest therefore that oral contact with hard food produces sensory inputs that will add to the rhythmic activity produced by the central programme so causing the large amplitude EMG signals seen when closing upon such food (*Fig.* 25.5). Conversely it is suggested that oral contact with *soft* (or masticated) food will tend to suppress elevator muscle activity and activate a transport mechanism carrying food back into the pharynx thus terminating mastication.

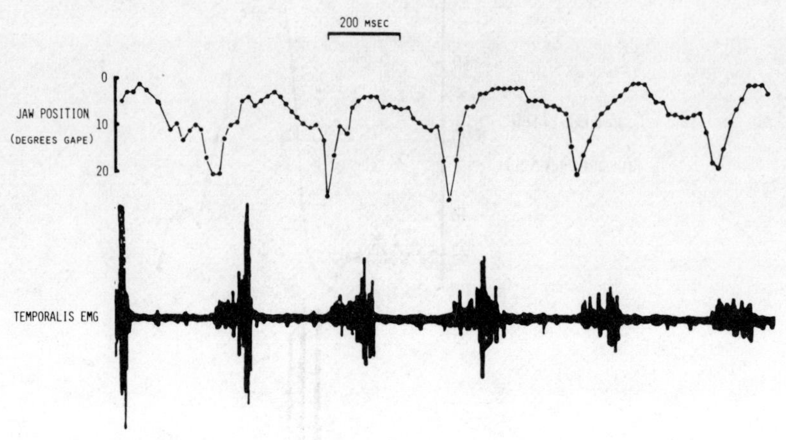

Fig. 25.5. Temporalis EMG and jaw movement profile (obtained from cineradiographs) of an opussum during chewing in which it breaks down and softens a small piece of dry bread. The section of record shown covers the transition from high amplitude temporalis EMG activity, which occurred in the first 4 cycles (not shown), to the low amplitude EMG activity which occurred in the 4 cycles (not shown) preceding the swallowing sequence.

A POSSIBLE MECHANISM OF PERIPHERAL CONTROL

Finally the question arises as to how sensory inputs interact with pattern or rhythm generators of the type considered by Bremer (83) and Dellow and Lund (142). There is indirect evidence as to how this may occur. Since the elevator and digastric motoneurons do not exhibit direct reciprocal relationships (570, 577) whilst the disynaptic reflex arcs do show reciprocal swings of excitability (570, 574) (*Fig.* 25.3), the hypothesis has been advanced (570, 577) that the disynaptic excitatory reflex arcs to elevator and digastric motoneurons (321, 550) have mutually crossed inhibitory connections at the interneuron level so constituting the type of neuronal network found in the spinal cord (287). In the type of network shown in *Fig.* 25.6, constant common drive to the two excitatory interneurons will cause alternately one and then the other side to be active whilst the opposing side in turn becomes refractory and is inhibited (245, 246, 276, 479). Since the excitatory reflex arcs have neurons in common with the oscillator group, it follows that the thresholds of those reflexes can be expected to vary with the state of the oscillator. For example, it might be predicted that when oscillator activity is driving closure, activation of the elevator reflex arc will summate with the activity in the oscillator and enhance the existing closing activity. The intra-oral forces produced during jaw closure will then generate sensory inputs which in principle are capable of activating the elevator reflex arc (575) so producing the positive feedback referred to previously. Overall negative feedback is then provided by the higher threshold reflex pathways constituting the jaw-opening reflex (575, 576).

It is not suggested that the operation of the neural network shown in *Fig.* 25.6 necessarily explains all aspects of mastication. The network is used to demonstrate that ideas of relative importance of 'central programme' and 'peripheral feedback' may be the result of the experimental approach to a system which can be made to function in a variety of ways. Experimentally one can arrange for a model of the type of network seen in *Fig.* 25.6 to function as a rhythm generator which is independent of feedback, as one which is capable of interacting with peripheral stimuli, as one which is highly dependent upon sensory feedback or as one simply demonstrating rebound phenomena. The examples given in this paper strongly suggest that the biological system controlling oral function can be made to operate in the same way.

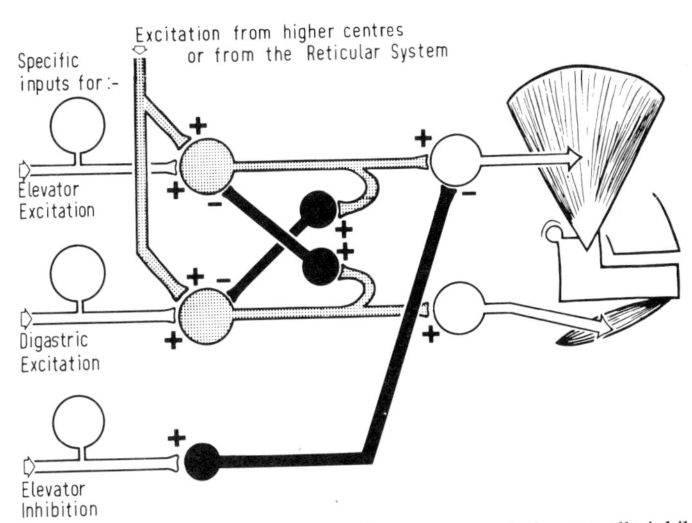

Fig. 25.6. A theoretical network showing an oscillator, composed of a mutually inhibitory pair of interneurons driven by a common input, combined with three known reflex pathways. Inhibitory cells are shown black and the excitatory interneuron system comprising the 'oscillator' is shown shaded. Collaterals from afferent neurons to the reticular system have been omitted for clarity.

Summary

In conclusion one can say that mastication is programmed in the sense that it represents the output of organized neural circuits but, apart from the initial stages of the first bite, it is subject to considerable modification by peripheral feedback. It appears that the sensory inputs generated during closure upon food may not only ensure maintenance of the rhythmic jaw activity by supplying 'drive' to oscillating neural networks but more significantly, in their modification of the central programme, those inputs ultimately define the activity as mastication.

Acknowledgements

Figs. 25.4, 25.5 and the related data, are part of unpublished work carried out jointly with Professor A. W. Crompton and Dr Karen Hiiemae under USPHS grant No. DE 03219.

Discussion CHAIRMAN: D. J. ANDERSON

LUND: You pointed out that there is a difference between the normal chewing rate of rabbits (5–6·3Hz) and the 4Hz output to the muscles which Dellow and I observed. The rate at which this rhythm is produced is very dependent on the state of the animal, on the state of anaesthesia or the state of the decerebrate preparation. In some animals rates above 5Hz are obtained. I do not claim that what we have been looking at is mastication as we know it in the intact animal, but it is the basic opening and closing cycle on which mastication is based.

THEXTON: I think that mastication in the intact animal is the result of peripheral sensory input summing with the activity of this oscillator.

MATTHEWS: You implied that, in the gnashing opossum, there was not enough time available for the elevator muscles to be switched off by the peripheral input and that the movement was therefore pre-programmed. Could you give a few more details of the timing of these events.

THEXTON: Let us put it this way. If the elevators are switched off at the moment that an overload state is produced the jaw still has inertia and mechanical contraction will still continue for a short time. The half relaxation time is about 20 msec.

MATTHEWS: Yes, but what was the interval between the teeth meeting resistance and the elevator muscle activity stopping ?

THEXTON: If the animal is smashing his jaws together—you can actually hear the crash together—then it occurs at or before contact.

MATTHEWS: I wanted to know what evidence you have that there was no time for peripheral feedback to occur.

THEXTON: If the teeth have already crashed together, and you can hear them, then it is already too late to have an effect on the muscle activity producing that snap.

CHAIRMAN: I don't think there is any more time to go on crashing together.

Paper No. 26

Disturbances of Mastication associated with some Neurological Disorders

C. J. Earl

Information about abnormalities of chewing and biting in neurological disease is scattered throughout the neurological literature. I propose to consider the subject under three headings.

1. Weakness of chewing movements and other movements of the jaw.
2. Conditions in which there is continuous spasm involving the muscles of mastication.
3. Abnormal movements of the jaw, which are usually associated with similar activity in the muscles of the face and tongue.

1. Unilateral masticatory muscle weakness causes deviation of the open jaw to the affected side on account of weakness of the pterygoid muscles, but in my experience patients do not make any complaint of disability and the weakness is found during routine examination of the cranial nerves in the investigation of other symptoms. Presumably these patients may be subject to the effects of abnormal strains on the temporomandibular joints if the condition lasts for very long. Any lesion which affects the motor division of the fifth nerve or its cells of origin in the brain stem will produce weakness of this sort. Milder unilateral weakness is seen in patients with hemiplegia, although on account of the facial weakness present at the same time, the degree of deviation of the open jaw may be difficult to assess.

When bilateral weakness is present, the situation is quite different and there is difficulty in biting and chewing and where weakness is severe, even of keeping the mouth closed. Weakness of this sort results from bilateral involvement of the fifth nerve in such pathological conditions as invasive tumours of the base of the skull or of the pons itself, and in these circumstances, of course, the weakness of jaw movement will not be an isolated phenomenon.

Generalized muscle disease may cause weakness of the jaw muscles. Difficulties in chewing are evident in dystrophia myotonica, but the most frequent cause in my experience is myasthenia gravis. In the early stages of this disease, the weakness is characteristically present only after chewing and in particular after chewing hard food, but power is recovered after a few minutes rest. However, in severe myasthenia gravis the masseter and temporal muscles quite frequently show persistent weakness and some wasting. Wasting in myasthenia is associated with structural abnormalities in the motor end plate and of muscle fibres (438).The weakness is usually more marked in the masseter and temporal muscles than in the pterygoids, but I have seen the situation reversed, with greater difficulty in maintaining the mouth open against resistance than in closing it.

In motoneuron disease, weakness of jaw muscles is common and is associated with weakness of the tongue, pharyngeal muscles and of the face. There are two lesions responsible for this, one affecting the motor cells in the fifth nerve nucleus and the other affecting cells in the motor cortex, producing in the jaw muscles, a combination of weakness and wasting with spasticity such as is seen elsewhere in the body in motoneuron disease. The spasticity of the jaw muscles finds its expression in a hyperactive jaw jerk, which can also be elicited in other conditions in which there is bilateral involvement of corticobulbar pathways.

2. Persistent spasm of the jaw muscles usually shows itself by forced closure of the mouth, although occasionally the mouth may be held forcibly in the open position, perhaps deviated to one side. Inability to open the mouth is a well known symptom of tetanus and is often one of the earliest of the patients' complaints. Similar spasm of the jaw muscles is also described in structural lesions of the pons. Gowers (215) described cases of this sort and spoke of the disease 'irritating' but not destroying the motor nucleus of the fifth nerve. According to him the spasm may be unilateral but is more often bilateral even with a unilateral lesion. He quoted a case described by Wernicke in which a tuberculoma occupying 'almost the whole of the vertical extent of the left half of the pons' produced a syndrome of loss of conjugate deviation of the eyes to the left, paralysis of the left facial nerve and what he described as 'extreme tension of the left masseter'. Another case of his own had similar signs but the pathology was not verified. Similar spasms may occur in encephalitis. Encephalitis lethargica was a common cause and a good example of such spasm of the jaw was given by Wilson (628) in his account of the disease. Similar spasms may occur in other types of encephalitis involving the brain stem, and trismus is mentioned as being frequently seen in post vaccinial encephalitis (76) although in a recent series of cases described by Spillane and Wells (671) it was not recorded. Similar continuous spasm in the facial muscles occurs also in pontine lesions and is known as facial myokymia. The most likely explanation of such phenomena seems to be that they are due to lack of some inhibitory effect on the motor cells in the nucleus. Similar muscular spasm has been described in spinal cord lesions as α-motoneuron rigidity (489) and may be due to 'isolation' of anterior horn cells.

However, at the present time, the commonest cause of such spasm is pharmacological rather than structural and 'spasms' of the jaw muscles occur as a result of administration of phenothiazine drugs or haloperidol. These reactions are acute and appear to be an idiosyncratic response to a normal dose of the drug and are unpredictable. The symptoms usually develop within a short time of administration of the drug although sometimes there is a long interval which is difficult to explain. The spasm may take various forms but the jaw muscles are very frequently involved together with the facial musculature, the tongue and neck muscles. The jaw may be held open, deviated to one side or firmly closed. The disturbance passes off in a few hours provided that no further doses of the drug are given. This is an important proviso since the condition is sometimes regarded as a hysterical disturbance for which a second dose of phenothiazine is given. The patient may be rapidly restored to normal after a dose of benztropine (Cogentin) administered by intravenous or intramuscular injection.

3. Abnormal movements of the jaws may be associated with fixed abnormal postures of the type already described, but they often occur alone. Spontaneous chewing movements are commonly associated with dementia in the elderly and occur in association with other types of mental illness. Their exact significance is uncertain. They occur in Huntington's chorea as a part of a more widespread

picture, and in the young they may be present as part of the clinical picture of athetosis or Sydenham's chorea. The abnormal facial movements which occur in patients who have taken large doses of phenothiazine drugs over long periods of time may be very striking (280). The face, tongue and jaws are all involved, and although the distribution resembles that of the acute spasms already described, which may follow a single dose of phenothiazine, there are important differences. Fixed abnormal postures are not usually associated with these movements, which show a pattern of continuous restless and often repetitive activity. The other important difference is that stopping the drug may make no difference to the movements or may in fact aggravate them. Treatment with L-dopa is occasionally effective, although this may be surprising in view of its own side effects.

Similar movements may occur as the result of natural disease processes although their exact nature remains obscure. The movements are frequently, and quite wrongly, regarded as not being organic in origin. One patient under my own care experienced difficulty mainly in keeping the mouth closed, but this became much easier if a sweet was sucked. Presumably, stimulation of the mucosa was the important factor in modifying the movement.

Lastly, and in quite a different context, there are the chewing movements which occur in attacks of temporal lobe epilepsy and the chewing movements of coma. These movements are not, of course a reflection of the site of origin of the epileptic discharge in the same sense that unilateral facial twitching or twitching of one hand betray the site of origin of a focal motor seizure. They are part of the automatic behaviour seen in temporal lobe attacks and are probably related to the hallucination of taste which is also a common feature of such episodes. One interesting characteristic which Penfield and Jasper (449) point out is that they always occur during the period of amnesia.

Patients in coma, whatever the cause, may develop chewing movements presumably similar to those seen in normal people during sleep. When coma is prolonged, these chewing movements may be so powerful and persistent, as to lead to grinding away of the teeth.

Discussion CHAIRMAN: D. J. ANDERSON

HANNAM: In patients showing muscle spasms resulting from phenothiazine drugs, are the spasms limited to the head and neck ?

EARL: Not solely, but almost entirely to the head and neck. This is true both in the acute phenothiazine spasms which I described first and in the more chronic spasms. The face, head, neck and jaw are predominantly involved.

KARLSSON: The trigeminal ganglion has often been implicated in the aetiology of trigeminal neuralgia; people refer to irregular myelin figures in the ganglion as one sign of the disease. Now the trigeminal motor root goes right through the ganglion and I wondered if you had seen any evidence of incoordination of jaw movements in these patients.

EARL: No. Of course their jaws may be held in very abnormal positions to prevent the occurrence of pain or when the pain occurs they may get a muscle spasm. As far as I know there is no abnormality on the motor side in trigeminal neuralgia. As you say, there may be areas of demyelination, possibly with false synapses between one myelinated fibre and another, and this is one of the theories as to how the pain arises, so that a touch stimulus evokes pain. But I do not think that there are any motor abnormalities.

KAWAMURA: Have you information on the chewing ability and chewing patterns in Parkinson's disease, because I am very interested in the functions of the basal ganglia in chewing movements ?
EARL: Chewing is abnormal in patients with Parkinsonism. I decided to leave this condition out because I do not think there is very much that I can say about it except that the abnormalities of movement of the jaw are very similar to the other abnormalities of movement that one sees in Parkinsonism. There is a slowness and loss of range of movement and some patients show an abnormal tongue posture. The tongue tends to push forward against the backs of the upper incisor teeth, but I am afraid I cannot say anything more than that.
SCHER: I should be grateful if you would pass an opinion on a problem associated with Parkinson's disease. The prosthodontist often has the problem of coping with the tremor of the mandible that these patients display and it is a curious fact that when the patient is asked to concentrate on something, say looking at a clock, the tremor increases. I have come across this in several instances and I wonder why the tremor is potentiated when the patient is asked to concentrate on something ?
EARL: I am afraid I cannot answer that. All I can say is that it also applies to the tremulous manifestations of Parkinsonism elsewhere and one of the tricks that some neurosurgeons employ when doing stereotactic operations, for bringing out a tremor so that they know when they have produced a big enough lesion to stop it, is to get the patient to count backwards, which takes a little more concentration than counting forwards. This must be related to the other observation which patients with Parkinsonism often make themselves, that the tremor, in the mild stages at least, can be abolished by concentrating on it and it is present only when concentrating on something else. The earliest complaint in Parkinsonism may be that the patient sitting watching television or reading a book suddenly realizes that his hand is shaking. Almost as soon as he realizes it, it stops.

Paper No. 27

Adaptation of the Masticatory System following Surgical Repositioning of the Jaws

I. H. Heslop and G. Wreakes

As practical clinicians we might be excused from saying, when faced with a masticatory apparatus like that in *Fig.* 27.1 or *Fig.* 27.2. 'Let us not talk so much about it, but rather let us put it right!' At the same time, we realize that the success of treatment depends upon an understanding of the physiological background and that clarification of imperfectly understood aspects of the physiology of mastication will help to prevent failures of today and will increase the chances of success in the future.

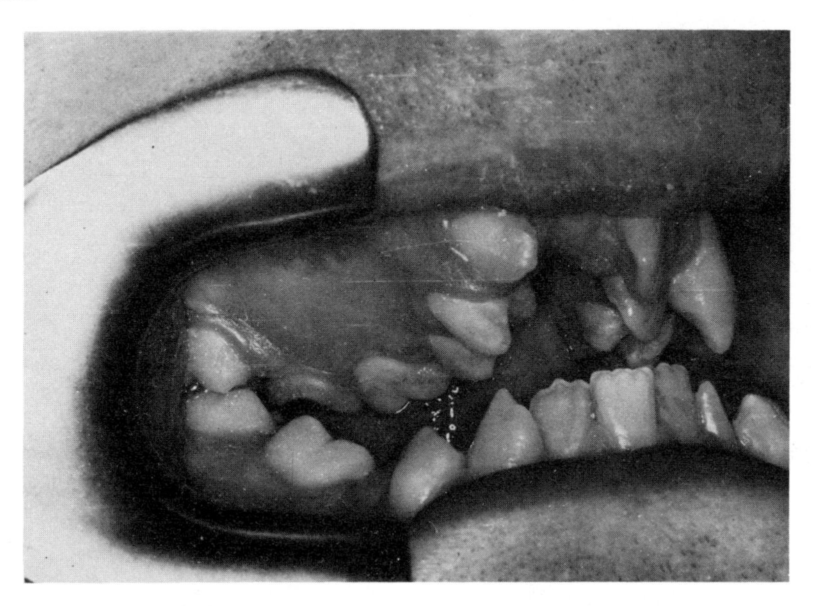

Fig, 27.1. Gross malocclusion.

The purpose of corrective jaw surgery is to improve the facial appearance and dental function in all types of congenital, developmental and acquired deformity. In this sense, it is an extension of orthodontics and in a large measure the analysis of the aetiological factors and the criteria for success are common to both disciplines. In this paper we confine ourselves to cases within the extremes of normal variation.

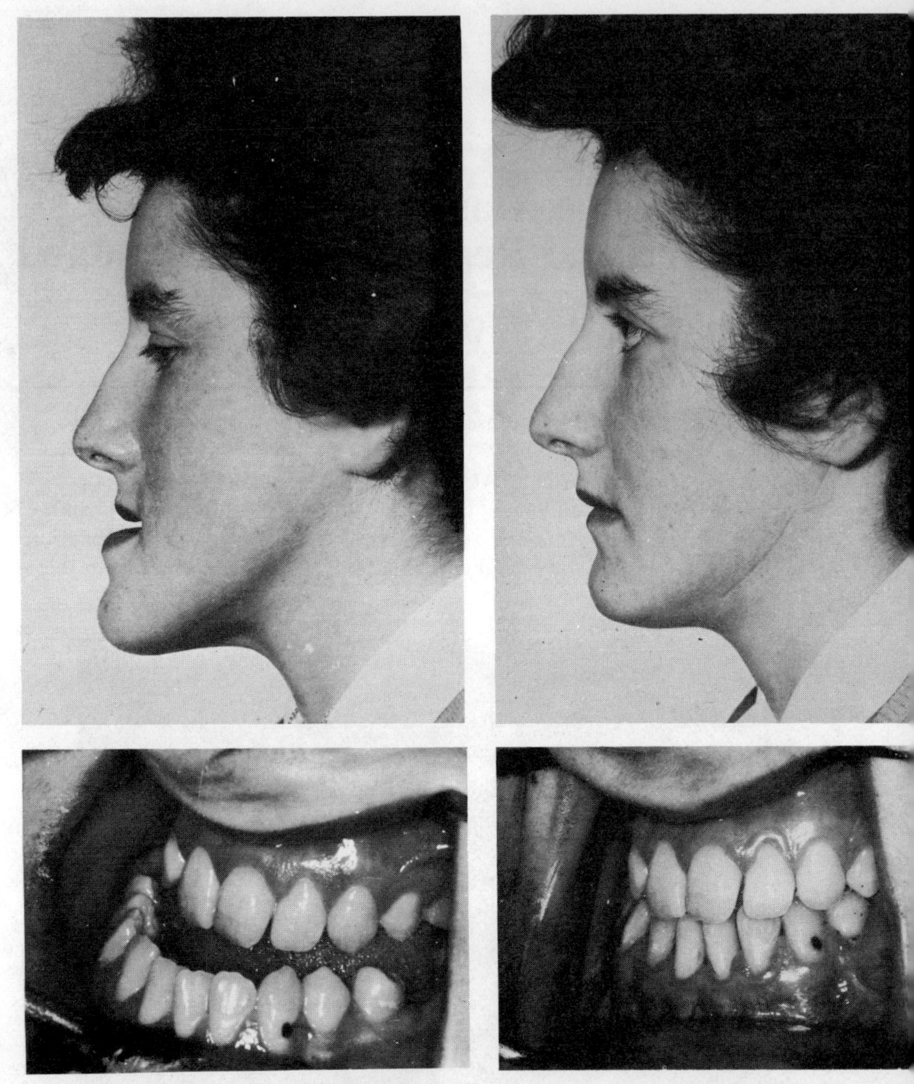

Fig. 27.2. Correction of facial and jaw deformity by ostectomy at the angles of the mandible.

The Oxford English Dictionary defines 'adapt' as 'to fit; make suitable; modify or alter'. It is adaptation in the sense of the first definition which produces the ideal occlusion from two major variable factors. These are:

1. The basic skeletal pattern, that is the morphology and spatial three-dimensional orientation of the maxilla and mandible excluding the teeth.

2. The soft tissue and muscles covering and attached to the bones. Traditionally, skeletal development has been regarded as being genetically determined, but at present much credence is given to the Moss functional matrix theory (415) which suggests that muscle activity may determine bone form, i.e. a type of adaptive response. The bones and muscles produce, by their interaction, a situation in which the teeth and their supporting bone, collectively known as the dento-alveolar

process, come to occupy the zone of neutrality between opposing muscle forces. In the vertical dimension the dento-alveolar processes appear to have an inherent ability, within limits, to establish occlusal contact. Depending on whether or not this is successful the result is a normal occlusion, an open bite, or a deep incisor overbite (*Fig.* 27.3). The last two are seen in extreme forms when the lower third of the face is either very deep or very shallow (*Fig.* 27.4). The adaptive potential will

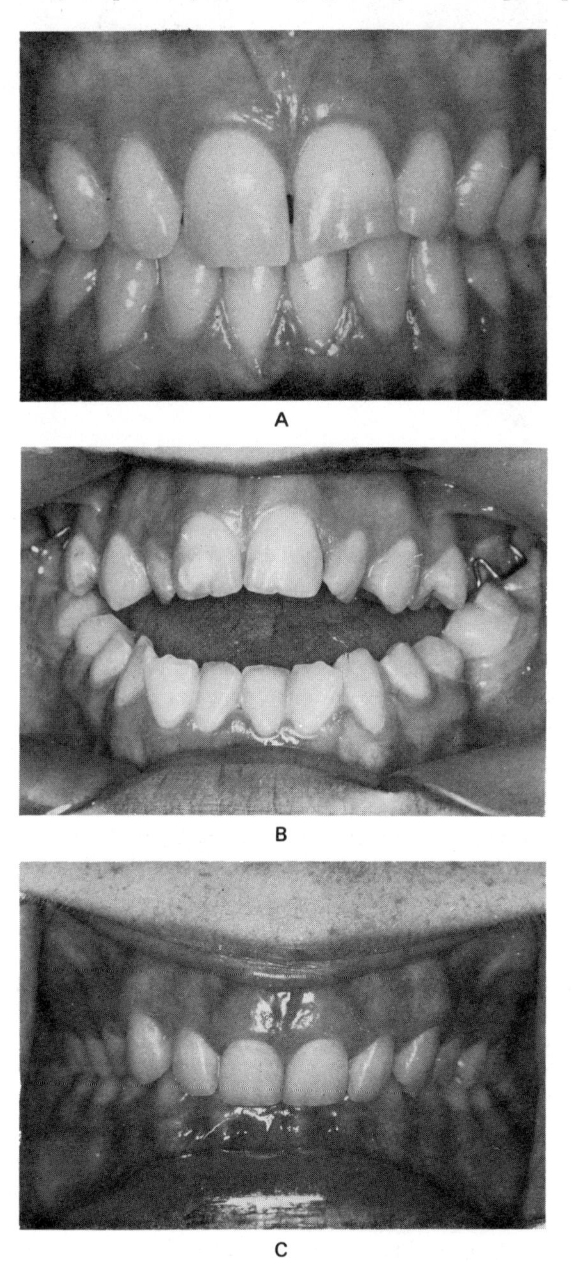

Fig. 27.3. A, Normal occlusion. B, Anterior open bite. C, Deep incisor overbite.

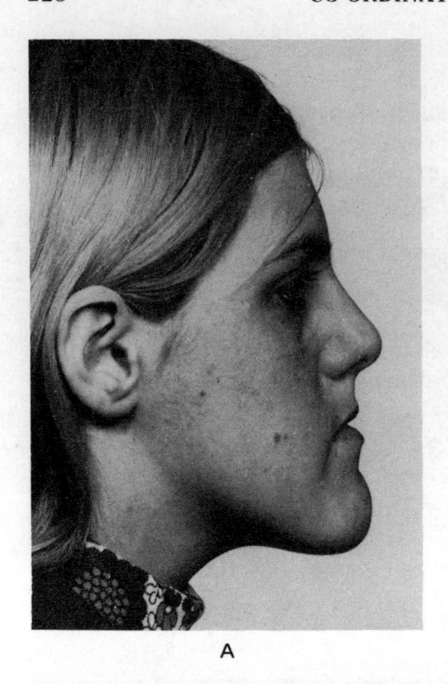
A

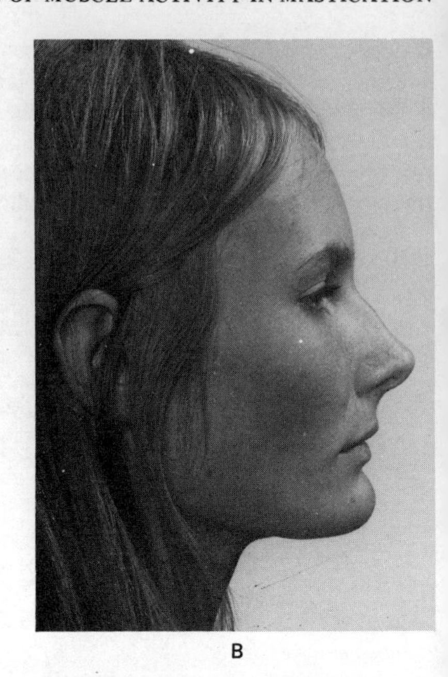
B

Fig. 27.4. Profiles of patients shown in *Fig.* 27.3. A, Same patient as in *Fig.* 27.3B. B, Same patient as in *Fig.* 27.3C.

in most people produce an occlusion which is both functionally and aesthetically acceptable. However, the adaptive response of the dento-alveolar tissues may sometimes be unacceptable to the patient and in the more severe form may require surgical intervention. The stability of the alterations produced by surgery depends on the creation of a new dynamic environment akin to that which exists in patients with a normal occlusion.

The role of surgery is in line with the second definition of adaptation, namely to alter or modify. Surgical repositioning of the mandible described by Kostecka (331) was one of the first steps in this direction. He described a blind operation on the ascending ramus to correct mandibular prognathism. Today, techniques are much much more sophisticated and the surgeon has at his disposal up to twenty-five operative techniques. Some patients may respond to one operation on one jaw; some will need operation on both jaws; others may need a combination of three or four operative techniques. Many require pre and/or postoperative orthodontics, usually as a means of ensuring that the dental arches will fit together well post-operatively, but also as a means of correcting the exuberant dento-alveolar development which was part of the original adaptive response (*Fig.* 27.5). Both types of orthodontic correction can be carried out concurrently and the fixed orthodontic appliances can sometimes be used for postoperative jaw fixation.

Since we are considering adaptation after surgery, it is important to realize that besides changing the position of the major part of the maxilla and/or mandible, we may in some cases secondarily change the attachments of the masticatory muscles; disturb the temporomandibular joints and upset the muscular attachments and relative positions of the lips, tongue and soft palate, and the diaphragm of the floor of the mouth. We will now consider the problems of adaptation in these areas.

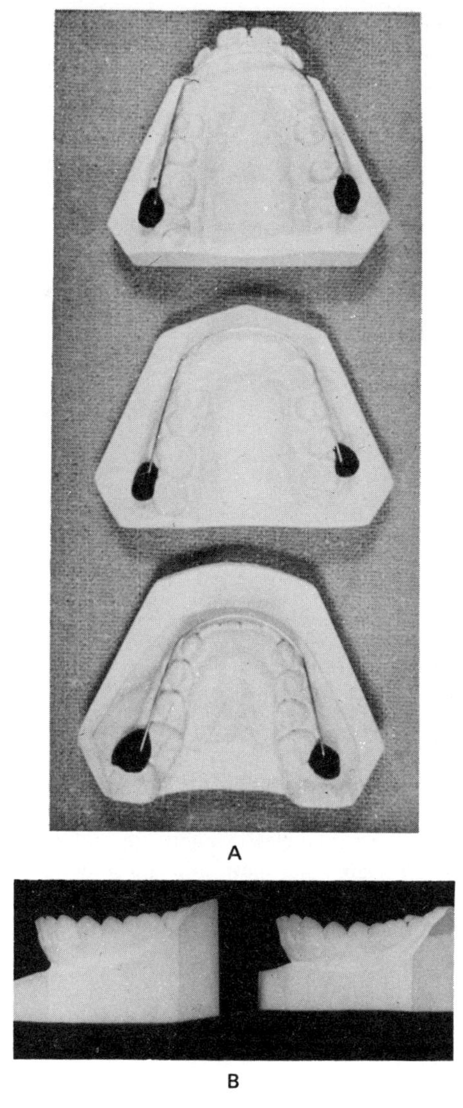

Fig. 27.5A. Movement of the upper incisors to produce an arch form coincident with the shape of the lower arch represented by the superimposed wire on the model. B, Levelling of the excessive curve of Spee.

TEETH AND ALVEOLAR BONE

When we consider how easy it is to detect a piece of paper or metal foil between the teeth, we realize how sensitive are the receptor mechanisms in the tooth-supporting tissues. By these standards, even orthodontic correction of the alignment of the teeth is a gross disturbance, but it is a gradual change and the body adapts to it and accepts it. Surgical repositioning of blocks of teeth and bone, however, produces an instant change and the body has to adapt first to the bite gagging and reduction of freeway space caused by the splint fixation for a period of four to six weeks and then has to readapt to the new dento-alveolar relationship. These changes mainly involve the lips and tongue, and clinical experience indicates that rapid adaptation occurs. It is important to realize that in all these anterior teeth and alveolar bone changes, whether produced orthodontically or surgically or by both means, no change is produced in the rest position of the mandible and so none of the muscles of mastication is disturbed. This undoubtedly accounts for the rapid adaptation.

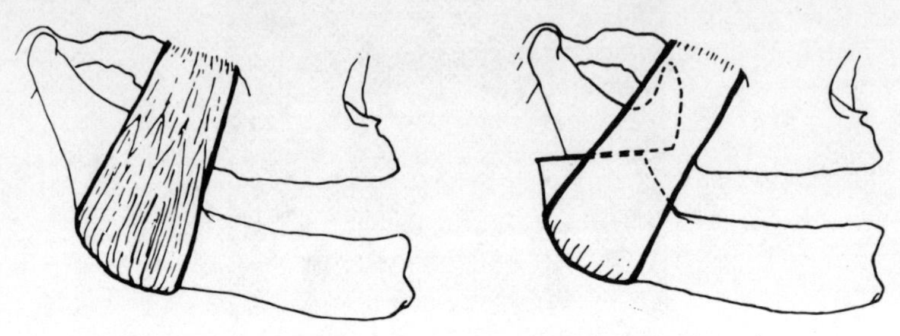

Fig. 27.6. Alterations produced to the pterygomasseteric sling in the Kostecka osteotomy.

MANDIBLE

When we look at surgical procedures, in which the mandible is divided and the whole body of the bone repositioned, the potential physiological upset is much greater and here we are mainly concerned with the muscles of mastication, in particular with the resting length and axis of the 'pterygomasseteric sling'. If we look first at the Kostecka operation, which left the muscles untouched, we see that the sling is elongated and its axis is altered (*Fig.* 27.6). Once the jaw splinting was removed the sling returned to its normal elastic length and, as Yemm has explained earlier in this Symposium this is an elastic length and not a matter of resting muscle tone. With this operation relapse was inevitable and sometimes it was complete. However, if the attachments of the sling to the mandible are detached in an open operation and the mandible is slid through the sling which is then allowed to reattach to the jaw in the new position, relapse does not occur, provided that one

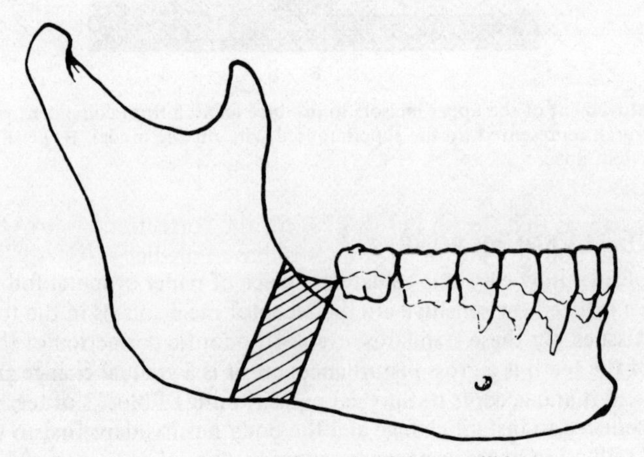

Fig. 27.7. Mandibular ostectomy at or near the angle of the mandible.

does not attempt large corrections of prognathism by these means. Large corrections are best achieved by ostectomy just anterior to the muscle sling or within the sling so that the ascending ramus and the sling length are unchanged (*Fig.* 27.7).

Increase or decrease in the length of the mandibular body is accomplished by sagittal splitting osteotomy at the angle (133, 436) (*Fig.* 27.8). Careful detachment of the medial pterygoid muscle and the sphenomandibular and stylomandibular ligaments is essential to prevent an increase of tension in the sling and consequent relapse. If the operation is used as a means of closing an anterior open bite, rotation of the mandible about a transverse axis will occur, with potential lengthening of the pterygomasseteric sling which must be divided at its lower border to allow the mandible to slip through and reattachment to occur. Although forward movement of the body of the mandible lengthens the depressor muscles of the mandible, these are small muscles and do not appear to be a significant cause of relapse. They seem to adapt readily by stretching during the period of fixation. However, premature release of jaw fixation can result in a small degree of relapse attributable to the elastic pull of these muscles. Although the plane of the diaphragm of the floor of the mouth may be altered by surgery, the actual muscle attachments are not.

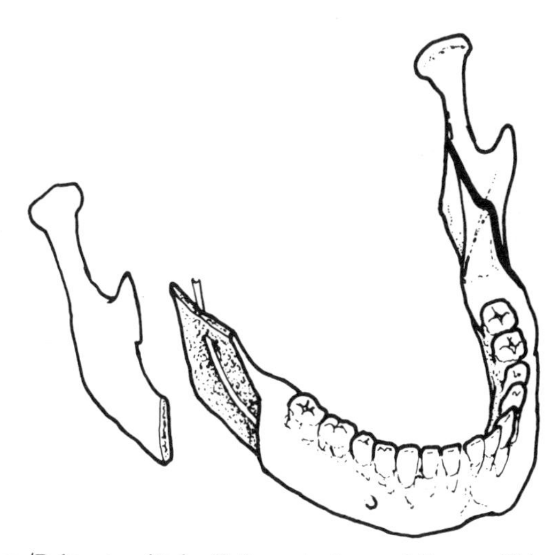

Fig. 27.8 Obwegeser/Dalpont sagittal splitting osteotomy of the mandible.

TONGUE AND LIPS

Operations such as that shown in *Fig.* 27.9 for the correction of cross-bite, in which the mandibular body on one or both sides is moved medially, restrict the tongue space a little, as do all backward movements of the body of the mandible, but the tongue musculature adapts well to these changes within the period of fixation. Only in rare cases in which the tongue is very large or the required posterior movement of the mandible is very great, is surgical reduction of tongue size necessary. Although surgical lengthening of short lips is not possible, correction of the dento-alveolar relationships and in some cases adjustment of the muscle attachments will improve their functional environment.

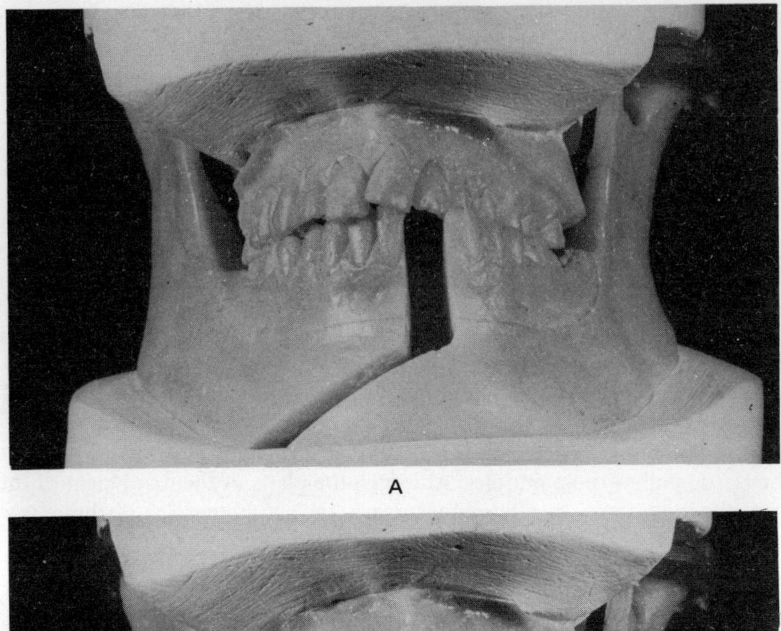

A

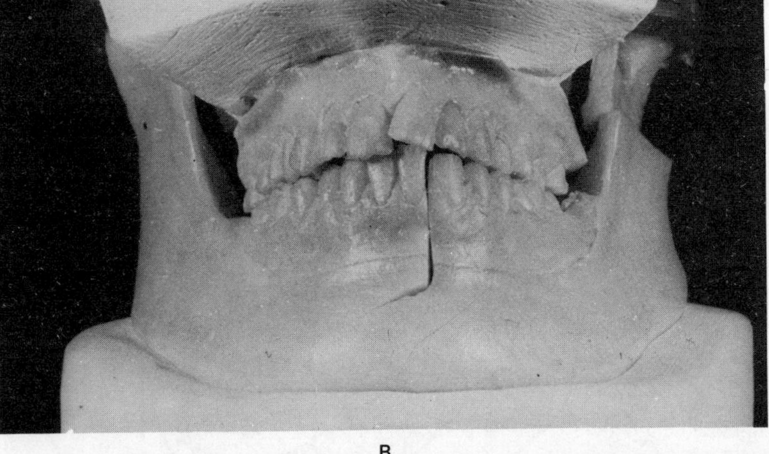

B

Fig. 27.9. Operation for correction of cross bite on the posterior teeth and asymmetry of the chin.

MAXILLA

Forward movement of the maxilla within the limit of 1–1·5 cm (*Fig.* 27.10) produces no problems in the masticatory system but it is important when moving the maxilla downwards not to go beyond the limits of the freeway space. Excessive downward movement of the maxilla results in relapse because the mandible pushes the maxilla up again to a point within the freeway space. This phenomenon has been seen in reverse in Schuchardt's operation (509) in which the posterior dento-alveolar segments are moved upwards into the antrum to close an anterior open bite. It is in total maxillary movement that the soft palate and its attachments may be disturbed and some temporary upset of speech may occur, but our patients have not experienced any difficulty in swallowing [associated with these operations].

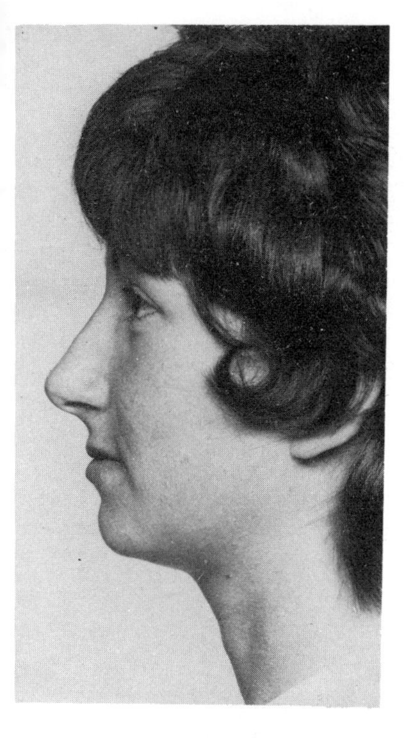

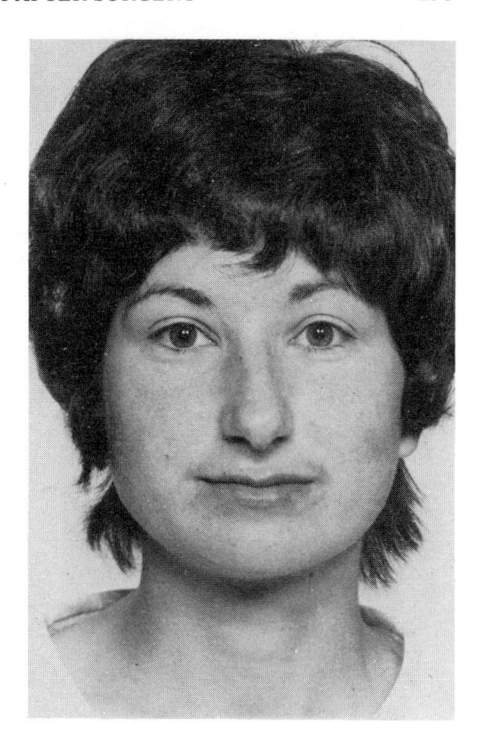

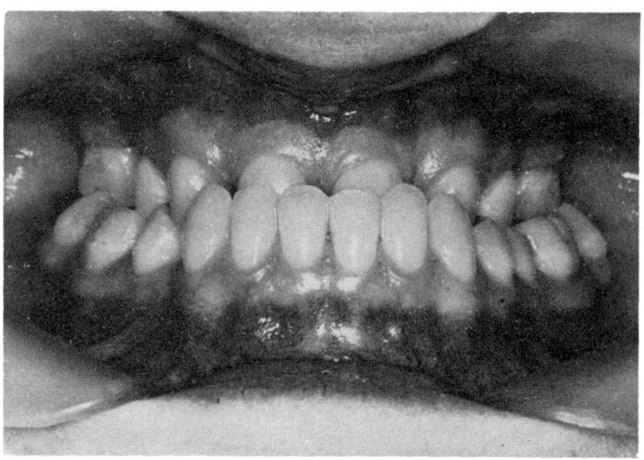

A

Fig. 27.10. Correction of facial and jaw deformity by maxillary osteotomy and bone graft with forward movement of the whole maxilla. A, Pre-operative series (*continued on* p. 234).

TEMPOROMANDIBULAR JOINTS

Medial rotation of the condylar heads in the Heslop/Sowray type of operation and similar procedures, might be expected to cause temporomandibular joint disturbance, but in fact only 10–15° rotation occurs with the removal of 1 cm at the symphysis and the operation is well tolerated by patients. Mandibular operations which result in rotation of the head of the condyle in its normal axis do not produce clinical problems in the joint, although it is important to preserve

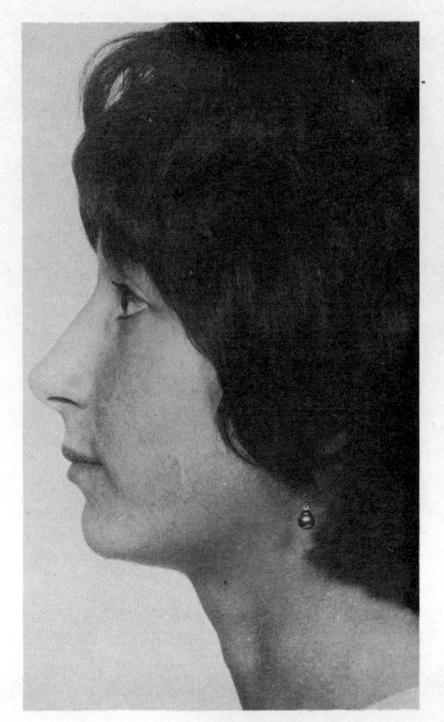

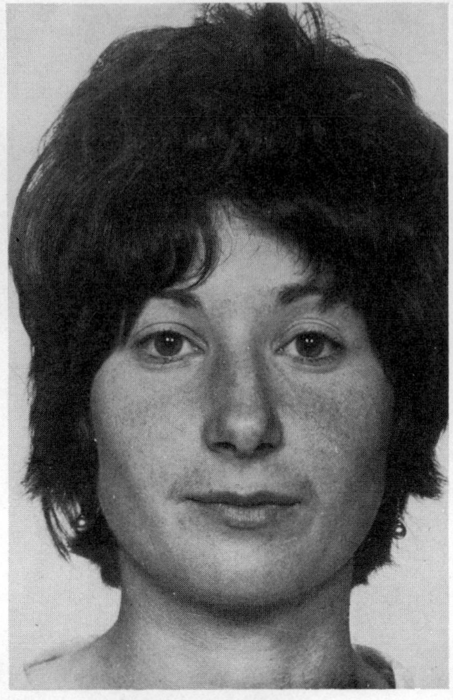

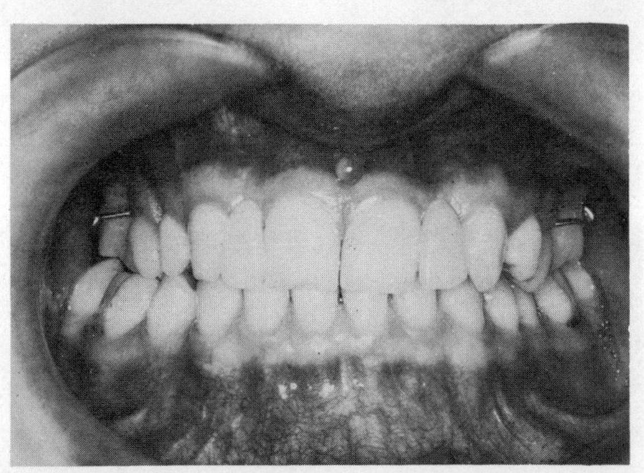

Fig. 27.10B. Post-operative series (*continued from* p. 233).

the correct position of the condylar head in the glenoid fossa after this part of the mandible has been mobilized at operation, otherwise postoperative upset of the joint and of the position of the jaw and the occlusion of the teeth may occur.

Conclusion

Surgical repositioning of the jaws for occlusal or facial dysharmony carried out within physiological limits results in rapid adaptation of the masticatory system and the success and stability of the treatment is assured. If we exceed the elastic limit of the pterygomasseteric sling, or unduly encroach on the freeway space, or the tongue space, adaptation is more difficult and relapse is likely to occur.

Discussion CHAIRMAN: Y. KAWAMURA

STOREY: I realize that you probably very rarely operate on prepubertal patients but I wonder whether your statement about the pterygomasseteric sling being non-adaptable in the adult would be true for the prepubertal patient.
HESLOP: We have not operated very much on prepubertal patients. On the whole we like to wait until growth has ceased or almost ceased, otherwise some relapse may occur. But the more work I do with Glyn Wreakes, the earlier I tend to start on the main correction of bone and tooth masses, because this allows the orthodontist to do the final correction of the teeth while the patient is still young. I am not sure whether there is much difference in the tendency to relapse in the young as compared with the adult patient. With regard to lengthening of the pterygomasseteric sling I have come more firmly to the conclusion, as a result of listening to the papers during the last two days, that the reason for relapse is not so much a question of lengthening of muscle particularly in view of what Dr Goldspink has told us. I think it is the fibrous tissue, the muscle sheath, ligaments, connective tissue and fascia which cause the relapse rather than the muscle fibres. I think that in the child there is just as much likelihood of relapse if you do not strip the muscle attachment.
GREENFIELD: I believe that the temporomandibular joint is a sensitive area to regard with great care in these operations. If you disturb the position of this joint during operation I think you will have a relapse. I notice that after operation many of these patients show very low activity of the masseter muscle and I believe that this is due to some inhibition from the receptors in the joint due to the fact that the joint has been pushed to an abnormal position in the fossa. This can be very easily tested without any elaborate apparatus. After operation you find that masseter activity is very much lower during biting in habitual occlusion than it was before operation, and if you ask the patient to protrude the mandible and bite, you can test whether the masseter is capable of activity or not, and confirm that you have not disturbed the masseteric nerve in your operation. This is a very good clinical test. The activity of the muscle can be tested with your fingers or with electromyography, but your fingers are just as good as anything else. If you detect a great difference in activity between the habitual and protrusive positions you can be pretty certain that you have placed the condyle in an abnormal retrusive position in the fossa.
HESLOP: After what type of operation have you noticed this ?
GREENFIELD: Any operation in which you have produced a new overbite and in so doing pushed the condyle into a retrusive position.
HESLOP: I think that one has to guard very carefully against pushing the condylar head back excessively during a period of fixation, because when you release the fixation, the condylar head returns to its normal position in the middle of the glenoid fossa and you will get relapse of the incisal relationship.
GREENFIELD: That is the patients' guarding response. They relapse because the position is uncomfortable and they cannot contract their masseter muscle. It is not due to muscle lengthening, I think it is due to direct inhibition from the joint and they automatically relapse to get into a position in which they can contract that masseter muscle again.
HESLOP: I think you are quite right that if one does alter the position of the condylar head in this way then you will get relapse.
GREENFIELD: I do not think that changing the length of the masseter is so important.

HESLOP: But if you change the length of the masseteric sling you do get relapse unless you split it and allow the jaw to go through. I did not realize that I was lengthening the sling when I first corrected an anterior open bite with the sagittal splitting technique, and it relapsed very rapidly.

CHAIRMAN: You did not mention age. Are there age limitations in these operations ?

HESLOP: There are age limitations for some of these operations, particularly the sagittal splitting operation and this is a technical problem. As people get older, the bone gets harder and the operation becomes more difficult. I recently did a sagittal split on a woman of 50, it was not easy as the bone did not split nicely. In general these operations are done on late teenagers or patients in their early twenties.

WREAKES: We do a lot of orthodontic preparation on our patients and I think by doing so we probably reduce the postoperative adaptation that is required. When Mr Heslop does the operation things just slot in very nicely and I think this probably limits the amount of condylar disturbance which would occur if we did not do fairly extensive preparation. Perhaps this is why we have not had as much of a problem as Mr Greenfield suggested.

Paper No. 28

The Incidence of Tooth Contacts in Normal Mastication and the Part they play in guiding the Final Stage of Mandibular Closure

D. J. Anderson

In spite of earlier claims to the contrary, the subjective impression of most people has been confirmed scientifically, and it has been possible for some years to assert with confidence that teeth make contact during chewing. The frequency of tooth contact during a chewing cycle increases towards the end of the cycle, but with some food materials contact may occur with nearly every chewing stroke (3, 17, 217, 404). It has also been established that although some individuals indulge in rather a chopping action of the jaws, with little lateral movement, after the style of carnivores, for most people, tooth contact is not a static position. During contact, the lower jaw moves along a pathway determined by the shape of the tooth cusps (1). This is referred to as the occlusal phase of the chewing cycle (4) and during this phase the teeth spend a short time in the position of maximum intercuspal contact and the force exerted by the elevator muscles is distributed over a greater tooth surface area than at any other stage in the chewing cycle. Throughout the period of tooth contact the elevator muscles are working isometrically or very nearly so.

No other movement in the body terminates with contact between two hard surfaces, and therefore tooth contact makes masticatory movements unique. A second unique feature is the restriction imposed on the final stages of the closure movement by the shape of the tooth cusps. Although the periodontal ligament cushions the shock of tooth contact, the maximum cushioning effect is probably no more than about 0·05 mm (455).

Since tooth contact makes masticatory movements unique, it is appropriate to attempt to discover if there is any event in the masticatory cycle which might be reflexly evoked as a result of signals set off at the moment of contact. This possibility calls for a renewal of interest in tooth contact, in view of what we now know of the characteristics of periodontal mechanoreceptors (16) and their effects on masticatory muscles (199, 518). If we consider the teeth actually dealing with the food mass, the movement of the teeth towards contact and the achievement of contact are unlikely to be associated with dramatic changes in the discharge of the periodontal mechanoreceptors of these teeth, except possibly to bring in high threshold receptors at the moment of contact. As soon as lateral and forward gliding in contact begins, receptors sensitive to horizontal movement will fire, and we know from animal experiments that directional sensitivity is a property of periodontal mechanoreceptors (431). The dramatic change in periodontal input comes, however, from those teeth which were not load-bearing during the initial stages of the movement, but make contact in the terminal phase of the movement. This sudden increase in total mechanoreceptor discharge distinguishes contact from the early stages of closure. Although it is tempting to assume that tooth contact is signalled solely by a change in periodontal mechanoreceptor discharge, this is not true. Hannam et al. (243) showed that muscle spindles can be involved and it is likely that joint receptors are also. Attempts to relate contact to some event in the chewing cycle can be based only on precise data and these are difficult to obtain. The first problem is to agree on what is meant by tooth contact. Is it contact between teeth on the working side, or the whole arch; on molars or incisors ? It is very difficult to compare in valid terms, work which has emerged from different laboratories. Different test materials have been used and since the behaviour differs with different food materials, there are risks in interchanging data between experiments when attempting to build a composite picture of performance.

The more serious difficulty in attempting to obtain a picture of events during chewing lies in the different methods of recording tooth contact. Existing metal fillings have been used in some experiments to complete electrical circuits between upper and lower teeth and metal bands have been attached to teeth (404). In other experiments, radio transmitters inserted into artificial dentures provide detailed evidence about the path taken by the lower jaw after tooth contact has been reached (308, 445). There are photographic and photoelectric techniques, which in general provide less precise and less detailed information than intra-oral recording methods (4, 8). Watt (608) and Hannam et al. (242) used the vibrations set up when the teeth come into contact to signal the event. In recent experiments Öwall and Elmqvist (440) have used an accelerometer to indicate the cessation of jaw closure and therefore tooth contact. In our experiments (17) we embedded robust stainless-steel wires into existing restorations and extended the wires into the buccal sulcus, looping them so that they made end-to-end contact with the teeth in the position of maximal intercuspation. Since the wires were only 0·5 mm in diameter, slight lateral or forward movement of the lower jaw separated the wires and broke an electrical circuit. Simply because the food became disastrously intermingled with the circuit leads which passed along the buccal sulcus to the exterior, we always made our recordings while chewing on the opposite side of the mouth. We looked specifically in static tests at the question of tilting. None of our subjects was able to clench on one side without making contact on the other, and furthermore when we separated the molars on one side with paper, contact on the opposite side was broken in spite of attempts to maintain it, in some subjects with paper only 0·04 mm thick, although in others the required thickness was 0·08 mm. So in our limited static tests, we showed that tilting is only slight. However, Møller (404) has shown that during unilateral gum chewing, tilting can occur, with the result that contact is achieved first on the molars of the non-working or balancing side. In his very careful studies, he found that molar contact precedes incisor contact and of course is broken before incisor contact is broken. Møller remarks, and perhaps he would like to comment on this, that 'during "natural" chewing contact occurs simultaneously on right and left sides, whereas during unilateral chewing it occurs earlier on the balancing side than on the chewing side'. My question really concerns the use of the term 'natural' chewing.

A number of studies, including those of Møller (404) have investigated the relation between the moment and duration of tooth contact, and EMG activity (4, 8, 242). In the study by Ahlgren and Öwall (8), force was also determined, and the question that this kind of experiment might answer is whether the change in receptor discharge taking place at the moment of tooth contact could play a part in regulating masticatory forces.

My attempts to correlate these data have run into difficulties, but it has been possible to show the time relationship between peak EMG activity and other events in three studies: those of Møller (404), Ahlgren (4), Ahlgren and Öwall (8). I have attempted to use data which are strictly comparable, but this attempt has not been wholly successful. In the three lower sets of data in *Fig.* 28.1 the food materials were apple (M), carrot (A) and peanuts (A & Ö). However, the interval of 50 msec between peak EMG and the onset of the occlusal phase in record (A) from Ahlgren (4) was determined with gum and not carrot. The cross-hatched rectangle in the lower record M from Møller (404) represents the period of molar contact and the solid rectangle the period of incisor contact. Not too much importance should be attached to the interval between peak EMG and moment of molar contact, since this is my estimate from Møller's results. EMG activity ceased approximately 60

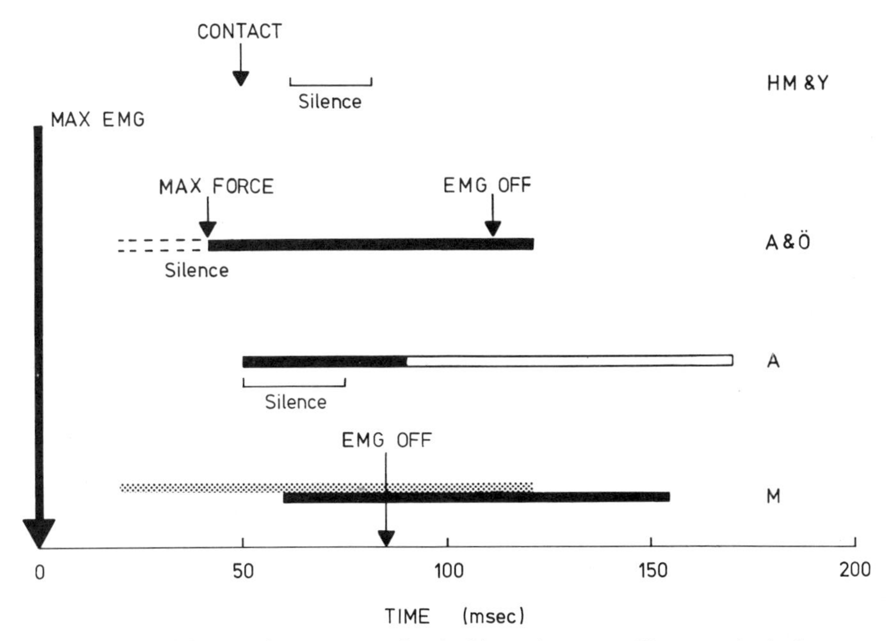

Fig. 28.1. Collated data on the events associated with tooth contact. The events in the lower three records have been related to the maximum EMG. In the top record (H, M and Y), the sequence of contact and silence has been positioned arbitrarily in relation to the other records.

msec after molar contact was made. In record A the duration of the occlusal phase (there was no differentiation between molar and incisor contact) was found to be 40–120 msec and the limits of the range are indicated by the solid, and the solid plus open bars. Although Ahlgren and Öwall (8) obtained more information by recording force with muscle activity and tooth contact (record A & Ö), they used the photoelectric method for recording jaw movements previously described by Ahlgren (4) which though simple, was less precise than the cinematographic technique also used by Ahlgren in the earlier study. Ahlgren and Öwall did not give figures for the interval between the maximum EMG and tooth contact, but referred to the figure of 50 msec previously obtained by Ahlgren. However, they found that the maximum force was developed approximately 40 msec after the peak EMG and that opening occurred about 80 msec after the peak force had been reached. If, as the authors say, the peak force is reached about the middle of the occlusal phase, it follows that the peak force must have occurred after tooth contact had been achieved, and this is indicated in *Fig.* 28.1.

Hannam et al. (242) (record HM & Y in *Fig.* 28.1) simply recorded the interval between tooth contact signalled by a crystal microphone on the forehead and the onset and duration of the silent period. This interval, of approximately 12 msec, is compatible with the hypothesis that contact, whether by exciting periodontal or other receptors, is responsible for switching off the muscle activity and therefore limiting the force applied to the teeth. However, since the mechanoreceptors in the periodontal tissue of food-bearing teeth are already being stimulated it seems more reasonable to attribute the switch-off to the sudden discharge of mechanoreceptor activity in non load-bearing teeth making contact on the balancing side. Møller found molar contact on the contra-lateral side preceded contact on the ipsilateral side by something like 30 msec, and since Hannam et al. found the latency between contact and silence to be about 12 msec, it would seem reasonable to ask whether silence might precede ipsilateral contact.

Very recent experiments by Öwall and Elmqvist (440) have challenged the view that tooth contact has any causal relationship with EMG silent periods—referred to by them as 'motor pauses'. The challenge comes from the observation that motor pauses are more frequent in the early part of a chewing cycle, whereas Öwall and Møller (441) and others have found that tooth contacts are more frequent towards the end of a chewing cycle. Furthermore, since motor pauses occurred in denture wearers and in subjects with a natural dentition, this further weakens support for the view that if tooth contact sets off the signals which are responsible for the silent period, their origin is from periodontal receptors.

How far have the experiments on tooth contact advanced our understanding of the control of masticatory movements ? As yet, not very far; partly for the reason that bedevils all research on masticatory activity in man, that mastication is not wholly a reflex activity, but is accessible to conscious control which is very likely to be of more importance in the experimental situation than in natural everyday private chewing. Clearly there are satisfactory methods available for the study of tooth contact and other events such as peak EMG, silent period, and peak force; what we lack so far is the combination of the best methods for the study of all these events in one and the same experiment.

Discussion CHAIRMAN: Y. KAWAMURA

MØLLER: I think I must be given an opportunity to comment on your question concerning tooth contact during natural chewing and unilateral chewing. You have quoted from my thesis (404) that during natural chewing, molar contact was made and broken before incisor contact and that on average, contact occurred simultaneously on both sides, whereas during unilateral chewing it occurred earlier on the balancing side than on the chewing side. These points have been clarified in a recent publication (664).

First, by natural chewing, I mean chewing without instruction (i.e. the way the bite is taken, its size and the manner by which it is chewed are left entirely to the subject). Natural chewing may be characterized as a series of unilateral strokes in which the bolus is shifted at random between the right and the left side. Therefore, if you average muscle activity and tooth contact from a number of consecutive strokes during natural chewing you will be constructing a stroke that never existed. It would be like counting how many people walk on the right side of the road and how many on the left and then putting the pavement in the middle of the street. This does not make sense; you have to consider each stroke separately.

Secondly, deliberate unilateral chewing of chewing gum represents an experimental situation in which you can study events related to the individual stroke during natural chewing, e.g. the occurrence of molar and incisor contact. The reason for using chewing gum is to avoid a change of the strokes by progressive reduction of the bolus. Thus one obtains a repeated and consistent chewing stroke on one side corresponding to 'freezing' conditions during a single stroke of a sequence in natural chewing. Such recordings indicate that contact is made first on the molars of the balancing side and then on the molars of the chewing side and the incisors; the mandible tilts, first around the bolus and then around the first contact on the balancing side. If now you average a number of strokes during natural chewing, an artefact is constructed indicating that contact occurs simultaneously on both sides, first on the molars and then on the incisors.

WATT: I have taken high speed cinematographic films of chewing and tried to relate them to synchronous EMGs to examine the silent periods that occur during mastication. In most of the sequences in which peanuts were chewed there were more silent periods in the first 10 strokes than in the second 10 strokes. With apple, however, the silent periods were about equally divided between the first and second 10 strokes. On the other hand, perhaps through inaccuracies in our film analysis, it was not always possible to relate the arrests of movement that occurred during chewing as seen on the film to silent periods that we observed in the EMGs. There seemed to be some odd relationship, but the error of the method seemed to be too great for us to elucidate this. But I should like your comment on the observation that silent periods are much more frequent at the beginning of chewing hard foods and much less frequent towards the end.

ANDERSON: I have no comment to make based on our own work, but this was an observation also made by Öwall and Elmqvist (440). Perhaps Dr Öwall would like to comment.

HANNAM: One has to be very careful with the silent period, because, as many people have shown already, it has a variety of possible causes, one of which is vibration and one would expect to see it at the beginning of a chewing cycle with hard foods. It may be important in the chewing cycle but some of us tend to believe that it is a biological artefact; I think we should not overrate its importance, it is such a fleeting pause. It is obviously reflex, it is obviously real, but it is multicausal. In terms of the duration of the chewing cycle, and the power which is developed in the chewing cycle, I do not think it is terribly significant.

ÖWALL: I have just investigated the frequency of silent periods during a chewing sequence and I quite agree with Professor Watt that they are more frequent in the early part of a chewing sequence than at the end of it. The distribution is the same for persons with their own teeth and full denture wearers. I found a silent period in about 75 per cent of chewing strokes in the first part of a sequence.

KAWAMURA: What test food did you use ?

ÖWALL: Peanuts.

WATT: With hard sugar-coated chewing gum there are initially frequent silent periods and, as the gum gets softer, they practically disappear.

Paper No. 29

The Incidence of Abnormal Tooth Contacts and their Detection

D. M. Watt

DEFINITIONS

Studies with intra-oral occlusal telemetry show that the position of maximum intercuspation or habitual occlusion is the position of contact most frequently achieved in chewing and swallowing (90, 193, 194, 217, 308, 443, 444, 445, 446).

If a patient can bring his teeth into contact in this position without muscle pain, periodontal or joint trauma, we can consider this to be a normal occlusion. We must, however, make a clear distinction between a morphological malocclusion and a functional malocclusion (*Fig.* 29.1). In a morphological malocclusion the arch form and/or jaw relationship is abnormal, but in spite of this the teeth might occlude in a stable manner in the position of maximum intercuspation so that no slide occurs after the first contact. In a functional malocclusion the first contact is unstable and a slide occurs into the position of maximum intercuspation. In such cases the dental arch may or may not be morphologically normal. The ideal occlusion might be defined as a morphologically normal occlusion in which 'muscle position' and 'tooth position' coincide (88) but there is some difficulty in defining these entities separately since one is dependent on the other. Tooth contact

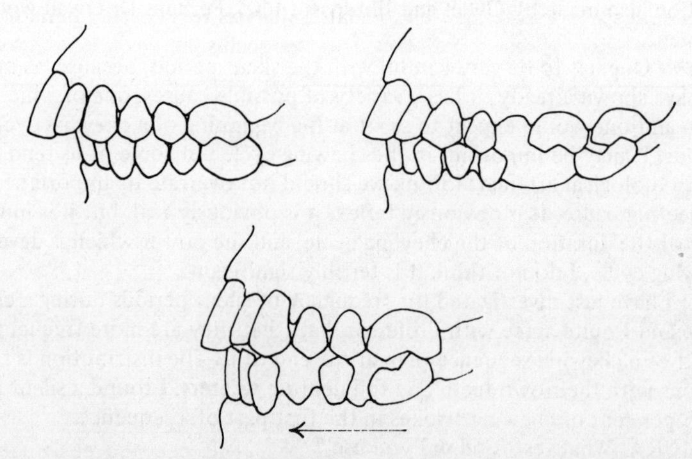

Fig. 29.1. The top diagrams represent morphological malocclusions which are functionally stable. The lower illustrates a morphological and functional malocclusion. Muscle position and tooth position do not coincide and a forward slide occurs into the position of maximum intercuspation.

modifies muscle activity and muscle activity modifies tooth position (501, 504, 625). We can define the muscle position as 'the position to which the mandible closes when the muscles are acting in an optimum manner' and the tooth position as 'the position of maximum intercuspation of the teeth', but we must recognize that there is a wide range of differences between muscle position and tooth position. At one end of the range are minute discrepancies to which the patient readily adapts, at the other end, major functional malocclusions to which the patient cannot adapt, and in which the mandible may slide several millimetres after first tooth contact. It seems unreasonable to consider the ideal occlusion 'normal' since we suspect that the majority of people have some slight discrepancy to which they easily adapt. It must also be accepted that the muscle position will alter with general factors such as fatigue, nervous tension, and bodily posture. Thus it is very difficult to describe 'normal' tooth contacts without defining a number of ambient conditions within which the description is valid but beyond which it is not.

Any definition of an abnormal tooth contact must therefore be rather flexible and I offer the following: An abnormal tooth contact is a contact to which the patient is unable to adapt without symptoms or obvious pathological change.

METHODS OF DETECTION OF ABNORMAL CONTACTS

Articulator Techniques

It is clear that any mechanical device (such as an articulator) puts such constraints upon its user that it is an invalid method of diagnosing abnormal contacts (55, 123, 177, 461, 520). The constraints are: (1) The mechanical limitations of even the best instrument introduce errors which are well beyond acceptable limits of occlusal physiology. It is possible, for example to detect separation of the teeth of as little as 0·02 mm (89, 590). (2) The techniques used in transferring the data on mandibular movement to the instrument are also subject to error which is unacceptably high (476, 587, 613). (3) It is necessary to judge, at a particular moment, a mandibular position which is taken to be a reference position from which the occlusion is assessed. Incidentally this constraint is equally applicable to some intra-oral telemetric studies. (4) The absence of a time base on which to judge occlusal performance makes articulator analysis very limited in its scope for assessing anything but the grossest parameter of occlusion.

If we adopt a classic gnathological approach to the diagnosis of abnormal tooth contacts, using the most retruded terminal hinge axis position as representing normality, we will be able to detect apparently abnormal tooth contacts in the majority of patients and the use of this position as a reference for the diagnosis of abnormal tooth contacts is unacceptable (28, 193, 587).

Direct Techniques

When we remove the hinge axis, what have we left ? Some may feel that the very foundation of our clinical method has been taken away without any reasonable substitute. How can you correct occlusion if you have no set point from which to make measurements ? In every occlusal situation there are three factors operating: the joints, the muscles, and the teeth, and all interact to a greater or lesser extent. Thus the final analysis must be carried out in the mouth. There are several methods available for locating contacts with paper, tape, wax and lacquer (465, 589, 607). These methods of course require a clinical decision that the position of tooth contact registered is the correct one, and that the contacts recorded are abnormal and should be removed. It is important to realize that the alterations are accompanied by adaptive responses which change the functional activity of the mandible. These adaptive responses, however, can be monitored by gnathosonic techniques whereby the sounds of occlusion are recorded.

Gnathosonic Techniques

When muscle position and tooth position coincide, the snapping contact of teeth produces an occlusal sound of short duration. This is picked up by a vibration transducer placed on the face or forehead and the jolt makes the transducer vibrate (85, 251, 608, 609, 610, 611, 612, 614, 615, 616, 617). The vibrations are of short duration and fade exponentially (*Fig.* 29.2). The characteristics of the transducer will of course influence the duration of the sounds and the results will vary to a certain extent with the type of transducer and the frequency response of the recording apparatus, which should be about 1kHz. However, this does not invalidate the findings, because the nature of the tooth contact is the main factor

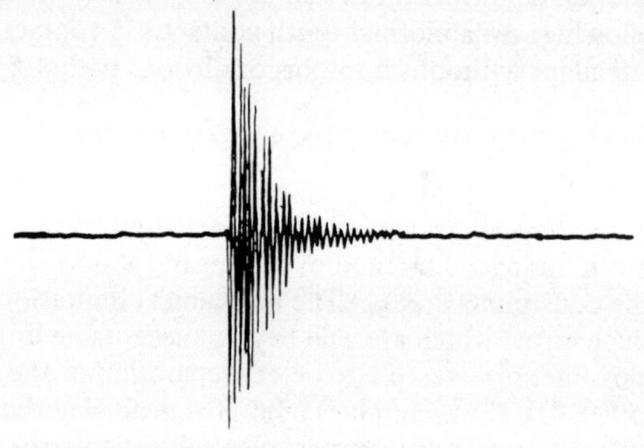

Fig. 29.2. A stable tooth contact jolts the vibration transducer once and makes it vibrate. The vibrations decay exponentially.

which affects the record. Sliding contacts of teeth cause lower amplitude vibrations, the duration of which is related to the length of the slide (*Fig.* 29.3). If there is an abnormal interceptive contact the transducer is jolted more than once in a single closure of the jaws and a double sound is produced (*Fig.* 29.4). The time interval between the peaks can be measured and the dimension of the interceptive contact can be deduced, provided that the mandibular velocity is known. Theoretically deflective contacts of about 0·02 mm can be measured by this means, but difficulties arise as the mandibular velocity changes during mandibular closure (*Fig.* 29.5) with the result that errors occur in the calculation of the dimension of the deflective contact. The adaptive activity of the jaws after the first contact causes additional difficulties in the measurement of the dimensions of interceptive contacts. For instance, an interceptive contact on the right premolar results in the mandible being swung slightly towards the left at the next closure in order to make more room for the interceptive contact which will apparently be

Fig. 29.3. A sliding contact of teeth produces lower amplitude vibrations the duration of which is related to the length of the slide.

reduced in size when measured by the sonic method. In addition, this adaptive activity may cause abnormal contacts elsewhere in the mouth and uncoordinated activity occurs (*Fig.* 29.6).

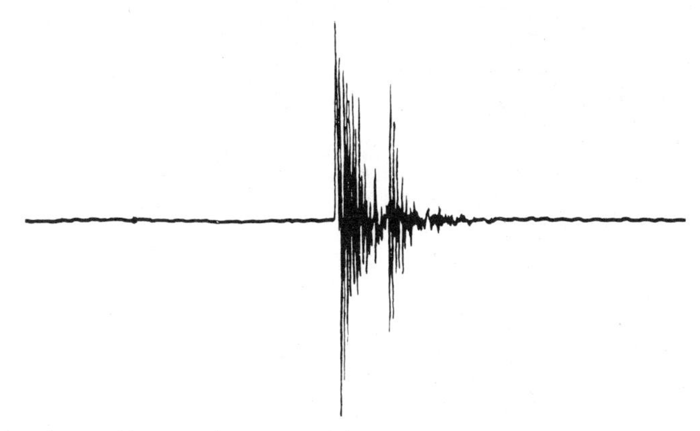

Fig. 29.4. An abnormal interceptive contact jolts the transducer more than once in a single closure of the jaws and a double sound is recorded.

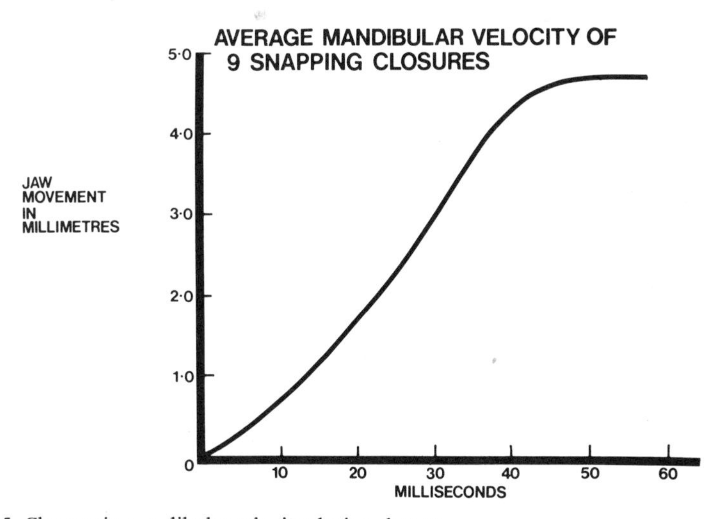

Fig. 29.5. Changes in mandibular velocity during closure.

On examination of large numbers of gnathosonic records it was found that they fell into three classes. In Class A the sounds of tooth contact are impact sounds of short duration indicating a stable occlusion. In Class B some sounds are short and some are prolonged, indicating some stable and some unstable contacts, and in Class C all the sounds of occlusion are prolonged, indicating unstable contacts. Examples of gnathosonic records from patients in each of these three classes are shown in *Fig.* 29.7. It can be seen that the sounds of occlusion in the Class C records are interspersed by vibration patterns of a different nature. These are caused by sliding contacts of teeth as they disarticulate on opening. The sounds have been called

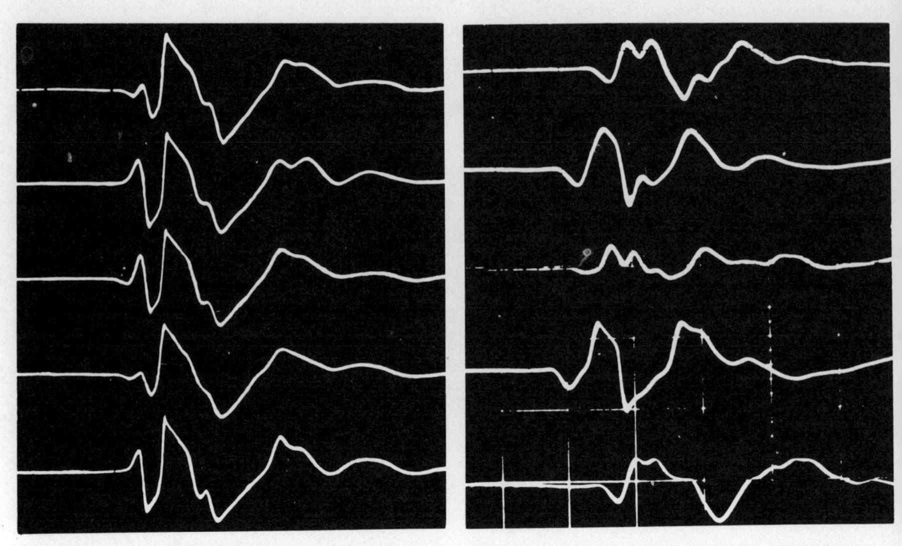

Fig. 29.6. *Left:* Oscilloscope records of five successive occlusal sounds of patient with stable occlusion. Note the similarity of the wave form and amplitude. A fast time-base (5 msec per cm) has been used to aid comparison. *Right:* Records from the same subject after placement of a 0·06 mm artificial premature contact.

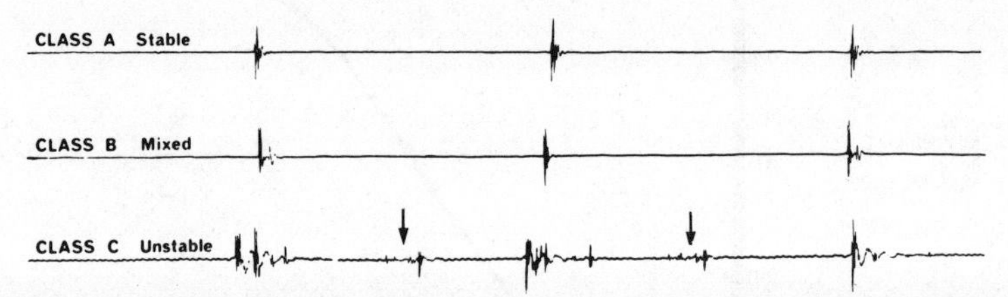

Fig. 29.7. Classification of occlusal sounds. Class A. All the sounds are impacts of short duration. Class B. Some sounds are short single impacts and some are prolonged. Class C. All the sounds are prolonged. In this case 'separation noise' (arrowed) is also present, indicating a contact slide on disarticulation.

separation noise (608).*Table* 29.1 shows the distribution of separation noise in equal samples of the three classes of gnathosonic records. It appears that when the occlusion is unstable (Class C) and the teeth slide into the position of maximum intercuspation on closing, they also slide out of that position on opening. This finding of course might be expected in patients whose muscle position and tooth position do not coincide and suggests that there is stress in the position of maximum intercuspation. The presence of occlusal creaks in the squeeze phase of occlusion lends further support to this concept. After the preliminary contact of teeth there appears to be increased muscle activity during the pause in occlusion before opening. I therefore call the pause the squeeze phase of occlusion. Slight discrepancies in the intercuspation of teeth during this phase give rise to creaking sounds which are usually followed by modification of the closure and consequent

Table 29.1 Distribution of separation noise between equal samples of the three classes of gnathosonic records

	Separation Noise Present (% Cases)
Class A	6
Class B	73
Class C	89

change in the preliminary occlusal sounds at the next contact (*Fig.* 29.8). There is some evidence that the creaks are related to peaks of increased muscle activity, but this is still under investigation. The creaking sound on the gnathosonic record looks like a number of rapid impacts and is produced by starting and stopping of the teeth as they slide jerkily over each other. The creaks may start quickly and gradually slow with an increase in amplitude or may start slowly and gradually speed up with a decrease in amplitude.

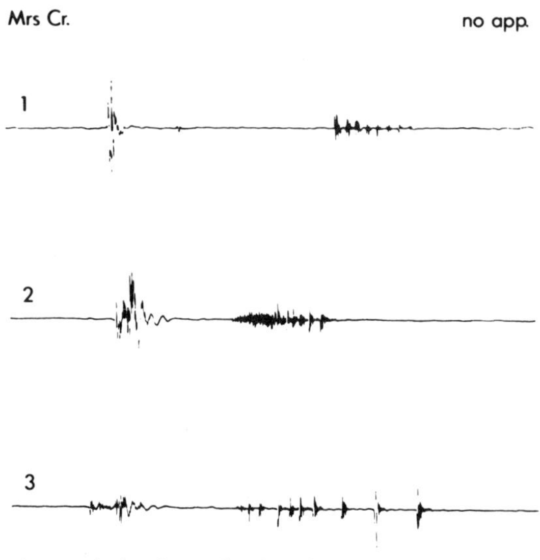

Fig. 29.8. Gnathosonic records showing occlusal creaks following three successive tooth contacts. The preliminary tooth contact sounds are on the left and the creaks, which look like a succession of impacts, are on the right. 1. Note that the preliminary sound of the first occlusion is comparatively stable and is followed 150 msec later by the creak. 2. On the next closure the patient has postured towards the creak, resulting in an unstable preliminary contact only 50 msec before the creak. 3. This traumatic contact is followed on the third closure by a low amplitude preliminary sound indicating a more gentle contact. The slow creak which follows suggests a less rapid application of pressure during the squeeze phase.

INCIDENCE OF ABNORMAL CONTACTS

The distribution between the three classes is shown in *Table* 29.2. The control group comprised 200 individuals between the ages of 17 and 60, who had natural teeth and no symptoms of gnathic dysfunctions. Wide disparity of occlusal morphology was to be found in the control group. Many of them had lost several teeth and the only criteria for selection was the presence of some occluding natural

Table 29.2 Percentage distribution between the three functional occlusal classes of a control group of 200 adults and a group of 160 adult patients under treatment for gnathic dysfunctions

	n	*A*	*B*	*C*
Control	200	59%	27%	14%
Patients	160	20%	29%	51%

teeth and the absence of joint or muscle pain. In the group of patients, all had symptoms of gnathic dysfunctions, they all had natural teeth and were over 17 years of age. It is interesting to note that only 14 per cent of the control group were in Class C, but 51 per cent of patients with muscle or joint symptoms were in this class. In a sample of children ranging from 2½ to 16 years, the distribution between the three classes was about equal, so it was decided to investigate this further by studying the duration of occlusal sounds in a large sample of children in different age groups. The results of this survey are shown in *Table* 29.3. The differences between the average duration of sounds in 5-year-old and 8–11-year-old children were highly significant ($P < 0{\cdot}001$), but the differences between the 5-year-old and 16-year-old children were not significant. Thus it seems that what is functionally abnormal occlusion in one age group is functionally normal in another age group and in assessing the incidence of abnormal contacts one must first specify the age of the sample investigated. Some may argue that empty occlusal contacts are irrelevant as far as mastication is concerned, but there is little doubt that the reference position for both masticatory activity and empty contacts is the position of maximum intercuspation. Ultra-high-speed film of mastication illustrates that invariably a long pause occurs in this position and I suggest that it is during this pause that the information from periodontal, mucosal, muscle and joint receptors is processed so that appropriate action can be taken at the next chewing stroke. I suggest that a recognizable and reproducible pattern in this feedback provides the reference for masticatory activity and the method of store is possibly 'holophonic' on the lines suggested by Higgins (256).

Conclusion and Summary

Occlusion is essentially a dynamic activity involving adaptive interaction of muscles,

Table 29.3 Mean occlusal sound duration in msec of 1584 children in different age groups

	Age of Children in Years						
	<5	5	6	7	8	9	10
Number in sample	35	164	142	158	149	181	151
Mean	27·54	34·71	41·42	41·59	53·87	50·72	47·23
Standard error	2·52	1·62	1·81	1·67	2·25	1·95	2·05

	Age of Children in Years						
	11	12	13	14	15	16	17
Number in sample	144	92	108	101	87	49	23
Mean	47·92	36·71	37·15	33·29	32·01	26·96	32·04
Standard error	2·11	1·92	2·08	2·08	2·16	1·75	3·00

joints and teeth. High fillings and slight changes following extraction of teeth can easily be accommodated by tooth movement and adaptive postural activity of muscle. Contacts within the range of the adaptive capacity can be considered normal, but when beyond the adaptive capacity, i.e. producing symptoms or pathology, may be considered abnormal. The ability to occlude the teeth successively into a stable intercuspal position without symptoms, indicates adaptation to contacts within normal range and the presence of a reference position for occlusal and masticatory activity. If the patient is unable to bring the teeth together without sliding this indicates the presence of abnormal contacts, the absence of a stable reference position, and uncoordinated masticatory activity. At different ages during growth this uncoordinated activity is present and was also found in 14 per cent of a sample of 200 adults who had no symptoms of muscle pain.

Discussion CHAIRMAN: B. MATTHEWS

AHLGREN: In your clinical work did you use gum or other test materials and if so did you find differences in the first sound and in the creak sound ?
WATT: The creak sounds can only be heard in empty movements because we are dealing with teeth moving over each other. When food is present there is so much noise from peanuts cracking and gum sucking that you cannot record much of value. *Fig.* 29.9 shows the sounds of 6 successive chews with gum and demonstrates their complexity. The mark S is a switch noise that we made to indicate the start of the contact. As the teeth go through the chewing gum you can see a noise. The peak C is tooth contact and then there is a pause in contact. The odd noise at the end marked O is a sucking noise as the teeth move out of the gum.

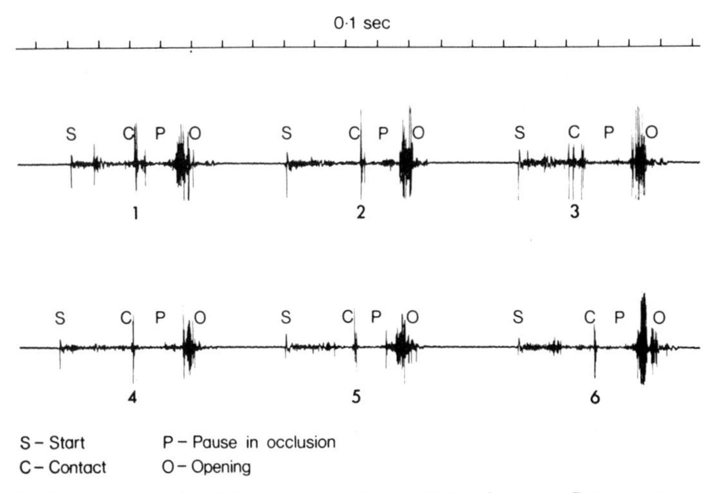

Fig. 29.9. Gnathosonic records of six successive chews of chewing-gum. Between S and C is the sound of the teeth penetrating the gum. The sound marked O is the sucking noise as the teeth move out of the gum. P is the pause in occlusion when there is little or no sound.

INGERVALL: Do you think it is important to standardize the position of the head when attempting to detect abnormal contacts, because one would think that the path of closure varies with the inclination of the head ?

WATT: That is true, and if you snap your teeth together with the head to one side and then put your head to the other side you realize that tooth contacts change. I do not like to clamp the head. I ask the patient to sit with the head upright in the normal posture.
NAIRN: How do you go about correcting abnormal tooth contacts ?
WATT: This is a good question. What I do is to ask the patient to put the tongue back and close the mouth gently, and I feel the condyles. When they seem to disappear under my fingers I know that they are in the retruded position. But I am not trying to force a retrusion.
NAIRN: Then what do you do ?
WATT: Then I remove the most obvious contacts I can see.
NAIRN: How do you see them ?
WATT: I use either wax or paper or tape.
NAIRN: Do you think that you get less inaccuracy that way than you would get by mounting models on an articulator ?
WATT: Yes. Having recorded the occlusion before a correction, I record again after the correction. *Fig.* 29.10 shows the sort of thing we monitor. This shows occlusal sounds of a patient who before a correction was in Class C and who was in Class A afterwards. Two and a quarter years later the occlusion is still stable.

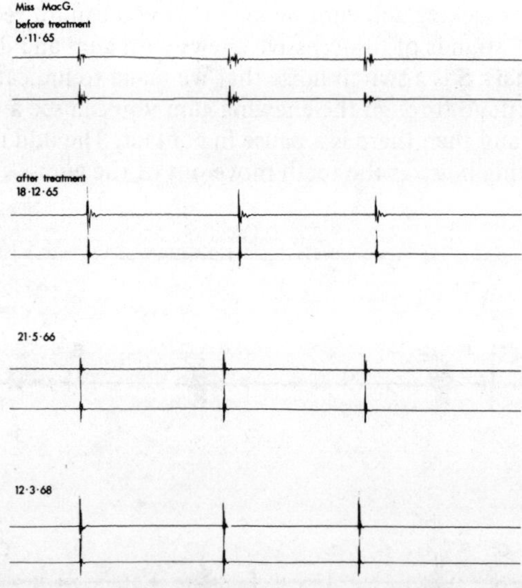

Fig. 29.10. Occlusal sounds are prolonged before correction and indicate instability of the occlusion. After correction the short impact sounds indicate a stable occlusion which has remained unchanged 2¼ years later.

SOLBERG: Have you noticed any differences between the sounds in males and females ? We tested 800 young men and women and found that there was a slightly significant difference, with women having a slightly less good sound in the age group 18–22 years.
WATT: That is very interesting because in a paper which will be published shortly we report a similar observation and I have attributed this to the earlier eruption of the 3rd molars in women. We found differences at a 5 per cent level of significance between males and females in different age groups.

Paper No. 30

Experimental Evidence on the Role of Abnormal Contacts in the Aetiology of Periodontal Disease

D. C. A. Picton

INTRODUCTION

It is now well established that contact between opposing teeth occurs during normal mastication and swallowing (1, 17, 217) while description of the response of the supporting tissues to the normal range of chewing forces and other physiological activities is to be found in the histology books. Occlusal interferences and abnormal or parafunctional habits are seen in many people in whom the teeth may be pressed together with considerable force, possibly for long periods and in an eccentric position (466). The objective of the present review is to examine the experimental evidence on the response of the supporting tissues to premature or eccentric contacts with particular regard to the onset or deterioration of chronic marginal periodontitis.

Although there have been a number of studies on human subjects in which the periodontal health of the supporting tissues has been related to more or less severe degrees of trauma from the occlusion (195, 467), controlled experiments in which there was full evaluation of the periodontium are restricted, with few exceptions, to animal studies. In these experiments the position of the teeth or their morphology was altered, so that premature contacts of one or more teeth were produced. The studies were designed to produce:

1. A simple displacing force or morphological addition in such a way that migration of the tooth away from the force could occur freely.
2. Repeated displacing or 'jiggling' forces applied to a tooth which could not migrate freely from them.
3. Devices of types used in (1) or (2) superimposed on teeth with established periodontal disease.

1. SIMPLE PREMATURE OCCLUSAL CONTACTS

A considerable number of studies have been carried out with various animals including rats (375), monkeys (59, 546), and dogs (214, 602). Two principal methods have been used to alter the occlusion. These were by propping the bite by the addition to the occlusal surface of high amalgam fillings (546), or gold inlays or crowns etc. (59) and secondly by displacing adjacent teeth by inserting a latex band between them (375). In the former the bite was propped open initially by amounts ranging from 1·0 to 7·0 mm so that the test and opposing teeth were the

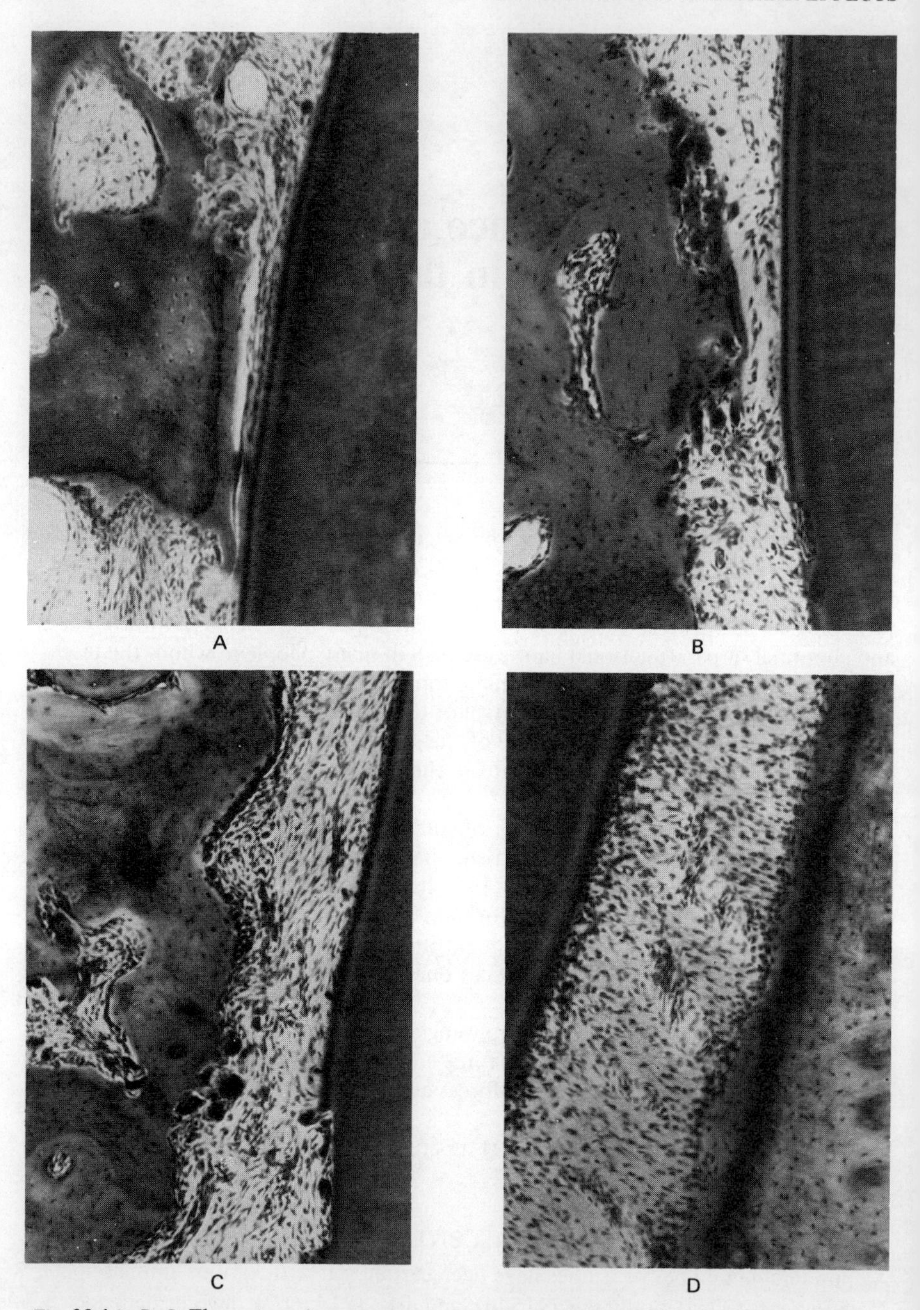

Fig. 30.1 A, B, C. Three stages in aseptic necrosis and remodelling of the periodontium of ferrets following the insertion of a maxillary splint which propped the bite on the opposing canines. The mesial surface of the root of the mandibular canine near to the alveolar crest is shown 10 days, 3 weeks and 6 weeks after the appliance was fitted. Little difference is apparent between A and B as the canine was being progressively displaced mesially. In C some resorption is taking place but no necrotic tissue is present in the young cellular periodontal membrane and the phase of remodelling is under way. D, From the distal aspect of the canine in A from the crestal region, showing the greatly widened membrane with proliferation of bone and periodontal membrane. Stained with haematoxylin, eosin and Biebrich Scarlet, × 5.

only teeth in occlusion. In each case the test animals were sacrificed after varying periods and the periodontal state was assessed on histological sections; the contralateral teeth, initially and for several days or weeks out of occlusion and for several studies the tissues of unoperated animals, were used for comparison.

Results

Remarkable uniformity is apparent in the gross and histological findings in these studies. Test and opposing teeth were progressively displaced horizontally and/or intruded into the alveolar bone until occlusion of the other teeth was re-established. Inspection of the teeth for 'mobility' if it was noted, revealed a marked increase in horizontal mobility (602). These changes are almost identical with those reported for teeth subjected to heavy orthodontic force in a horizontal direction with narrowing of the periodontal space in regions of compression while widening was noted in areas of tension (*Fig.* 30.1). Within a short period, i.e. 3 days (197), loss of cellular identity was apparent in areas of compression, and necrosis due to prolonged ischemia was followed by autolysis. Within a similar period osteoclasts appeared in the marrow spaces adjacent to the alveolar wall, setting up undermining resorption of bone in the region of necrosis. Where less severe compression was produced, resorption of the alveolus from the periodontal aspect was apparent, together with removal of small areas of cementum. In the tissue spaces created by the osteoclasts, proliferation of young connective tissue elements was seen, with gradual transition from the phase of resorption to one of repair with reformation of new principal fibres, cementum and alveolar bone. The duration of the former phase was dependant on the distance the test and opposing teeth had to be displaced to allow the occlusion to approach a normal state. The striking lack of leucocytes is noted by most workers, and therefore the lesions can be described as aseptic necrosis, followed by reformation of the periodontium.

The changes in the supracrestal regions are of particular importance. Thus, almost without exception, the epithelial attachment on the test or opposing teeth is described as being either on enamel only (59, 602) (*Fig.* 30.2) or, if the junctional zone was found to be on cementum, a similar extension was noted on the contralateral or unoperated control teeth (197). Thus no pocketing or increasing rate of recession had been produced by these workers. In several studies the degree of infiltration of the crestal tissues by inflammatory cells was within normal limits or at least similar to that in the control region. In other studies greater inflammation was seen, but this can be attributed to the raised level of stagnation at the gingiva around the margins of crowns, etc. (602). Stones (546) however, noted more epithelial downgrowth on to cementum and greater infiltration of inflammatory cells around the test teeth but it seems possible that this was due to the marked intrusion and tilting of the test teeth and consequent stagnation which occurred. Another consistent finding was the integrity of the trans-septal fibres which were considered by Glickman (192) to act as a physical barrier to the spread of irritants from the interdental soft tissues into the periodontal space. The angulation of this normally horizontal band was changed to become deeply inclined towards the intruded tooth (197). Signs of necrosis above the alveolar margin were always absent despite, at times, severe levels of necrosis and resorption of root and bone in the periodontium. Removal of the occlusal interferences was followed by return of the test teeth towards the original position while repeated replacement and removal, two to five times, created no new histological features but rather an increasing degree of resorption of dentine and cementum (602).

Concerning the forces involved, from limited evidence in man, it seems probable that the test and opposing teeth were subjected to increased occlusal loads (18) although nothing is known of the duration or frequency of these forces. Despite this, high crowns worn continuously by young adults did not cause pain from their insertion until the occlusion was re-established (15, 432).

A

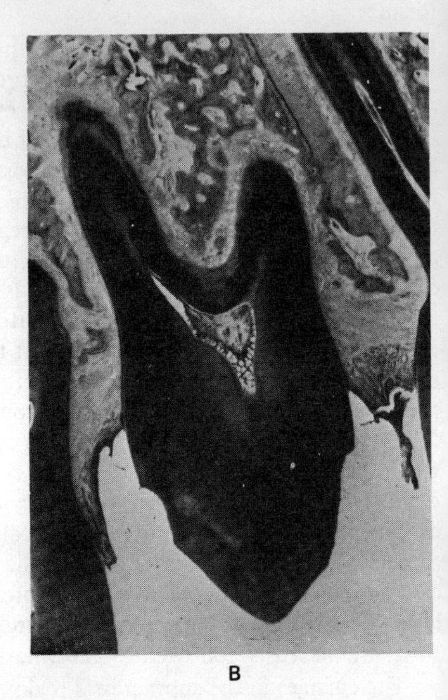

B

Fig. 30.2 A, B, Histological record of the periodontal state of a monkey 6 months after a 'high' crown was fitted to B while A remained on the control side of the maxillary arch. The lack of epithelial downgrowth on to the cementum and of inflammatory cell infiltration in the periodontal membrane or gingiva is apparent in both A and B. (*Reproduced by kind permission of Dr Glickman and Dr Smulow and the Editor of the Journal of Periodontology.)*

2. 'JIGGLING' PREMATURE CONTACTS

The relevance of the above studies to man is said to be of limited value in that a stressed tooth in parafunctional habits is not in a 'high' position and can generally not be displaced into a less vulnerable position. Accordingly, a modification of the experimental design was introduced to cause the test tooth to be displaced in one direction on closure of the arches but in another when the teeth were not in occlusion. Thus the test and opposing teeth were subjected to a heavy displacing force in centric occlusion together with an additional opposing force which tended to return the test tooth to the original position when the teeth were apart. Studies of this type in monkeys (196, 621) and in children (475) have been reported.

The results of this regimen were similar in many features to those in (1) above. Although most of the test and opposing teeth did not adopt a more intruded position, a high degree of horizontal mobility was seen with marked widening of the periodontal space and angular resorption of the alveolar crestal region (196). The histological changes were almost identical to those in (1) except that the alveolus was reformed to accommodate a much greater horizontal displacement of the tooth from the rest position rather than a total change of position in the alveolar process (475, 621).

Histologically, the epithelial tissues, round-cell infiltration and supracrestal fibres were indistinguishable from those in (1) in quality, although the degree of gingival inflammation was more severe due, probably, to the higher level of stagnation around the orthodontic bands and springs (196). Thus the adaptation of the periodontium to the premature contacts here also seems to have been aseptic remodelling, independent from the changes in the gingiva.

3. 'JIGGLING' PREMATURE CONTACT SUPERIMPOSED ON ESTABLISHED MARGINAL PERIODONTITIS

Although there was marked gingival inflammation in several of the studies reported above and some of the species chosen, notably monkey, normally have quite severe gingivitis, there is only one study in which the 'jiggling' design has been set up on a deliberately created severe level of periodontitis. The regimen for this has been described recently (353, 554). Following controlled surgical trauma to the mesial surface of the root of the fourth mandibular premolar of six dogs, copper bands extending into the lesion were cemented on the teeth for three weeks. On removal of the bands, plaque-retaining rubber bands were placed on the teeth for four weeks, at the end of which time severe periodontitis was established. On the test side, a device was fitted which caused the test tooth to be displaced mesially on closure while another appliance was fitted to force the test tooth distally when the teeth were apart. The contralateral tooth was treated in the same way except that no 'jiggling' forces were created. During the next six months the vascular reactions (555) and periodontal condition were assessed by means of inspection, radiography, gingival exudate and tooth mobility estimates and at death on histological examination (353).

The findings, in brief, were that gingival inflammation and plaque scores were similar on the two sides but there was greater horizontal bone loss on the test side with angular bony defects, significantly increased mobility and greater proliferation of the epithelial attachment over the cementum. In short, there was a marked worsening of the degree of periodontitis on the test tooth as compared with the control tooth following the introduction of trauma from the premature occlusal contacts.

The fundamental difference between this study and those mentioned previously is the extent of the inflammatory processes. For, at the time Lindhe and Svanberg introduced the trauma from occlusion, inflammatory invasion of the periodontal space had occurred far beyond the alveolar crest so that dead tissue in regions of compression would have tended to become infected with little chance of normal repair. Thus if a physical barrier is presented, by the trans-septal fibres for example, to invasion from the inflamed gingiva then it seems possible that necrosis in the membrane in regions of compression is aseptic and repair, with remodelling of the periodontium and the inflammation, remains restricted to the supracrestal regions.

It would be interesting to know whether the alternative regimen of superimposing periodontitis on the occlusal interferences would cause the same affect, since the supporting tissues would have adapted aseptically to the occlusal forces by that time and in the absence of necrotic areas in the newly reformed periodontium the spread of the inflammation might well be contained. The importance of the trans-septal fibres as stressed by Glickman (192) in this context could also be explored with this approach. Thus it is confirmed that these fibres are reformed after excision within a few weeks in adult monkeys (456) and across extraction sockets in a similar period (422, 447). If an interval of a month was allowed, therefore, between the trauma described by Lindhe and Svanberg (353) and the placement of latex bands, it seems possible that the trans-septal fibres would reform apical to the surgical lesion and the inflammatory lesion would not spread.

In conclusion, it seems clear from the evidence that by establishing occlusal interferences or premature contacts experimentally in the presence of healthy supporting tissues or gingivitis, aseptic necrosis in the periodontal membrane is followed by remodelling of the periodontium to accommodate the new position of the tooth or the increased masticatory loads without inflammatory changes. The same occlusal alterations imposed on teeth with active inflammatory lesions in the periodontal ligament appear likely to cause progression of that lesion. The relevance of these findings to man has yet to be confirmed since the effect of parafunctional habits may not correspond with the experimentally created 'high' crowns or fillings but, if shown to be correct, then it seems probable that treatment of the marginal periodontitis by elimination of stagnation is of more immediate importance than correction of the parafunctional habit.

Acknowledgement
I am grateful to Dr J. P. Moss of University College Hospital Dental School for permission to use the histological material from ferrets.

Discussion CHAIRMAN: B. MATTHEWS

GRAF: I should like to show some data from our measurements of loads on a single molar crown, which might shed some light on occlusal trauma and also perhaps on the neurophysiological aspects of masticatory control. The method has been reported by Graf et al. (216). It employs a piezo-electric transducer which can be incorporated in the crown of a root canal treated molar and measures masticatory forces simultaneously in the axial, orofacial and anteroposterior directions. *Fig.* 30.3 shows the forces developed in the lower first molar during chewing on the same side as the transducer, with the tooth in normal occlusion, and with the height of the crown raised by about 0·2 mm. Increasing the height of the crown results in an increase in the amplitude of the force in all three directions. Loads are recorded on the balancing side (*Fig.* 30.4) when the experiment is repeated with the subject chewing on the opposite side to the transducer. When the crown height is increased, the forces on the balancing side decrease, especially in an orofacial direction. Perhaps the physiologists can tell us if there is some inhibition or facilitation—some peripheral input perhaps from periodontal or other receptors which could influence the loads and control them.
HANNAM: There is one observation which has been made recently, with regard to the human ability to discriminate forces applied to teeth during orthodontic treatment. This is a long way from loading teeth in the way just referred to, but if you look at teeth which are being treated orthodontically then the discriminating power decreases markedly about the middle of the treatment and then returns to normal again at the end of treatment. Presumably signals from around teeth which are being moved are quite drastically altered.
PICTON: This of course raises the question of what changes occur in the supporting tissues during normal function. During the day, the tooth adopts a progressively

more intruded position in the socket. That means, presumably, that the receptors are in a different position at the end of the day from earlier on. So how do you

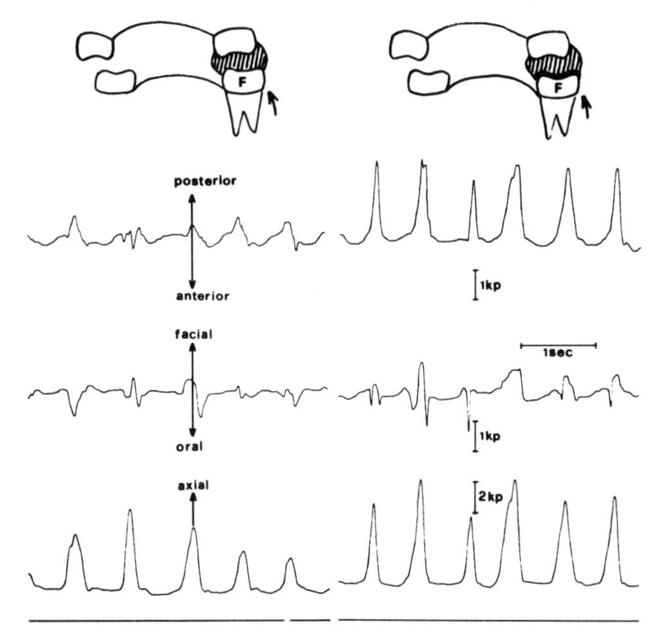

Fig. 30.3. The vectorial load pattern on a lower first molar during chewing of Swiss air dried beef on the working side. *Left*, tooth in 'normal' occlusion; *right*, crown about 0·2 mm too high.

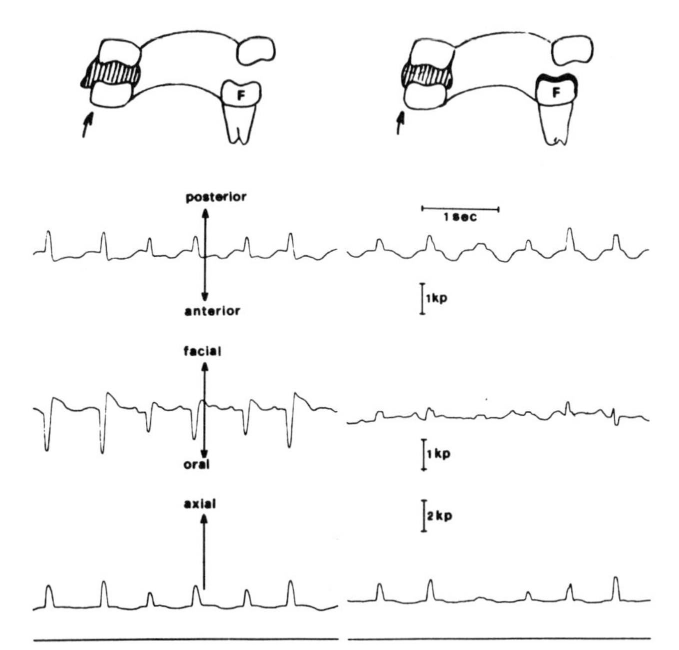

Fig. 30.4. Force patterns on the same tooth, when the food was chewed on the opposite side. *Left*, tooth in 'normal' occlusion; *right*, crown about 0·2 mm too high.

conceive the orthodontic situation ? Are the endings still there, or are they destroyed or non-functional ?

HANNAM: I cannot answer that. I think that in the example I quoted, the teeth were being rotated and this would produce profound alterations in periodontal membrane structure. In experiments on the cat in which we were stimulating once a second for long periods of time, the signalling ability at the end of 30 minutes was about the same as it was at the beginning.

CHAIRMAN: We can confirm that. Hilton in our laboratory found similar results, although some units showed a substantial decline in their response with repeated stimulation (266).

Paper No. 31

Clinical Observations on the Effects of Abnormal Occlusion on Muscle and Joint Function

H. Thomson

The contact relationships between opposing teeth are probably no more nor less harmonious than those between other opposing members of the human body. The hands can be well or ill related to each other in the performance of operations requiring manipulative skill. Examples include sewing a fine seam, or shaping a vase on a potter's wheel. Similarly, the feet do not always perform according to intention and have been known to exhibit premature contacts against each other or against various unexpected interferences; sometimes causing disorders of joint and muscle. These members can, of course, be trained to function more efficiently; and learned muscle activities are a well-known feature of neuromuscular function. In these activities some people exhibit greater skill than others. In general, it can be said that these members perform most efficiently when not in contact, while maintaining well trained relationships to each other.

This principle can be said to apply to opposing teeth in relationship to each other where they are often found to be too much in contact and not always beneficially.

The mandibular teeth move in relationship to the maxillary teeth in the performance of various natural functions such as mastication, swallowing, speech, facial expression and of various parafunctions such as clenching the teeth and grinding them, opening of hair grips and sometimes beer bottles, supporting pipes and pencils and sometimes other people, as in acts of daring far above the circus ring, or acts of violence. In many, if not most people, the various and sometimes bizarre uses to which the teeth are put, by present-day standards, result in adaptation by the tissues of the masticatory system. But not in all people. The variability of responses to the demands made on these tissues may provide a useful source for investigating the problems of joint and muscle disturbances.

The teeth, together with the mandibular muscles and joints, form a triad of tissues which perform the functions of the masticatory system under the control of the neuromuscular system. These functions are controlled both at reflex level and by impulses relayed through the various higher centres of the brain.

DEFINITIONS

What constitutes abnormal occlusion is not easy to define, since what may be considered an abnormality often results in adequate adaptation. It is no easier to define normal occlusion where apparent normality is sometimes associated with various disorders of joint or muscle function, as will be seen in the examples to be shown.

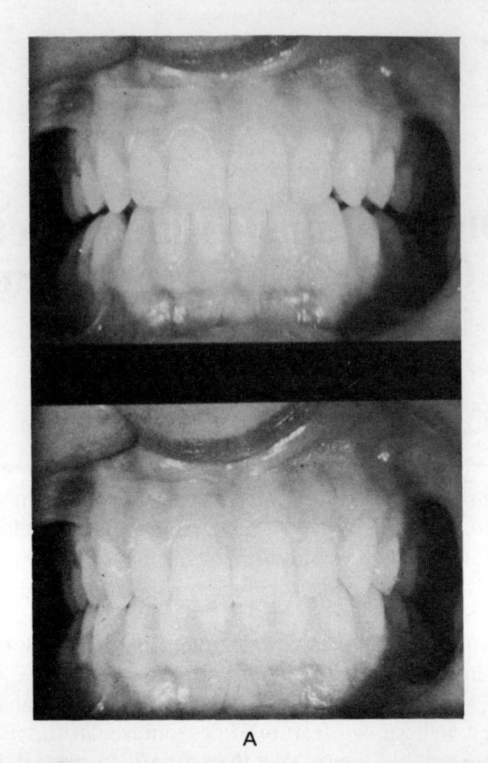

A

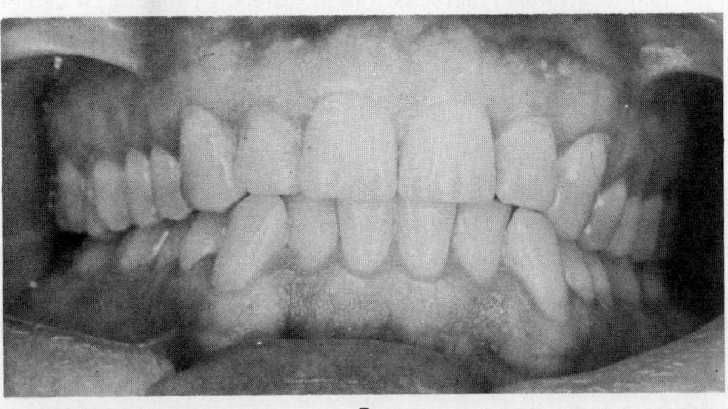

B

Fig. 31.1A, Normal closure. No symptoms. B, Normal closure. Severe unilateral pain.

Occlusion, strictly defined, is contact between opposing teeth while the mandible is stationary. Occlusion has been much studied under this definition and the study of of it has suffered as a result, since it is during the movement of the mandibular teeth in contact with the food bolus or the maxillary teeth that the problems, the possible disorders, occur. As is well known the term 'articulation' refers to contact between opposing teeth while the mandible is in motion and this is not always a desirable feature of mandibular function. It might be helpful to use the term 'occlusal function' to describe the contact between teeth and food bolus and between opposing teeth during the functions of mastication and swallowing. On the other hand, parafunction refers to contact between opposing teeth other than during these two functions. Examples have already been mentioned.

What is implied in the title of this paper, therefore, is the effects of, or the association between, abnormal occlusal function (including parafunction) and muscle and joint function. Optimal occlusal function can be assessed on the basis

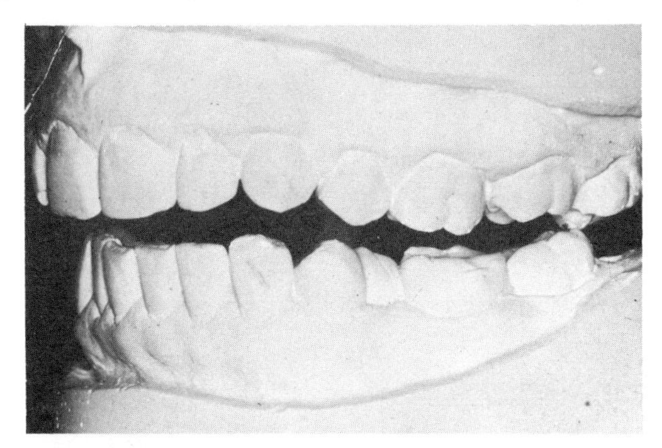

Fig. 31.2. Balancing interference. Severe unilateral pain.

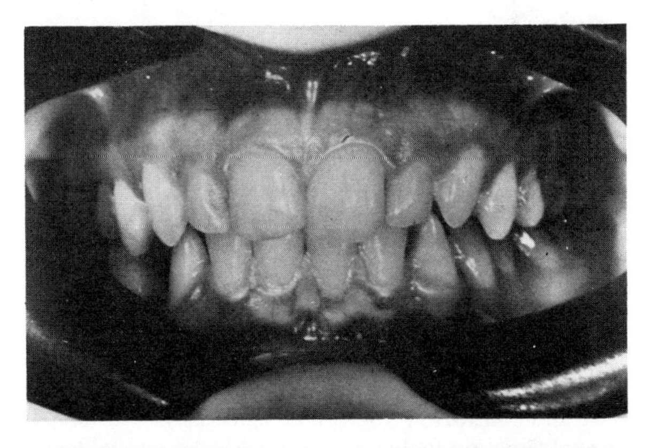

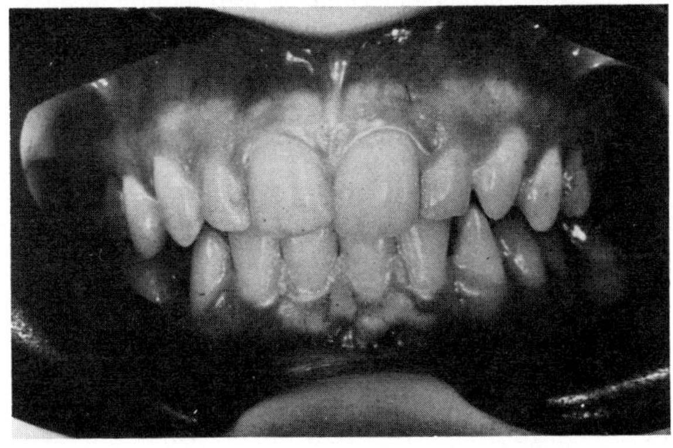

A

Fig. 31.3A, Minor displacing activity to left. Unilateral pain (*continued on* p. 262).

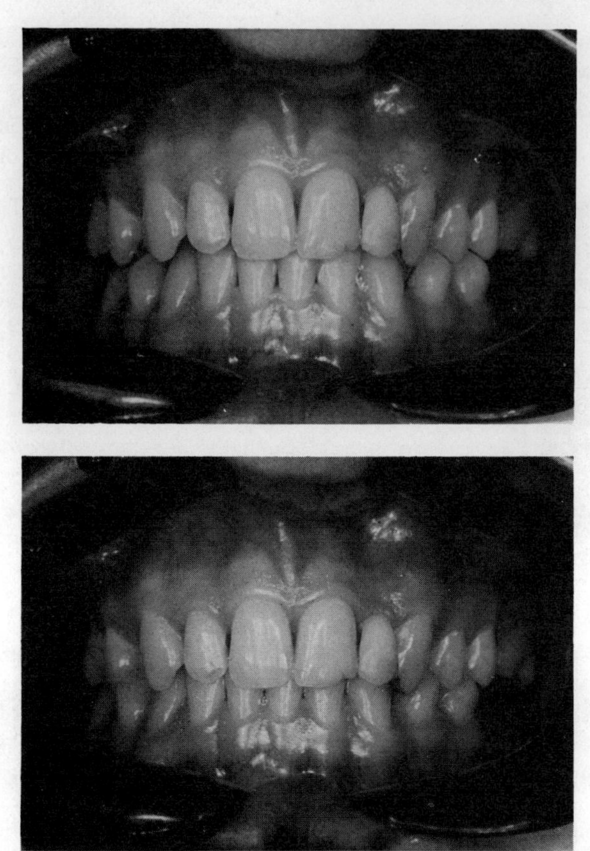

Fig. 31.3B, **Displacing activity to right. No symptoms** (*continued from* **p. 261).**

of the patient's symptoms in eating, speaking, laughing, yawning and at rest and on the basis of the following clinical signs: stable rest and intercuspal positions, intercuspal position only on swallowing, bilateral mastication, comfort on firm intercuspal closure and comfort in both joint regions.

Signs of abnormal function include: unilateral function, balancing interferences, displacing activities, mandibular overclosure and wear facets.

MUSCLE AND JOINT DYSFUNCTION

The symptoms, signs and sounds of muscle and joint dysfunction which are sometimes associated with abnormal occlusal function are well known and include: pain in the joint region during mastication, intermittent dull ache in the joint region, discomfort on opening the mandible, deviation on opening, stiffness in the mandibular musculature on waking, tenderness in the joint region, and finally, click or crepitus on jaw movement. These features are generally unilateral and constitute the mandibular dysfunction syndrome. It is not the purpose of this paper to discuss the pathological changes in the tissues of the masticatory system which account for these symptoms, except to suggest that they are probably more muscular in origin than arthritic. Radiographic findings generally prove negative and there is usually nothing in the patient's medical history of any value in making a diagnosis.

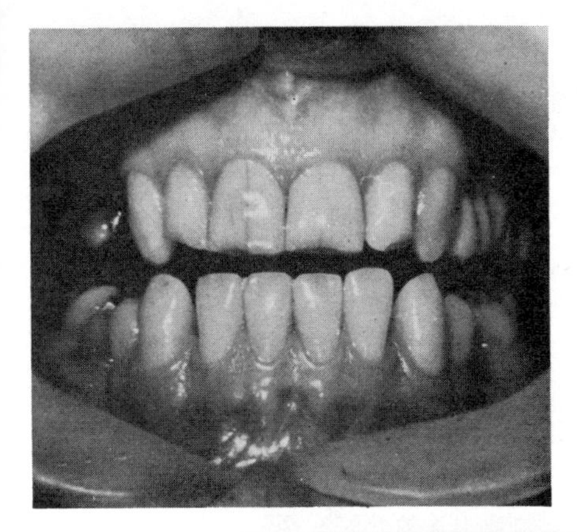

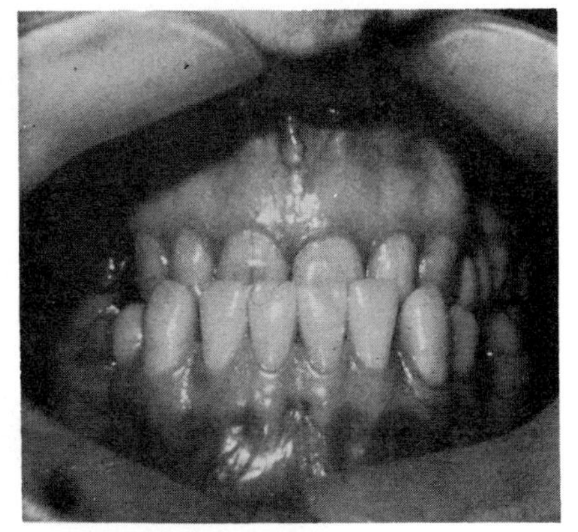

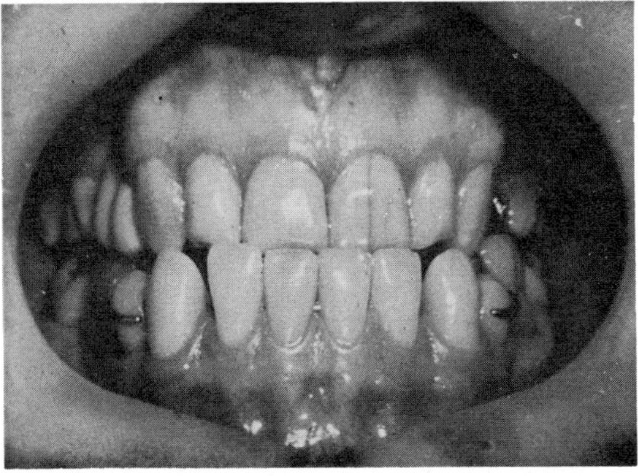

A

Fig. 31.4A, Mandibular overclosure. Unilateral pain. Treated (*continued on* p. 264).

The association between these symptoms and abnormal occlusal function has been established and is illustrated in *Figs.* 31.1–31.5.

Similar occlusal and jaw relationships result in variable tissue responses. The variable factor may be found in the phenomenon known as adaptation. This may act in one or both of two ways. First, the stimuli provided by altered jaw positions and movements may result in impulses for movements which are within the endogenous pattern of joint and muscle activity and which occur without injury, fatigue or spasm of the muscle fibres or without injury to the tissues of the mandibular joints or periodontal membranes of the teeth. Secondly, impulses from the various higher centres of the brain commonly resulting in muscle hyperactivity are resisted or tolerated by the mandibular muscles without disorders. Conversely, these impulses may result in pathological responses which can include: torn muscle fibres which leave a scar which in turn will resist further stretching and may produce a painful sensation; fatigue in the muscles which is reversible; spasm of a muscle or part of a muscle.

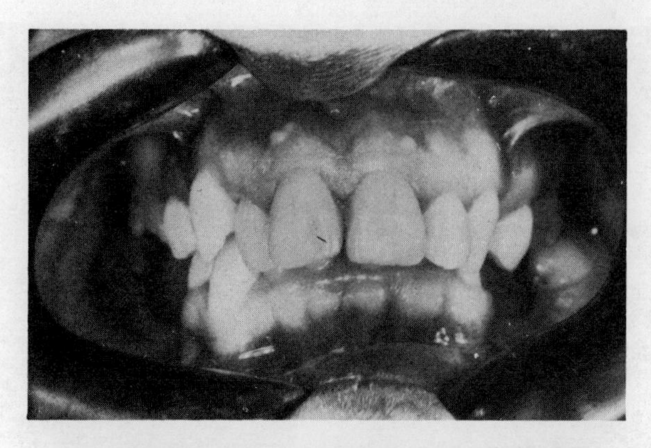

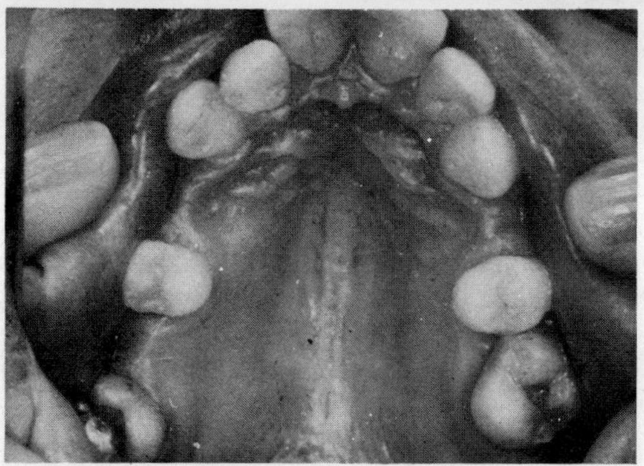

Fig. 31.4B. Mandibular overclosure. No symptoms but bruising of mucosa (*continued from* p. 263).

Discussion CHAIRMAN: B. MATTHEWS

SESSLE: In determining what might be the neural mechanisms involved in the temporomandibular dysfunction syndrome, we have to rely largely on animal studies. Our studies on jaw muscle function in animals indicate that peripheral noxious stimuli are inhibitory to elevator muscles, and the same is true of descending central influences. I wonder how you or other people might explain how the various peripheral and central influences which are largely inhibitory in animals are supposedly excitatory in man.
THOMSON: May I suggest that these are protective to the masticatory system in man and perhaps this protective mechanism can be disturbed by unexpected encounters with food or opposing cusps ?
SESSLE: Nociceptive inputs presumably have a protective function, but the inhibitory effects can also be produced by stimuli that are not at all painful. Moreover, the inhibition that we noted with central feedback probably does not serve a protective function. But the inhibitory effects can be overcome by central drive (643) and perhaps, as you indicated, by certain peripheral inputs. We noted that the periodontium is indeed a powerful excitatory input to elevator motoneurons, and perhaps the central excitatory state (e.g. stress versus relaxation) might determine the overall balance between inhibition and excitation.

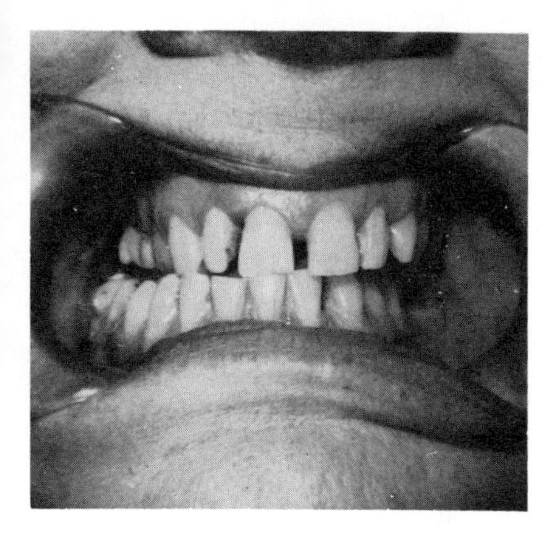

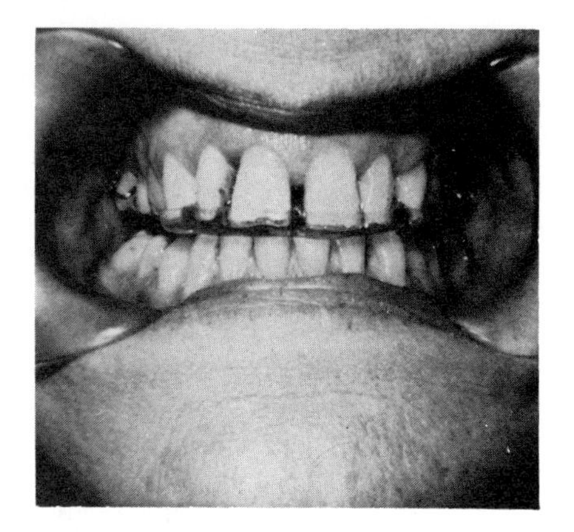

A

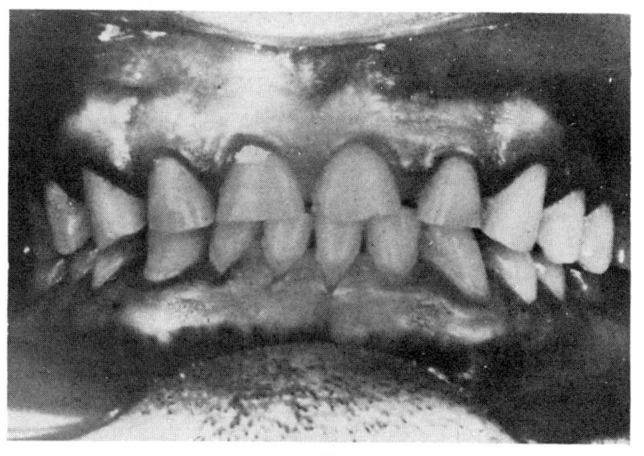

B

Fig. 31.5A, Parafunction. Unilateral pain. Treated. B, Parafunction. No symptoms.

WATT: I should like to suggest that there may be four different kinds of contact. One, unilateral and above occlusal level to which the patient will never adapt and which, when we make it artificially, will produce symptoms. Secondly, one about the occlusal level that is fairly large and to which the patient does not even attempt to adapt to bilaterally and to which the patient accommodates. In the third type, the contact is rather small but the patient, in adapting to it, produces a contact on the other side to which he then tries to adapt producing a sort of oscillatory, uncoordinated activity. This is the most dangerous type. The last type of contact is so small that the patient adapts to it and has no symptoms.
THOMSON: That makes some sense, except that it does not account for the overclosure which was acceptable until the age of 65 and then was not acceptable when treated according to our tenets, although it is an alteration in the occlusion.
WATT: I do not think overclosure is very important.
STOREY: I was intrigued by your observation that the patients often had a history of pain and your suggestion that there has to be a scar there. I think that there is some evidence now that there is such a thing as pain memory (662) and I wonder perhaps whether there may be a history of pain without an organic residue. There may be a memory residue and this may be a trigger.

THOMSON: I produced the scar theory very speculatively. One does know that muscle fibres do get torn but whether the pain memory acts as a trigger I do not know.
THEXTON: I should like to come back to something Dr Sessle said about protective reflexes and to whether the central state could alter them. We have reported (645) that stimulation of areas of the diencephalon which produce a defence reaction reverse the jaw-opening reflex into jaw closure. This would seem to be something which might be significant here.
FUNAKOSHI: We have evidence that periodontal mechanoreceptor stimulation increases masticatory muscle tone in experiments on humans and animals (179). These findings support your hypothesis.
GOLDBERG: You drew a parallel between the control of jaw and limb muscles. I think we have to keep in mind that, although jaws and limbs can be controlled voluntarily, there are, I believe, completely different systems operating in their control.

References

1. Adams S. H. and Zander H. A. (1964) Functional tooth contacts in lateral and in centric occlusion. *J. Am. Dent. Assoc.* **69,** 465–473.
2. Adatia A. K. and Gehring E. N. (1971) Proprioceptive innervation of the tongue. *J. Anat.* **110,** 215–220.
3. Ahlgren J. (1966) Mechanisms of Mastication. *Acta Odontol. Scand.* **24,** Suppl. 44.
4. Ahlgren J. (1967) Pattern of chewing and malocclusion of teeth. A clinical study. *Acta Odontol. Scand.* **25,** 3–13.
5. Ahlgren J. (1967) Kinesiology of the mandible. An EMG study. *Acta Odontol. Scand.* **25,** 593–611.
6. Ahlgren J. (1969) The silent period in the EMG of the jaw muscles during mastication and its relationship to tooth contact. *Acta Odontol. Scand.* **27,** 219–227.
7. Ahlgren J. (1976) Masticatory movements of the mandible in the sagittal plane. A cinematographic study. *Scand. J. Dent. Res.* (In press.)
8. Ahlgren J. and Öwall B. (1970) Muscular activity and chewing force: a polygraphic study of human mandibular movements. *Arch. Oral Biol.* **15,** 271–280.
9. Ainamo J. (1971) Prenatal occlusal wear in guinea-pig molars. *Scand. J. Dent. Res.* **79,** 69–71.
10. Ainamo J. and Talari A. (1976) Eruptive movement of teeth in human adults. In: Poole D. F. G. and Stack M. V. (ed.), *The Eruption and Occlusion of Teeth.* Colston Symposium, No. 27. London, Butterworths. (In press.)
11. Alder A. B., Crawford G. N. C. and Edwards R. G. (1959) The effect of denervation on the longitudinal growth of voluntary muscle. *Proc. R. Soc. Lond. B.* **150,** 554–562.
12. Allbrook D. B., Han M. F. and Hellmuth A. E. (1971) Populations of muscle satellite cells in relation to age and mitotic activity. *Pathology* **3,** 233–243.
13. Allen W. F. (1919) Application of the Marchi method to the study of the radix mesencephalica trigemini in the guinea pig. *J. Comp. Neurol.* **30,** 169–216.
14. Alley K. E. (1974) Morphogenesis of the trigeminal mesencephalic nucleus in the hamster: cytogenes and neurone death. *J. Embryol. Exp. Morphol.* **31,** 99–121.
15. Anderson D. J. (1967) Experimental malocclusion. In: Anderson D. J., Eastoe J. E., Melcher A. H. and Picton D. C. A. (ed.), *Mechanisms of Tooth Support. A Symposium.* Bristol, John Wright. pp. 126–130.
16. Anderson D. J., Hannam A. G. and Matthews B. (1970) Sensory mechanisms in mammalian teeth and their supporting structures. *Physiol. Rev.* **50,** 171–195.
17. Anderson D. J. and Picton D. C. A. (1957) Tooth contact during chewing. *J. Dent. Res.* **36,** 21–26.
18. Anderson D. J. and Picton D. C. A. (1958) Masticatory stresses in normal and modified occlusion. *J. Dent. Res.* **37,** 312–317.
19. Anderson K. V. and Mahan P. E. (1971) Interaction of tooth pulp and periodontal ligament receptors in a jaw-depression reflex. *Exp. Neurol.* **32,** 295–302.
20. Andrew L. (ed.) (1966) *Control and Innervation of Skeletal Muscle.* Thomas, Dundee.
21. Ardran G. M. and Kemp F. H. (1960) Biting and mastication. A cinéradiographic study. *Dent. Pract. Dent. Rec.* **11,** 23–36.
22. Ariano M. A., Armstrong R. B. and Edgerton V. R. (1973) Hindlimb muscle fibre populations of five mammals. *J. Histochem. Cytochem.* **21,** 51–55.
23. Ariëns Kappers C. U. (1936) *A Comparative Anatomy of the Nervous System of the Vertebrates including Man.* Vol. 1. New York, MacMillan, pp. 380–412.
24. Asada Y. and Bennett M. V. L. (1971) Experimental alteration of coupling resistance at an electrotonic synapse. *J. Cell Biol.* **49,** 159–172.
25. Atkinson H. F. and Shepherd R. W. (1961) Temporomandibular joint disturbances and the associated masticatory patterns. *Aust. Dent. J.* **6,** 219–222.
26. Atwood D. A. (1956) A cephalometric study of the clinical rest position of the mandible. Part I; The variability of the clinical rest position following the removal of occlusal contacts. *J. Prosthet. Dent.* **6,** 504–519.
27. Atwood D. A. (1966) A critique of research of the rest position of the mandible. *J. Prosthet. Dent.* **16,** 848–854.
28. Atwood D. A. (1968) A critique of research of the posterior limit of the mandibular position. *J. Prosthet. Dent.* **20,** 21–36.

29. Aziz-Ullah and Goldspink G. (1974) Distribution of mitotic nuclei in the biceps brachii of the mouse during post natal growth. *Anat. Rec.* **179,** 115–118.
30. Bähr U. and Schwindling R. (1974) Optische Untersuchungen zur Bestimmung der physiologischen Ruhelage. *Dtsch. Zahnaertzl. Z.* **29,** 1002–1007.
31. Baker R. and Llinás R. (1971) Electrotonic coupling between neurons in the rat mesencephalic nucleus. *J. Physiol. (Lond.)* **212,** 45–63.
32. Bando E., Fukushima S., Kawabata H. and Kohno S. (1972) Continuous observations of mandibular positions by telemetry. *J. Prosthet. Dent.* **28,** 485–490.
33. Banker, B. Q., Przybylski R. J., Van Der Meulen J. P. and Victor M. (ed.) (1972) *Research in Muscle Development and the Muscle-Spindle.* Excerpta Medica, Amsterdam.
34. Barker D. (1948) The innervation of the muscle spindles. *Q. J. Micr. Sci.* **89,** 143–186.
35. Barker D., Harker D., Stacye M. J. and Smith C. R. (1972) Fusimotor innervation. In: Banker, B. Q. et al. (ed.), *Research in Muscle Development and the Muscle Spindle.* Excerpta Medica, Amsterdam, pp. 227–250.
36. Barnard R. J., Edgerton V. R., Furukawa T. and Peter J. B. (1971) Histochemical, biochemical and contractile properties of red, white and intermediate fibres. *Am. J. Physiol.* **220,** 410–414.
37. Barnard R. J., Edgerton V. R. and Peter J. B. (1970) Effect of exercise on skeletal muscle. I Biochemical and histochemical properties. *J. Appl. Physiol.* **28,** 762–766.
38. Barnard R. J., Edgerton V. R. and Peter J. B. (1970) Effect of exercise on skeletal muscle. II Contractile properties. *J. Appl. Physiol.* **28,** 767–770.
39. Barron K. D. and Sklar S. (1961) Response of bulbospinal motoneurons to axon section. *Neurology (Minneap.)* **11,** 866–875.
40. Barron K. D. and Tuncbay T. O. (1964) Phosphatase histo-chemistry of feline cervical spinal cord after brachial plexotomy. *J. Neuropath. Exp. Neurol.* **23,** 368–386.
41. Basmajian J. V. (1967) *Muscles Alive: their Functions revealed by Electromyography.* 2nd ed. Ch. 4. Baltimore, Williams & Wilkins.
42. Baum J. (1900) Beiträge zur Kenntnis der Muskelspindeln. *Anat. Hefte* **13,** 249–305.
43. Bazett H. C. and Penfield W. G. (1922) A study of the Sherrington decerebrate animal in the chronic as well as acute condition. *Brain* **45,** 185–265.
44. Beaudreau D. E. and Jerge C. R. (1968) Somatotopic representation in the Gasserian ganglion of tactile peripheral fields in the cat. *Arch. Oral Biol.* **13,** 247–256.
45. Beck H. O. and Morrison W. E. (1962) A method for reproduction of movements of the mandible. *J. Prosthet. Dent.* **12,** 873–883.
46. Benedetti E. L. and Emmelot P. (1968) Hexagonal array of subunits in tight junctions separated from isolated rat liver plasma membranes. *J. Cell Biol.* **38,** 15–24.
47. Bennett M. V. L. (1966) Physiology of electrotonic junctions. *Ann. N.Y. Acad. Sci.* **137,** 509–539.
48. Bennett M. V. L. (1972) A comparison of electrically and chemically mediated transmission. In: Pappas G. D. and Purpura D. P. (ed.), *Structure and Function of Synapses.* New York, Raven Press, pp. 221–256.
49. Bennett M. V. L. (1973) Permeability and structure of electrotonic junctions and intercellular movements of tracers. In: Kater S. D. and Nicholson C. (ed.), *Intracellular Staining in Neurobiology.* New York, Elsevier, Chapter 8, pp. 115–134.
50. Bennett M. V. L. (1974) Flexibility and rigidity in electrotonically coupled systems. In: Bennett M. V. L. (ed.) *Synaptic Transmission and Neuronal Interaction.* Society of General Physiologists Series. Vol. 28. New York, Raven Press, pp. 153–178.
51. Bennett M. V. L., Nakajima Y. and Pappas G. D. (1967) Physiology and ultrastructure of electrotonic junctions. I. Supramedullary neurons. *J. Neurophysiol.* **30,** 161–179.
52. Bennett M. V. L., Nakajima Y. and Pappas G. D. (1967) Physiology and ultrastructure of electronic junctions. III. Giant electromotor neurons of *Malapterus electricus. J. Neurophysiol.* **30,** 209–235.
53. Bennett M. V. L., Pappas G. D., Aljure E. and Nakajima Y. (1967) Physiology and ultrastructure of electrotonic junctions. II. Spinal and medullary electromotor nuclei in mormyrid fish. *J. Neurophysiol.* **30,** 180–208.
54. Bennett M. V. L., Pappas G. D., Gimenez M. and Nakajima Y. (1967) Physiology and ultrastructure of electrotonic junctions. IV. Medullary electromotor nuclei in gymnotid fish. *J. Neurophysiol.* **30,** 236–300.
55. Bergström G. (1950) On the reproduction of dental articulation by means of articulators. *Acta Odontol. Scand.* **9,** Suppl. 4.

56. Berry D. C. and Poole D. F. G. (1974) Masticatory function and oral rehabilitation. *J. Oral Rehab.* **1**, 191–205.
57. Bessette R., Bishop B. and Mohl N. (1971) Duration of masseteric silent periods in patients with TMJ syndrome. *J. Appl. Physiol.* **30**, 864–869.
58. Beyron H. (1964) Occlusal relations and mastication in Australian aborigines. *Acta Odontol. Scand.* **22**, 597–678.
59. Bhaskar S. N. and Orban B. (1955) Experimental occlusal trauma. *J. Periodontol.* **26**, 270–284.
60. Bierman W. and Ralston H. J. (1965) Electromyographic study during passive and active flexion of the knee of the normal human subject. *Arch. Phys. Med. Rehabil.* **46**, 71–75.
61. Biscoe T. J. and Taylor A. (1967) The effect of admixture of fast and slow muscle in determining the form of the muscle twitch. *Med. Biol. Eng.* **5**, 473–479.
62. Blair C. A. (1973) Reflex excitability of the masseter muscle during learned movements in the monkey. (Ph.D. Thesis) Seattle, Washington.
63. Blom S. (1960) Afferent influences on tongue muscle activity. A morphological and physiological study in the cat. *Acta Physiol. Scand.* **49**, Suppl. 170, 1–97.
64. Bodden R., Housch T. and Koulourides T. (1975) Experimental evaluation of the cariogenic potential of foodstuffs in man. *J. Dent. Res.* **54**, (Special Issue A) 178, (Abstract).
65. Bodian D. (1966) Electron microscopy: two major synaptic types on spinal motoneurons. *Science* **151**, 1093–1094.
66. Bodian D. and Mellors R. C. (1945) The regenerative cycle of motoneurons, with special reference to phosphatase activity. *J. Exp. Med.* **81**, 469–488.
67. Bonaguro J. G., Dusza G. R. and Bowman D. C. (1969) Ability of human subjects to discriminate forces applied to certain teeth. *J. Dent. Res.* **48**, 236–241.
68. Boos R. H. (1940) Intermaxillary relation established by biting power. *J. Am. Dent. Assoc.* **27**, 1192–1199.
69. Bossy J. (1958) A propos de l'innervation proprioceptive du muscle stylohyoidien et du ventre postérieur du muscle gastric. *Arch. Anat. Histol. Embryol (Strasb.)* **41**, 37–50.
70. Boucher L. J., Zwemer T. J. and Pflughoeft F. (1959) Can biting force be used as a criterion for registering vertical dimension ? *J. Prosthet. Dent.* **9**, 594–599.
71. Bowman D. C. and Nakfoor P. M. (1968) Evaluation of the human subject's ability to differentiate intensity of forces applied to the maxillary central incisors. *J. Dent. Res.* **47**, 252–259.
72. Bowman J. P. (1968) Muscle spindles in the intrinsic and extrinsic muscles of the Rhesus monkey's (*Macaca mulatta*) tongue. *Anat. Rec.* **161**, 483–485.
73. Bowman J. P. and Combs C. M. (1968) The discharge patterns of lingual spindle afferent fibres in the hypoglossal nerve of the Rhesus monkey. *Exp. Neurol.* **21**, 105–119.
74. Bowman J. P. and Combs C. M. (1969) The cerebrocortical projection of hypoglossal afferents. *Exp. Neurol.* **23**, 291–301.
75. Bradley K., Easton D. M. and Eccles J. C. (1953) An investigation of primary or direct inhibition. *J. Physiol. (Lond.)* **122**, 474–488.
76. Brain W. R. and Walton J. N. (1969) *Diseases of the Nervous System.* p. 483. London, Oxford University Press.
77. Brashear A. D. (1936) The innervation of the teeth. *J. Comp. Neurol.* **64**, 169–185.
78. Bratzlavsky M. (1972) Reflexes with intraoral afferents in human lip musculature. *Exp. Neurol.* **37**, 179–187.
79. Bratzlavsky M. (1972) Pauses in activity of human jaw closing muscles. *Exp. Neurol.* **36**, 160–165.
80. Bratzlavsky M. (1976) Behaviour of human brainstem reflexes in muscles with antogonistic function. In: Shahani M. (ed.) *The Motor System—Neurophysiology and Muscle Mechanism.* Amsterdam, Elsevier. (In press.)
81. Bratzlavsky M. (1976) Human brainstem reflexes. In: Shahani M. *The Motor System—Neurophysiology and Muscle Mechanism.* Amsterdam, Elsevier. (In press.)
82. Bratzlavsky M. and Vander Eecken H. (1974) Afferent influences upon human genioglossal muscle. *J. Neurol.* **207**, 19–25.
83. Bremer F. (1923) Physiologie nerveuse de la mastication chez le chat et le lapin. *Arch. Int. Physiol.* **21**, 308–352.

84. Brenman H. S., Black M. A. and Coslet J. G. (1968) Interrelationship between the electromyographic silent period and dental occlusion. *J. Dent. Res.* **47**, 502.
85. Brenman H. S. and Millsap J. S. (1958) A 'sound' approach to occlusion. Bulletin of the Philadelphia County Dental Assoc.
86. Bridgman C. F. (1968) The structure of tendon organs in the cat: a proposed mechanism for responding to muscle tension. *Anat. Rec.* **162**, 209–220.
87. Brightman M. W. and Reese T. S. (1969) Junctions between intimately apposed cell membranes in the vertebrate brain. *J. Cell Biol.* **40**, 648–677.
88. Brill N., Lammie G. A., Osborne J. and Perry H. T. (1959) Mandibular positions and mandibular movements. *Br. Dent. J.* **106**, 391–400.
89. Brill N., Schübeler S. and Tryde G. (1962) Aspects of occlusal sense in natural and artificial teeth. *J. Prosthet. Dent.* **12**, 123–128.
90. Brion M. A. M., Pameijer J. H. N., Glickman I. and Roeber F. W. (1969) Recent intraoral telemetry findings and their developments in the study of occlusion. *Int. Dent. J.* **19**, 541–552.
91. Brown L. P. (1936) Appellations of the dental practitioner. *Dent. Cosmos* **78**, 246–258.
92. Bryan R. N., Trevino D. L. and Willis W. D. (1972) Evidence for a common location of alpha and gamma motoneurons. *Brain Res.* **38**, 193–196.
93. Buchtal F., Guld C. and Rosenfalck P. (1957) Volume conduction of the spike of the motor unit potential investigated with a new type of multielectrode. *Acta Physiol. Scand.* **38**, 331–354.
94. Buller A. J., Mommaerts W. F. H. M. and Seraydarian K. (1969) Enzyme properties of myosin in fast and slow twitch muscles of the cat following cross-innervation. *J. Physiol. (Lond.)* **205**, 581–597.
95. Bunt A. H., Lund R. D., and Lund J. S. (1974) Retrograde axonal transport of horseradish peroxidase by ganglion cells of the albino rat retina. *Brain Res.* **73**, 215–228.
96. Burke R. E. (1967) Motor units of cat triceps surae muscle. *J. Physiol. (Lond.)* **193**, 141–160.
97. Burke R. E., Levine D. N., Zajac F. E., Tsairis P. and Engel W. K. (1971) Mammalian motor units: physiological-histochemical correlation in three types in cat gastrocnemius. *Science* **174**, 709–712.
98. Butler P. M. (1973) Molar wear facets of early tertiary North American primates. Symposium of the 4th International Congress of Primatology **3**, 1.
99. Butler P. M. (1974) A zoologist looks at occlusion. *Br. J. Orthodont.* **1**, 205–212.
100. Butler P. M. and Mills J. R. E. (1959) A contribution to the odontology of *Oreopithecus. Bull. Brit. Mus. (Nat. Hist.)* Part A. Geology **4**, 1–26.
101. Caminos R. A. (1954) *Late – Egyptian Miscellanies.* London, Oxford University Press.
102. Carlsöö S. (1952) Nervous coordination and mechanical function of the mandibular elevators. *Acta Odontol. Scand.* **10**, Suppl. 11.
103. Carlsöö S. (1956) An electromyographic study of the activity of certain supra-hyoid muscles (mainly the anterior belly of the digastric muscle), and of reciprocal innervation of the elevator and depressor musculature of the mandible. *Acta Anat. (Basel)* **26**, 81–93.
104. Chalcroft J. P. and Bullivant S. (1970) An interpretation of liver cell membrane and junction structure based on observation of freeze fracture replicas of both sides of the fracture. *J. Cell Biol.* **47**, 49–60.
105. Chase M. H. and McGinty D. J. (1970). Modulation of spontaneous and reflex activity of the jaw musculature by orbital cortical stimulation in the freely-moving cat. *Brain Res.* **19**, 117–126.
106. Chase M. H., Sterman M. B., Kubota K. and Clemente C. D. (1973) Modulation of masseteric and digastric neural activity by stimulation of the dorso-lateral cerebral cortex in the squirrel monkey. *Exp. Neurol.* **41**, 277–289.
107. Christensen L. V. (1967) Muscle spindles in the lateral pterygoid muscle of miniature swine. *Arch. Oral Biol.* **12**, 1203–1204.
108. Christensen J. (1970) Effect of occlusion-raising procedures on the chewing system. *Dent. Pract. Dent. Rec.* **20**, 233–238.
109. Clark R. K. F. and Wyke B. D. (1974) Contributions of temporomandibular articular mechanoreceptors to the control of mandibular posture: an experimental study. *J. Dent.* **2**, 121–129.

110. Clarke S. L. (1926) Nissl granules of primary afferent neurons. *J. Comp. Neurol.* **41,** 423–451.
111. Clemmesen S. (1951) Some studies on muscle tone. *Proc. R. Soc. Med.* **44,** 637–646.
112. Close R. I. (1972) Dynamic properties of mammalian skeletal muscles. *Physiol. Rev.* **52,** 129–197.
113. Cody F. W. J., Harrison L. M., Taylor A. and Weghofer B. (1974) Distribution of tooth receptor afferents in the mesencephalic nucleus of the fifth cranial nerve. *J. Physiol. (Lond.)* **239,** 49–50 *P.*
114. Cody F. W. J., Lee R. W. H. and Taylor A. (1972) A functional analysis of the components of the mesencephalic nucleus of the fifth nerve in the cat. *J. Physiol. (Lond.)* **226,** 249–261.
115. Cody F. W. J. and Taylor A. (1973) The behaviour of spindles in the jaw closing muscles during eating and drinking in the cat. *J. Physiol. (Lond.)* **231,** 49–50 *P.*
116. Cooper S. (1953) Muscle spindles in the intrinsic muscles of the human tongue. *J. Physiol. (Lond.)* **122,** 193–202.
117. Cooper S. (1960) Muscle spindles and other muscle receptors. In: Bourne G. H. (ed.), *The Structure and Function of Muscle,* Vol. 1, New York, Academic Press, pp. 381–420.
118. Cooper S., Daniel P. D. and Whitteridge D. (1953) Nerve impulses in the brainstem of the goat. Short latency responses obtained by stretching the extrinsic eye muscles and jaw muscles. *J. Physiol. (Lond.)* **120,** 471–490.
119. Cooper S. and Eccles J. C. (1930) The isometric response of mammalian muscles. *J. Physiol. (Lond.)* **69,** 377–385.
120. Corbin K. B. (1940) Observations on the peripheral distribution of fibres arising in the mesencephalic nucleus of the fifth cranial nerve. *J. Comp. Neurol.* **73,** 153–177.
121. Corbin K. B. and Harrison F. (1939) The sensory innervation of the spinal accessory and tongue musculature in the rhesus monkey. *Brain* **62,** 191–197.
122. Corbin K. B. and Harrison F. (1940) Function of the mesencephalic root of the fifth cranial nerve. *J. Neurophysiol.* **3,** 423–435.
123. Craddock F. W. (1949) The accuracy and practical value of records of condyle path inclination. *J. Am. Dent. Assoc.* **38,** 697–710.
124. Creed R. S., Denny-Brown D., Eccles J. C., Liddell E. G. T. and Sherrington C. S. (1932) *Reflex Activity of the Spinal Cord.* Oxford, Clarendon.
125. Crompton A. W. (1971) The origin of the tribosphenic molar. In: Kermack D. M. and Kermack K. A. (ed.), *Early Mammals.* London, Academic Press, pp. 65–87.
126. Crompton A. W. and Hiiemae K. M. (1969) How mammalian molar teeth work. *Discovery* (Yale Peabody Museum) **5,** 23–34.
127. Crompton A. W. and Hiiemae K. M. (1970) Molar occlusion and mandibular movements during occlusion in the American opossum, *Didelphis marsupialis. Zool. J. Linn. Soc.* **49,** 21–47.
128. Crompton A. W. and Jenkins F. A. (1968) Molar occlusion in late Triassic mammals. *Biol. Rev.* **43,** 427–458.
129. Crompton A. W. and Jenkins F. A. (1973) Mammals from reptiles: A review of mammalian origins. In: Donath F. A., Stehli F. G. and Wetherill G. W. (ed.), *Annual Review of Earth and Planetary Science.* **1,** 131–155. Annual Reviews Inc., Palo Alto, California.
130. Crompton A. W., Thexton A. J., Hiiemae K. M. and Cook P. P. (1976) The activity of the hyoid and jaw muscles during the chewing of soft food in the American opossum. In: Gilmore D. and Robinson B. (ed.), *Biology and Environment.* Vol. II. The Biology of Marsupials. London, Macmillan. (In press.)
131. Curtis D. R. (1962) The depression of spinal inhibition by electrophoretically administrated strychnine. *Int. J. Neuropharmacol.* **1,** 239–250.
132. Curtis D. R. (1969) The pharmacology of spinal postsynaptic inhibition. In: Akert K. and Waser P. G. (ed.), *Progress in Brain Research,* Vol. 31. Amsterdam, Elsevier, pp. 171–189.
133. Dal Pont G. (1961) Retromolar osteotomy for the correction of prognathism. *J. Oral Surg.* **19,** 42–47.
134. Darling A. I. and Mendis B. R. R. N. (1975) Response of human dentine to attrition. *J. Dent. Res.* **54,** (Special Issue A), L439 (Abstract).
135. Davey M. R. and Taylor A. (1966) Activity of cat jaw muscle stretch receptors recorded

from their cell bodies in the mid-brain during spontaneous jaw movements. *J. Physiol. (Lond.)* **185,** 62 *P.*
136. Davey M. R. and Taylor A. (1967) The activity of jaw muscle spindles recorded together with active jaw movement in the cat recovering from anaesthesia. *J. Physiol. (Lond.)* **190,** 8–9 *P.*
137. Davis D. D. (1961) Origin of the mammalian feeding mechanism. *American Zoologist* **1,** 229–241.
138. De Boever J. (1969) Experimental occlusal balancing-contact interference and muscle activity. *Parodontologie* **23–24,** 59–69.
139. De Lange A., Hannam A. G. and Matthews B. (1969) The diameters and conduction velocities of fibres in the terminal branches of the inferior dental nerve. *Arch. Oral Biol,* **14,** 513–519.
140. De Long M. (1971) Central patterning of movement. *Neurosci. Res. Program. Bull.* **(9)** 10–30.
141. Dellow P. G. (1969) Control mechanisms of mastication. *Ann. Austr. Coll. Dent. Surg.* **2,** 81–95.
142. Dellow P. G. and Lund J. P. (1971) Evidence for central timing of rhythmical mastication. *J. Physiol. (Lond.)* **215,** 1–13.
143. Denny-Brown D. E. (1929) The histological features of striped muscle in relation to its functional activity. *Proc. R. Soc. (Lond.) B* **104,** 371–411.
144. Derry D. E. (1933) Incidence of dental disease in ancient Egypt. *Br. Med. J.* **1,** 112.
145. Dibdin G. H. and Griffiths M. J. (1975) An intra-oral telemetry system for the continuous recording of vertical jaw movement. *Phys. Med. Biol.* **20,** 355–365.
146. Dmytruk R. J. (1974) Neuromuscular spindles and depressor masticatory muscles of monkey. *Am. J. Anat.* **141,** 147–154.
147. Doggenweiler C. F. and Frenk S. (1965) Staining properties of lanthanum on cell membranes. *Proc. Natl Acad. Sci. USA* **53,** 425–430.
148. Dubowitz V. and Pearse A. G. E. (1960) A comparative histochemical study of oxidative enzyme and phosphorylase activity in skeletal muscle. *Histochemie* **2,** 105–117.
149. Dubowitz V. and Pearse A. G. E. (1960) Reciprocal relationships of phosphorylase and oxidative enzymes in skeletal muscle. *Nature (Lond.)* **185,** 701.
150. Duxbury A. J. and Rothwell P. S. (1973) A digital optoelectronic method for recording mandibular movement in association with oral electromyograms and temporomandibular joint noises. *J. Dent. Res.* **52,** 932 (Abstract).
151. Eccles J. C. (1964) *The Physiology of Synapses.* Berlin, Springer.
152. Eccles J. C., Eccles R. M., Iggo A. and Lundberg A. (1960) Electrophysiological studies on gamma motoneurones. *Acta Physiol. Scand.* **50,** 32–40.
153. Eccles J. C., Llinás R. and Sasaki K. (1966) Intracellularly recorded responses of the cerebellar Purkinje cells. *Exp. Brain Res.* **1,** 161–183.
154. Eccles J. C., Schmidt R. and Willis W. D. (1963) Pharmacological studies on presynaptic inhibition. *J. Physiol. (Lond.)* **168,** 500–530.
155. Eccles R. M. and Lundberg A. (1959) Synaptic actions in motoneurones by afferents which may evoke the flexion reflex. *Arch. Ital. Biol.* **97,** 199–221.
156. Edgerton V. R., Gerchman L. and Carrow R. (1969) Histochemical changes in rat skeletal muscle after exercise. *Exp. Neurol.* **24,** 110–123.
157. Edström L. and Kugelberg E. (1968) Histochemical composition, distribution of fibres and fatiguability of single motor units. *J. Neurol. Neurosurg. Psychiatry* **31,** 424–433.
158. Eklund G. and Hagbarth K. E. (1966) Normal variability of tonic vibration reflexes in man. *Exp. Neurol.* **16,** 80–92.
159. Elliot-Smith G. (1921) The Royal Mummies, Cairo, Catalogue Général des Antiquités Egyptiennes du Musée du Caire.
160. Elliot-Smith G. (1932) Incidence of dental disease in ancient Egypt. *Br. Med. J.* **2,** 760.
161. Emery W. B. (1962) *A Funerary Repast in an Egyptian Tomb.* Leiden, Nederl. Inst. voor het Nabuje Oosten.
162. Erichsen W. (1932) *Papyrus Harris.* Bruxelles, Editions de la Fondation Egyptologique Reine Elizabeth.
163. Erman A. (1894) *Life in Ancient Egypt.* Trans. H. M. Tivard. London, Macmillan.
164. Eversole L. R. and Standish S. M. (1970) Histochemical demonstration of muscle fibre types. *J. Histochem. Cytochem.* **18,** 591–593.

165. Eyzaguirre C. (1958) Modulation of sensory discharges by efferent spindle excitation. *J. Neurophysiol.* **21,** 465–480.
166. Faigenblum M. J. (1966) Negative oral pressures. A research report. *Dent. Pract. Dent. Rec.* **16,** 214–216.
167. Farias M. C. and Be Ment S. L. (1974) Dynamic characteristics of rapidly adapting knee joint receptors in cats. *Soc. Neurosc. Progr. and Abstr.* October 20–24, Abst. 195.
168. Farquhar M. G. and Palade G. E. (1963) Junctional complexes in various epithelia. *J. Cell Biol.* **17,** 375–412.
69. Fenichel G. M. and Engel W. K. (1963) Histochemistry of muscle in infantile spinal muscular atrophy. *Neurology* **13,** 1059–1066.
70. Ferrier D. (1886) *The Functions of the Brain.* 2nd ed., London, Smith, Elder. pp. 260–261.
71. Fex S. and Sonesson B. (1970) Histochemical observations after implantation of a 'fast' nerve into an innervated mammalian 'slow' skeletal muscle. *Acta Anat.* **77,** 1–10.
72. Fillenz M. (1965) Responses in the brain stem of the cat to stretch of extrinsic ocular muscles. *J. Physiol. (Lond.)* **128,** 182–199.
73. Franks A. S. T. (1964) Studies on the innervation of the temporomandibular joint and lateral pterygoid muscle in animals. *J. Dent. Res.* **43,** 947–948.
74. Franks A. S. T. (1965) De beheersing van de bewegingen in het kaakgewricht: onderzoekingen naar de innervatie van het gewricht en van de m. pterygoideus lateralis. *Ned. Tijdschr. Tandheelkd.* **72,** 605–619.
75. Freeman W. (1925) The relationship of the radix mesencephalica trigemini to the extraocular muscles. *Arch. Neurol. Psychiat.* **14,** 111–113.
76. Freimann R. (1954) Untersuchungen über Zahl und Anordnung der Muskelspindeln in den Kaumuskeln des Menschen. *Anat. Anz.* **100,** 258–264.
77. French F. A. (1929) The intra-oral versus the extra-oral method of recording mandibular measurements. *J. Am. Dent. Assoc.* **16,** 1100–1102.
78. Friend D. S. and Gilula N. B. (1972) Variations in tight and gap junctions in mammalian tissues. *J. Cell Biol.* **53,** 758–776.
79. Funakoshi M. and Amano N. (1974) Periodontal jaw muscle reflexes in the albino rat. *J. Dent. Res.* **53,** 598–605.
80. Gail P. de, Lance J. W. and Neilson P. D. (1966) Differential effects on tonic and phasic reflex mechanisms produced by vibration of muscles in man. *J. Neurol. Neurosurg. Psychiatry* **29,** 1–11.
81. Galambos R. (1956) Suppression of auditory nerve activity by stimulation of efferent fibres to cochlea. *J. Neurophysiol.* **19,** 424–437.
82. Garnick J. and Ramfjord S. P. (1962) Rest position: an electromyographic and clinical investigation. *J. Prosthet. Dent.* **12,** 895–911.
83. Garton C. P. (1968) A device for recording thresholds to mechanical stimulation of human teeth. *J. Dent. Res.* **47,** 974. (Abstract).
84. Gibbs C. H., Suit S. R. and Benz S. T. (1973) Masticatory movements of the jaw measured at angles of approach to the occlusal plane. *J. Prosthet. Dent.* **30,** 283–288.
85. Gill H. I. (1971) Neuromuscular spindles in human lateral pterygoid muscles. *J. Anat.* **109,** 157–167.
86. Gillespie C. A., Simpson D. R. and Edgerton V. R. (1970) High glycogen content of red as opposed to white skeletal muscle fibres of guinea pigs. *J. Histochem. Cytochem.* **18,** 552–556.
87. Gillings B. R. D. (1967) Photoelectric mandibulography: a technique for studying jaw movements. *J. Prosthet. Dent.* **17,** 109–121.
88. Gilula N. B., Reeves O. R. and Steinbach A. (1972) Metabolic coupling, ionic coupling and cell contacts. *Nature (Lond.)* **235,** 262–265.
89. Gilula N. B. and Satir P. (1971) Septate and gap junctions in molluscan gill epithelium. *J. Cell Biol.* **51,** 869–872.
90. Gladden M. H. (1975) Elastic fibres in human muscle spindles. *J. Anat.* **119,** 187–188 (Abstract).
91. Glees P. (1938) Neuroplasmatische Verbindungen zwischen Zellen des Mesencephalen Trigeminuskerne bei *Scyllium canicula. Koninkl. Ned. Akad. Wetenschap. Proc.* **41,** 426–430.
92. Glickman I. (1965) Clinical significance of trauma from occlusion. *J. Am. Dent. Assoc.* **70,** 607–618.

193. Glickman I., Haddad A. W., Martignoni M., Mehta N., Roeber F. W. and Clark R. E. (1974) Telemetric comparison of centric relation and central occlusion reconstructions. *J. Prosthet. Dent.* **31**, 527–536.
194. Glickman I., Pameijer J. H. N. and Roeber F. W. (1968) Intraoral occlusal telemetry. Part I. A multifrequency transmitter for registering tooth contacts in occlusion. *J. Prosthet. Dent.* **19**, 60–68.
195. Glickman I. and Smulow J. B. (1965) Effect of excessive occlusal forces upon the pathway of gingival inflammation in humans. *J. Periodontol.* **36**, 51–57.
196. Glickman I. and Smulow J. B. (1968) Adaptive alterations in the periodontium of the Rhesus monkey in chronic trauma from occlusion. *J. Periodontol.* **39**, 101–105.
197. Glickman I. and Weiss L. A. (1955) Role of trauma from occlusion in initiation of periodontal pocket formation in experimental animals. *J. Periodontol.* **26**, 14–20.
198. Gogan P., Gueritand J. P., Horcholle-Bossavit G. and Tyć-Dumont S. (1974) Electrotonic coupling between motoneurons in the abducens nucleus of the cat. *Exp. Brain Res.* **21**, 139–154.
199. Goldberg L. J. (1971) Masseter muscle excitation induced by stimulation of periodontal and gingival receptors in man. *Brain Res.* **32**, 369–381.
200. Goldberg L. J. (1972) The effect of jaw position on the excitability of two reflexes involving the masseter muscle in man. *Arch. Oral Biol.* **17**, 565–576.
201. Goldberg L. J. (1972) Excitatory and inhibitory effects of lingual nerve stimulation on reflexes controlling the activity of masseteric motoneurons. *Brain Res.* **39**, 95–108.
202. Goldberg L. J. (1972) An excitatory component of the jaw opening reflex in the temporal and masseter muscles of cats and monkeys. *Experientia* **28**, 44–46.
203. Goldberg L. J. (1976) Motoneurone mechanisms: reflex controls. In: Sessle B. J. and Hannam A. G. (ed.), *Mastication and Swallowing: Biological and Clinical Correlates.* Toronto, Univ. of Toronto Press. (In press.)
204. Goldberg L. J. and Browne P. A. (1974) Differences in the excitability of two populations of trigeminal primary afferent central terminals. *Brain Res.* **77**, 195–209.
205. Goldberg L. J. and Nakamura Y. (1968) Lingually induced inhibition of masseteric motoneurones. *Experientia* **24**, 371–373.
206. Goldspink G. (1968) Sarcomere length during the post-natal growth of mammalian muscle fibres. *J. Cell Sci.* **3**, 539–548.
207. Goldspink G., Tabary C., Tabary J. C., Tardieu C. and Tardieu G. (1974) Effect of denervation on the adaptation of sarcomere number and muscle extensibility to the functional length of the muscle. *J. Physiol. (Lond.)* **236**, 733–742.
208. Golgi C. (1893) Intorno all'origine del quarto nervo cerebrale. *Atti della reale Accad. dei Lincei.* Sér. V. Vol. II.
209. Goodenough D. A. and Revel J. P. (1970) A fine structural analysis of intercellular junctions in the mouse liver. *J. Cell Biol.* **45**, 272–290.
210. Goodenough D. A. and Stoeckenius W. (1972) The isolation of mouse hepatocyte gap junctions. *J. Cell Biol.* **54**, 646–656.
211. Goodwin G. M. and Luschei E. S. (1974) Effects of destroying spindle afferents from jaw muscles on mastication in monkeys. *J. Neurophysiol.* **37**, 967–981.
212. Goodwin G. M. and Luschei E. S. (1975) Discharge of spindle afferents from jaw closing muscles during chewing in alert monkeys. *J. Neurophysiol.* **38**, 560–571.
213. Gordon A. M., Huxley, A. F. and Julian F. J. (1966) The variation in isometric tension with sarcomere length in vertebrate muscle fibres. *J. Physiol. (Lond.)* **184**, 170–192.
214. Gottlieb B. and Orban B. (1931) Tissue changes in experimental traumic occlusion with special reference to age and constitution. *J. Dent. Res.* **11**, 505–510.
215. Gowers W. R. (1893) *A Manual of Diseases of the Nervous System.* Vol. 2. London, Churchill, p. 221.
216. Graf H., Grassl H. and Aeberhard H. J. (1974) A method for measurement of occlusal forces in three directions. *Helv. Odontol. Acta* **18**, 7–11.
217. Graf H. and Zander H. A. (1963) Tooth contact pattern in mastication. *J. Prosthet. Dent.* **13**, 1055–1066.
218. Granit R. (1955) *Receptors and Sensory Perception.* Yale University Press. New Haven.
219. Granit R. (1970) *The Basis of Motor Control.* London, Academic Press.
220. Gray E. G. (1959) Axosomatic and axodendritic synapses of the cerebral cortex: an electron microscope study. *J. Anat.* **93**, 420–433.

221. Green J. D., De Groot J. and Suttin J. (1957) Trigemino-bulbar reflex pathways. *Am. J. Physiol.* **189,** 384–388.
222. Greenfield B. E. and Wyke B. D. (1956) Electromyographic studies of some muscles of mastication. (I) Temporal and masseter activity in various jaw movements in normal subjects. *Br. Dent. J.* **100,** 129–143.
223. Greenfield B. E. and Wyke B. (1966) Reflex innervation of the temporomandibular joint. *Nature (Lond.)* **211,** 940–941.
224. Greenwood L. F. and Sessle B. J. (1976) Pain, brain stem mechanisms and motor function. In: Sessle B. J. and Hannam A. G. (ed.), *Mastication and Swallowing: Biological and Clinical Correlates.* Toronto, Univ. of Toronto Press. (In press.)
225. Griffin C. J. (1972) The fine structure of end-rings in human periodontal ligament. *Arch. Oral Biol.* **17,** 785–797.
226. Griffin C. J. and Malor R. (1974) An analysis of mandibular movement. In: Kawamura Y. (ed.), *Frontiers of Oral Physiology,* Vol. 1. Physiology of Mastication. Basel, Karger, pp. 159–198.
227. Griffin C. J. and Munro R. R. (1971) Electromyography of the masseter and anterior temporalis muscles in patients with temporomandibular dysfunction. *Arch. Oral Biol.* **16,** 929–949.
228. Griffin G. E., Williams P. E. and Goldspink G. (1973) Region of longitudinal growth in striated muscle fibres. *Nature (Lond.)* **232,** 28–29.
229. Griffiths M. J. and Dibdin G. H. (1973) Telemetry of mandibular posture. *J. Dent. Res.* **52,** 934. (Abstract).
230. Grigg P. (1974) Torque and angular dependence of discharge in joint afferent neurons in the cat. *Soc. Neurosc. Progr. and Abstr.* October 20–24, Abstr. 258.
231. Grossman R. C. and Hattis B. F. (1967) Oral mucosal sensory innervation and sensory experience. In: Bosma J. F. (ed.), *Symposium on Oral Sensation and Perception.* Springfield, Thomas, pp. 5–62.
232. Gura E. V., Limanskii Yu. P. and Pilyavskii A. I. (1969) Synaptic potentials of motoneurons of the masseter muscle. *Neirofiziologiia* **1,** 199–204.
233. Gura E. V., Limanskii Yu. P. and Pilyavskii A. I. (1971) Synaptic potentials of the motoneurons of the digastric muscle. *Neirofiziologiia* **3,** 52–57.
234. Hagbarth K. E. and Eklund G. (1966) Motor effects of vibratory stimuli in man. In: Granit R. (ed.), *Muscular Afferents and Motor Control.* Nobel Symposium I, Stockholm, Almqvist & Wiksell.
235. Hamburger V. (1961) Experimental analysis of the dual origin of the trigeminal ganglion in the chick embryo. *J. Exp. Zool.* **148,** 91–134.
236. Hannam A. G. (1969) The response of periodontal mechanoreceptors in the dog to controlled loading of the teeth. *Arch. Oral Biol.* **14,** 781–791.
237. Hannam A. G. (1970) Receptor fields of periodontal mechanosensitive units in the dog. *Arch. Oral Biol.* **15,** 971–978.
238. Hannam A. G. (1972) Effect of voluntary contraction of the masseter and other muscles upon the masseteric reflex in man. *J. Neurol. Neurosurg. Psychiatry* **35,** 66–71.
239. Hannam A. G. and Farnsworth T. (1973) Information transmission in periodontal mechanosensitive neurones. *J. Dent. Res.* **52,** (Special Issue) 102 (Abstract).
240. Hannam A. G. and Matthews B. (1969) Reflex jaw opening in response to stimulation of periodontal mechanoreceptors in the cat. *Arch. Oral Biol.* **14,** 415–419.
241. Hannam A. G., Matthews B. and Yemm R. (1968) The unloading reflex in masticatory muscles of man. *Arch. Oral Biol.* **13,** 361–364.
242. Hannam A. G., Matthews B. and Yemm R. (1969) Changes in the activity of the masseter muscle following tooth contact in man. *Arch. Oral Biol.* **14,** 1401–1406.
243. Hannam A. G., Matthews B. and Yemm R. (1970) Receptors involved in the response of the masseter muscle to tooth contact in man. *Arch. Oral Biol.* **15,** 17–24.
244. Hanson J. and Widen L. (1970) Afferent fibres in the hypoglossal nerve of cat. *Acta Physiol. Scand.* **79,** 24–36.
245. Harman L. (1964) Neuromines: action of a reciprocally inhibitory pair. *Science* **146,** 1323–1325.
246. Harman L. and Lewis E. (1966) Neural modelling. *Physiol. Rev.* **46,** 513–552.
247. Harris J. E. and Weeks K. R. (1973) *X-raying the Pharaohs.* New York, Scribner's.
248. Harrison F. and Corbin K. B. (1942) The central pathway for the jaw-jerk. *Am. J. Physiol.* **135,** 439–445.

249. Hayes W. C. (1951) Inscriptions from the palace of Amenhotep III. *J. Near East. Studies* **10,** 82–104.
250. Hedegård B., Lundberg M. and Wictorin L. (1970) Masticatory function—a cineradiographic study. IV. Duration of the masticatory cycle. *Acta Odontol. Scand.* **28,** 859–865.
251. Henatsch H. D., Meyr-Lohmann J., Schmidt J. and Windhorst U. (1975) Influence of inhomogeneous extrafusal mechanics upon the static and dynamic behaviour of primary muscle-spindle endings in the cat. *J. Anat.* **119,** 200–202. (Abstract).
252. Henneman E. and Olson C. B. (1965) Relations between structure and function in the design of skeletal muscle. *J. Neurophysiol.* **28,** 581–598.
253. Herrick C. J. (1914) The medulla oblongata of larval amblystoma. *J. Comp. Neurol.* **24,** 343–427.
254. Herring S. W. and Scapino R. P. (1973) Physiology of feeding in miniature pigs. *J. Morpho.* **141,** 427–460.
255. Hickey J. C., Allison M. L., Woelfel J. B., Boncher C. O. and Stacy R. W. (1963) Mandibular movements in three dimensions. *J. Prosthet. Dent.* **13,** 72–92.
256. Higgins H. C. (1968) The non-local storage of temporal information. *Proc. R. Soc. B* **171,** 327–334.
257. Hiiemae K. (1976) Mammalian Mastication. A Review. In: Butler P. M. and Joysey K. (ed. *Studies in the Development and Function of Teeth.* Academic Press. (In press.)
258. Hiiemae K. M. and Ardran G. M. (1968) A cineradiographic study of feeding in *Rattus Norvegicus. J. Zool. (Lond.)* **154,** 139–154.
259. Hiiemae K. M. and Crompton A. W. (1971) A cinefluorographic study of feeding in the American opossum, *Didelphis marsupialis.* In: Dahlberg A. A. (ed.), *Dental Morphology and Evolution.* Univ. of Chicago Press. pp. 299–334.
260. Hiiemae K. and Houston W. J. B. (1971) The structure and function of the jaw muscles of the rat. (*Rattus norvegicus L.*) II. *Zool. J. Linn. Soc.* **50,** 101–109.
261. Hiiemae K. M. and Jenkins F. A. (1969) The anatomy and internal architecture of the muscles of mastication in the American opossum, *Didelphis marsupialis. Postilla* **140,** 1–49.
262. Hiiemae K. and Kay R. F. (1973) *Evolutionary Trends in the Dynamics of Primate Mastication.* Symp. IVth Int. Congr. Primat. Vol. 3. Craniofacial Biology of Primates. Zingeser M. R. (ed.), Basle, Karger. pp. 28–64.
263. Hiiemae K. M. and Thexton A. J. (1975) Consistency and bite size as regulators of mastication in cats. *J. Dent. Res.* **54,** (Special Issue A) 194, (Abstract).
264. Hild W. (1957) Observations on neurons and neuroglia from the area of the mesencephalic fifth nucleus of the cat in vitro. *Z. Zellforsch.* **47,** 127–146.
265. Hildebrand G. Y. (1931) Studies in the masticatory movements of the human lower jaw. *Scand. Arch. Physiol.* Suppl. 61.
266. Hilton P. B. (1972) The effect of repeated stimulation on the response from slowly adapting periodontal mechanoreceptors. In: Emmelin N. and Zotterman Y. (ed.), *Oral Physiology.* Oxford, Pergamon, pp. 217–222.
267. Hinrichsen C. F. L. (1970) Coupling between cells of the trigeminal mesencephalic nucleu *J. Dent. Res.* **49,** 1369–1373.
268. Hinrichsen C. F. L. and Larramendi L. M. H. (1968) Synapses and cluster formation of th mouse mesencephalic fifth nucleus. *Brain Res.* **7,** 296–299.
269. Hinrichsen C. F. and Larramendi L. M. (1969) Features of trigeminal mesencephalic nucleus structure and organization. I. Light microscopy. *Am. J. Anat.* **126,** 497–505.
270. Hinrichsen C. F. L. and Larramendi L. M. H. (1970) The trigeminal mesencephalic nucleu II. Electron microscopy. *Am. J. Anat.* **127,** 303–320.
271. Hoefer P. F. A. (1941) Innervation and 'tonus' of striated muscle in man. *Arch. Neurol. Psychiatry* **46,** 947–972.
272. Hoffman P. and Tönnies J. F. (1948) Nachweis des völlig konstanten Vorkrommens des Zunger-Kieferreflexes beim Menschen. *Pfluegers Arch.* **250,** 103–108.
273. Holtzer H., Marshall J. and Finck H. (1957) An analysis of myogenesis by the use of fluorescent antimyosin. *J. Biophys. Biochem. Cytol.* **3,** 705–723.
274. Honée G. L. J. M. (1966) An investigation on the presence of muscle spindles in the human lateral pterygoid muscle. *Ned. Tijdschr. Tandheelkd.* **73,** 43–48 (Suppl. 3).
275. Hoy R. R. and Wilson D. M. (1969) Rhythmic motor output in leg motoneurons of milkweed bug, *Oncopeltus. Fed. Proc.* **28,** 588.
276. Hoyle G. (1964) Exploration of neuronal mechanisms underlying behaviour in insects.

In: R. F. Reiss (ed.), *Neural Theory and Modelling.* Stanford, Stanford University Press, p. 346.
277. Hufschmidt H. J. and Spuler H. (1962) Mono- and polysynaptic reflexes of the trigeminal muscles in human beings. *J. Neurol. Neurosurg. Psychiatry* **25,** 332–335.
278. Hugelin A. and Bonvallet M. (1956) Etude électrophysiologique d'un réflexe monosynaptique trigéminal. *C. R. Soc. Biol. (Paris)* **150,** 2067–2071.
279. Hunt C. C. and Ottoson D. (1975) Dynamic and static responses of primary and secondary endings of isolated mammalian muscle spindles. *J. Anat.* **119,** 194–196 (Abstract).
280. Hunter R., Earl C. J. and Jantz D. (1964) A syndrome of abnormal movements and dementia in leukotomised patients treated with phenothiazines. *J. Neurol. Neurosurg. Psychiatry* **27,** 219–223.
281. Imamoto K. and Shimizu N. (1970) Fine structure of the mesencephalic nucleus of the trigeminal nerve in the rat. *Arch. Histol. Jap.* **32,** 51–67.
282. Ishikawa H. (1965) The fine structure of the myo-tendon junction in some mammalian skeletal muscles. *Arch. Histol. Jap.* **25,** 275–296.
283. Ito M. (1957) The electrical activity of spinal ganglion cells investigated with intracellular microelectrodes. *Jap. J. Physiol.* **7,** 297–323.
284. Ito M. and Saiga M. (1959) The mode of impulse conduction through the spinal ganglion. *Jap. J. Physiol.* **9,** 33–42.
285. Ito M. and Takahashi I. (1960) Impulse conduction through spinal ganglion. In: Katsuki Y. (ed.), *Electrical Activity of Single Cells.* Tokyo, Igaku Shoin, pp. 159–179.
286. Iwata N., Sakai Y. and Deguchi T. (1971) Effects of physostigmine on the inhibition of trigeminal motoneurons by cutaneous impulses in the cat. *Exp. Brain Res.* **13,** 519–532.
287. Jankowska E., Jukes M. G., Lund S. and Lundberg A. (1967) The effect of DOPA on the spinal cord: reciprocal organisation of pathways transmitting excitatory action to alpha motoneurons of flexors and extensors. *Acta Physiol. Scand.* **70,** 369–388.
288. Jarabak J. R. (1957) An electromyographic analysis of muscular behaviour in mandibular movements from rest position. *J. Prosthet. Dent.* **7,** 682–710.
289. Jerge C. R. (1963) Organization and function of the trigeminal mesencephalic nucleus. *J. Neurophysiol.* **26,** 379–392.
290. Jerge C. R. (1963) Function of the nucleus supratrigeminalis. *J. Neurophysiol.* **26,** 393–402.
291. Jerge C. R. (1964) The neurological mechanism underlying cyclic jaw movements. *J. Prosthet. Dent.* **14,** 667–681.
292. Jerge C. R. (1967) The neural substratum of oral sensation. In: Bosma J. F. (ed.), *Symposium on Oral Sensation and Perception.* Springfield, Thomas, pp. 63–83.
293. Johnson R. G. and Sheridan J. D. (1971) Junctions between cancer cells in culture: ultrastructure and permeability. *Science* **174,** 717–719.
294. Johnston J. B. (1909) The radix mesencaphalica trigemini. *J. Comp. Neurol.* **19,** 593–664.
295. Joniot B. (1974) Physiologic mandibular resting posture. *J. Prosthet. Dent.* **31,** 4–9.
296. Joseph J. (1960) *Man's Posture.* Springfield, Ill. Thomas.
297. Joseph J., Nightingale A. and Williams P. L. (1955) A detailed study of the electric potentials recorded over some postural muscles while relaxed and standing. *J. Physiol. (Lond.)* **127,** 617–625.
298. Kallen F. C. and Gans C. (1972) Mastication in the little brown bat (*Myotis lucifugus*). *J. Morphol.* **136,** 385–420.
299. Karlsen K. (1965) The location of motor end plates and the distribution and histological structure of muscle spindles in the jaw muscles of the cat. *Acta Odontol. Scand.* **23,** 521–547.
300. Karlsen K. (1969) Muscle spindles in the lateral pterygoid muscle of a monkey. *Arch. Oral. Biol.* **14,** 1111–1112.
301. Karlsson U. and Andersson-Cedergren E. (1966) Motor myoneural junctions on frog intrafusal muscle fibres. *J. Ultrastruct. Res.* **14,** 191–211.
302. Karlsson U. and Andersson- Cedergren E. (1971) Satellite cells of the frog muscle spindle as revealed by electron microscopy. *J. Ultrastruct. Res.* **34,** 426–438.
303. Karlsson U., Andersson-Cedergren E. and Ottoson D. (1966) Cellular organisation of the frog muscle spindle as revealed by serial sections for electron microscopy. *J. Ultrastruct. Res.* **14,** 1–35.
304. Karlsson U. and Hawks A. T. (1974) Sensory deformation in the primate muscle spindle: preliminary observations with quantitative electron microscopy. *J. Ultrastruct. Res.* **50,** 384–385 (Abstract).

305. Karlsson U., Hooker W. and Bendeich E. (1971) Quantitative changes in the frog muscle spindle with stretch. *J. Ultrastruct. Res.* **36,** 743–756.
306. Karlsson U. and Schultz R. L. (1965) Fixation of the central system for electron microscopy by aldehyde perfusion. I. Preservation with aldehyde perfusates versus direct perfusion with osmium tetroxide with special reference to membranes and the extracellular space. *J. Ultrastruct. Res.* **12,** 160–186.
307. Karlsson U. and Schultz R. L. (1965) Fixation of the central nervous system for electronmicroscopy by aldehyde perfusion. II. Effect on osmolarity, pH of perfusate and fixative concentration. *J. Ultrastruct. Res.* **12,** 187–206.
308. Kavanagh D. and Zander H. A. (1965) A versatile recording system for studies of mastication. *Med. Biol. Eng.* **3,** 291–300.
309. Kawai H. (1963) Cytoplasmic granules in nerve cells of the spinal cord following transection of sciatic nerve in guinea pig. *Exp. Neurol.* **7,** 457–463.
310. Kawamura Y. (1963) Recent concepts in the physiology of mastication. In: Staple P. H. (ed.), *Advances in Oral Biology.* Vol. 1, New York, Academic Press, pp. 77–109.
311. Kawamura Y. (1967) Neurophysiologic background of occlusion. *Periodontics* **5,** 175–183
312. Kawamura Y. (1974) Neurogenesis of mastication. In: Kawamura Y. (ed.), *Frontiers of Oral Physiology.* Vol. 1, Basel, Karger, pp. 77–120.
313. Kawamura Y. (1974) Responses of stomatognathic structures to noxious stimuli. *Adv. Neurol.* **4,** 351–356.
314. Kawamura Y. and Abe K. (1974) Role of sensory information from temporomandibular joint. *Bull. Tokyo Med. Dent. Univ.* **21,** Suppl. 78–82.
315. Kawamura Y. and Fujimoto J. (1957) Some physiologic considerations on measuring rest position of the mandible. *Med. J. Osaka Univ.* **8,** 247–255.
316. Kawamura Y. and Fujimoto J. (1958) A study of the jaw opening reflex. *Med. J. Osaka Univ.* **9,** 377–387.
317. Kawamura Y., Funakoshi M. and Tsukamoto S. (1958) Brain-stem representation of jaw muscle activities of the dog. *Jap. J. Physiol.* **8,** 292–304.
318. Kawamura Y., Majima T. and Kato I. (1967) Physiologic role of deep mechanoreceptor in temporomandibular joint capsule. *J. Osaka Univ. Dent. Sch.* **7,** 63–76.
319. Kay R. F. and Hiiemae K. (1974) Jaw movement and tooth use in recent and fossil primates. *Am. J. Phys. Anthropol.* **40,** 227–256.
320. Kennedy D. and Calabrese R. L. (1974) Presynaptic inhibition: primary afferent depolarization in crayfish neurons. *Science* **186,** 451–454.
321. Kidokoro Y., Kubota K., Shuto S. and Sumino R. (1968) Reflex organization of cat masticatory muscles. *J. Neurophysiol.* **31,** 695–708.
322. Kidokoro Y., Kubota K., Shuto S. and Sumino R. (1968) Possible interneurons responsible for reflex inhibition of motoneurons of jaw-closing muscles from the inferior dental nerve. *J. Neurophysiol.* **31,** 709–716.
323. Kirkwood P. A. and Sears T. A. (1974) Monosynaptic excitation of motor neurones from secondary endings of muscle spindles. *Nature (Lond.)* **232,** 243–244.
324. Kitiyakara A. and Angevine D. M. (1963) A study of the pattern of post-embryonic growth of M. gracilis in mice. *Dev. Biol.* **8,** 322–340.
325. Klatsky M. (1940) A cinefluorographic study of the human masticatory apparatus in function. *Am. J. Orthod.* **26,** 664–670.
326. Klineberg I. (1971) Structure and function of temporomandibular joint innervation. *Ann. R. Coll. Surg. Engl.* **49,** 268–288.
327. Koivumaa K. K. (1961) Cinefluorographic analysis of the masticatory movements of the mandible. *Suom. Hammaslääk. Toim.* **57,** 306–336.
328. Korn H. and Bennett M. V. L. (1972) Electrotonic coupling between teleost oculomotor neurons; restriction to somatic regions and relation to function of somatic and dendritic sites of impulse initiation. *Brain Res.* **38,** 433–439.
329. Korn H., Sotelo C. and Crepel F. (1973) Electrotonic coupling between neurons in the rat lateral vestibular nucleus. *Exp. Brain Res.* **16,** 255–275.
330. Kosaka K. (1912) Zur Frage der physiologischen Natur der cerebralen Trigeminuswurzel. *Folia Neurobiol.* **6,** 1–16.
331. Kostecka F. (1931) Die Chirurgische Therapie der Progenie. *Zahnärztl. Rundsch.* **40,** 669–688.
332. Krasne F. B. and Bryan J. S. (1973) Habituation: regulation through presynaptic inhibition. *Science* **182,** 590–592.
333. Kreibel M. E., Bennett M. V. L., Waxman S. G. and Pappas G. D. (1969) Oculomotor

neurons in fish: electrotonic coupling and multiple sites of impulse initiation. *Science* **166,** 520–524.

334. Kristensson K., Olsson Y. and Sjöstrand J. (1971) Axonal uptake and retrograde transport of exogenous proteins in the hypoglossal nerve. *Brain Res.* **32,** 399–406.
335. Kubota K., Kidokoro Y. and Suzuki J. (1968) Postsynaptic inhibition of trigeminal and lumbar motoneurons from the superficial radial nerve in the cat. *Jap. J. Physiol.* **18,** 198–215.
336. Kubota K. and Masegi T. (1972) Muscle spindle distribution in the masticatory muscle of the Japanese shrew-mole. *J. Dent. Res.* **51,** 1080–1091.
337. Kugelberg E. (1952) Facial reflexes. *Brain* **75,** 385–396.
338. Kurth L. E. (1942) Mandibular movements in mastication. *Am. J. Orthod.* **29,** 1769–1790.
339. Kuypers H. G. J. M. (1958) An anatomical analysis of corticobulbar connections to the pons and lower brain stem in the cat. *J. Anat.* **92,** 198–218.
340. Kydd W. L., Harrold W. and Smith D. E. (1967) A technique for continuously monitoring the interocclusal distance. *J. Prosthet. Dent.* **18,** 308–310.
341. Laird W. R. E., Davies E. H., Manson G. and von Fraunhofer J. A. (1971) Measurement of occlusal tooth separation by means of electrical field variations. *Biomed. Eng.* **6,** 504–508.
342. Larramendi L. M. H., Lemkey N. and Fickenscher L. (1967) Synaptic vesicles of inhibitory and excitatory terminals in the cerebellum. *Science* **156,** 967–969.
343. Laursen A. M. (1972) Static and phasic muscle activity of monkeys with pyramidal lesions. *Brain Res.* **40,** 125–126.
344. LaVail J. H. and LaVail M. M. (1972) Retrograde axonal transport in the central nervous system. *Science* **176,** 1416–1417.
345. Law E. (1954) Lingual proprioception in pig, dog and cat. *Nature (Lond.)* **174,** 1107–1108.
346. Leek F. F. (1966) Observations on the dental pathology seen in ancient Egyptian skulls. *J. Egypt. Archaeol.* **52,** 59–64.
347. Leek. F. F. (1972) Teeth and bread in ancient Egypt. *J. Egypt. Archaeol.* **58,** 126–132.
348. Leksell L. (1945) The action potential and excitatory effects of the small ventral root fibres to skeletal muscle. *Acta Physiol. Scand.* **10,** Suppl. 31.
349. Lewinsky W. and Stewart D. (1937) A comparative study of the innervation of the periodontal membrane. *Proc. R. Soc. Med.* **30,** 73–87.
350. Lewis D. M. (1969) The response of primary afferent fibres from cat spindles to single volleys in sensory nerves. *J. Physiol. (Lond.)* **203,** 22–24 *P.*
351. Lindblom G. (1960) On the anatomy and function of the temporomandibular joint. *Acta Odontol. Scand.* **17,** Suppl. 28.
352. Lindblom G. (1971) A longitudinal research of dysfunctional disturbances (arthrosis) in the temporomandibular joint—their diagnosis and treatment results up to (1969). *Swed. Dent. J.* **64,** 559–584.
353. Lindhe J. and Svanberg G. (1974) Influence of trauma from occlusion on progression of experimental periodontitis in the beagle dog. *J. Clin. Periodontol.* **1,** 3–14.
354. Llinás R. (1964) Mechanisms of supraspinal actions upon spinal cord activities. Pharmacological studies on reticular inhibition of alpha extensor motoneurons. *J. Neurophysiol.* **27,** 1127–1137.
355. Llinás R., Baker R. and Sotelo C. (1974) Electrotonic coupling between neurons in cat inferior olive. *J. Neurophysiol.* **37,** 560–571.
356. Lodge D., Duggan A. W., Biscoe T. J. and Caddy K. W. T. (1973) Concerning recurrent collaterals and afferent fibers in the hypoglossal nerve of the cat. *Exp. Neurol.* **41,** 63–75.
357. Long C., Thomas D. and Crochetiere W. J. (1964) Objective measurements of muscle tone in the hand. *Clin. Pharmacol. Therap.* **5,** 907–917.
358. Lorente de Nó R. (1922) Contribución al conocimiento del nervo trigemino. *Libro en honor de Ramón y Cajal* **2,** 13–30.
359. Lorente de Nó R. (1933) Vestibulo-occular reflex arc. *Arch. Neurol. Psychiat. Chicago* **30,** 245–291.
360. Lorente de Nó R. (1947) Action potentials of the motoneurons of the hypoglossal nucleus. *J. Cell. Comp. Physiol.* **29,** 207–288.
361. Lous I., Sheikholeslam A. and Moller E. (1970) Postural activity in subjects with functional disorders of the chewing apparatus. *Scand. J. Dent. Res.* **78,** 404–410.
362. Lowe A. A. and Sessle B. J. (1973) Tongue activity during respiration, jaw opening, and swallowing in cat. *Can. J. Physiol. Pharmacol.* **51,** 1009–1011.

363. Lowe A. A. and Sessle B. J. (1975) Genioglossus activity during respiration, jaw opening and swallowing in the cat and monkey. *J. Dent. Res.* **53**, 201, 587. (Abstract.)
364. Lucas A. (1962) *Ancient Egyptian Materials and Industries,* 4th ed. London, Arnold.
365. Lund J. P. (1971) Experiments on the neural control of mastication. Ph.D. Thesis. Univ. of Western Ontario.
366. Lund J. P. (1976) Oral-facial sensation in the control of mastication and voluntary movements of the jaw. In: Sessle B. and Hannam A. (ed.), *Mastication and Swallowing: Biological and Clinical Correlates.* Toronto, Univ. of Toronto Press. (In press.)
367. Lund J. P. and Dellow P. G. (1971) The influence of interactive stimuli on rhythmical masticatory movements in rabbits. *Arch. Oral Biol.* **16**, 215–223.
368. Lund J. P. and Dellow P. G. (1973) Rhythmic masticatory activity of hypoglossal motoneurons responding to an oral stimulus. *Exp. Neurol.* **40**, 243–246.
369. Lund J. P. and Lamarre Y. (1973) The importance of positive feedback from periodontal pressoreceptors for voluntary isometric contraction of jaw closing muscles in man. *J. Biol. Buccale* **1**, 345–351.
370. Lund J. P., McLachlan R. S. and Dellow P. G. (1971) A lateral jaw movement reflex. *Exp. Neurol.* **31**, 189–199.
371. Lund J. P. and Sessle B. J. (1974) Oral-facial and jaw muscle afferent projections to neurones in cat frontal cortex. *Exp. Neurol.* **45**, 314–331.
372. Lund P., Nishiyama T. and Møller E. (1970) Postural activity in the muscles of mastication with the subject upright, inclined and supine. *Scand. J. Dent. Res.* **78**, 417–424.
373. Luschei E. S. and Goodwin G. M. (1974) Patterns of mandibular movement and jaw muscle activity during mastication in monkeys. *J. Neurophysiol.* **37**, 954–966.
374. Lynn A. M. J. and Yemm R. (1971) External forces required to move the mandible of relaxed human subjects. *Arch. Oral Biol.* **16**, 1443–1447.
375. Macapanpan L. C. and Weinmann J. P. (1954) The influence of injury to the periodontal membrane on the spread of gingival inflammation. *J. Dent. Res.* **33**, 263–272.
376. McIntyre A. K. (1951) Afferent limb of the myotatic reflex arc. *Nature (Lond.)* **168**, 168–169.
377. McIntyre A. K. and Robinson R. G. (1959) Pathway for the jaw jerk in man. *Brain* **82**, 468–474.
378. Mackay B., Harrop T. J. and Muir A. R. (1969) The fine structure of the muscle tendon junction in the rat. *Acta Anat.* **73**, 588–604.
379. McLean L. F., Brenman H. S. and Friedman M. G. F. (1973) Effects of changing body position on dental occlusion. *J. Dent. Res.* **52**, 1041–1045.
380. McNamara J. A. jun. (1972) Neuromuscular and skeletal adaptations to altered oro-facial function. Ann Arbor, Univ. of Michigan.
381 Magoun H., Ranson S. W. and Fisher C. (1933) Corticifugal pathways for mastication. Lapping and other motor functions in the cat. *Arch. Neurol. Psychiat. Chicago* **30**, 292–308.
382. Mahan P. E. and Anderson K. V. (1970) Jaw depression elicited by tooth pulp stimulation. *Exp. Neurol.* **29**, 439–448.
383. Manly R. S. C., Pfaffmann D. D., Lathrop D. and Keyser J. (1952) Oral sensory thresholds of persons with natural and artificial dentitions. *J. Dent. Res.* **31**, 305–312.
384. Manni E., Bortolami R. and Azzena G. B. (1965) Jaw muscle proprioception and mesencephalic trigeminal cells in birds. *Exp. Neurol.* **12**, 320–328.
385. Marey E. (1887) Recherches experimentales sur la morphologie des muscles. *C. R. hebd. Seanc. Acad. Sci. (Paris)* **105**, 446–451.
386. Marsden C. D., Merton P. A. and Morton H. B. (1972) Servo action in human voluntary movements. *Nature (Lond.)* **238**, 140–143.
387. Matsunami K. and Kubota K. (1972) Muscle afferents of trigeminal mesencephalic tract nucleus and mastication in chronic monkeys. *Jap. J. Physiol.* **22**, 545–555.
388. Matthews B. (1975) Mastication. In: Lavelle C. (ed.), *Applied Physiology of the Mouth.* Bristol, Wright, Ch. 10.
389. Matthews P. B. C. (1969) Evidence that the secondary as well as the primary endings of the muscle spindles may be responsible for the tonic stretch reflex of the decerebrate cat. *J. Physiol. (Lond.)* **204**, 365–393.
390. Matthews P. B. C. (1972) *Mammalian Muscle Receptors and their Central Actions.* London, Arnold.

391. Mauro A., Shafiq S. A. and Milhorat A. T. (1970) Regeneration of striated muscle and myogenesis. Excerpta Medica International Congress Series No. 218.
392. Meier-Ewert K., Gleitsmann K. and Reiter F. (1974) Acoustic jaw reflex in man: its relationship to brain stem and other micro-reflexes. *Electroencephogr. Clin. Neurophysiol.* **36**, 629–637.
393. Merrillees N. C. R. (1962) Some observations on the fine structure of a Golgi tendon organ of a rat. In: Barker D. (ed.), *Symposium of Muscle Receptors.* Hong Kong, Hong Kong University Press. p. 199.
394. Merton P. A. (1953) Speculations on the servo control of movement. In: Wolstenholme G. E. W. (ed.), *The Spinal Cord.* London, Churchill. pp. 247–255.
395. Meyer E. (1875) Uber rothe und blasse quegestreifte Muskeln. *Arch. Anat. Physiol.* 217–232.
396. Miles A. E. W. (1961) Assessment of ages of a population of Anglo-Saxons from their dentitions. *Proc. R. Soc. Med.* **25**, 881–886.
397. Mills J. R. E. (1955) Ideal dental occlusion in the primates. *Dent. Pract. Dent. Rec.* **6**, 47–63.
398. Mills J. R. E. (1976) Attrition in animals. In: Poole D. F. G. and Stack M. V. (ed.), *The Eruption and Occlusion of Teeth.* Colston Symposium No. 27. London, Butterworths. (In press.)
399. Milner-Brown H. S., Stein R. B. and Yemm R. (1973) Contractile properties of human motor units during voluntary isometric contractions. *J. Physiol. (Lond.)* **228**, 285–306.
400. Milner-Brown H. S., Stein R. B. and Yemm R. (1973) The orderly recruitment of human motor units during voluntary isometric contractions. *J. Physiol. (Lond.)* **230**, 359–370.
401. Milner-Brown H. S., Stein R. B. and Yemm R. (1973) Changes in firing rate of human motor units during linearly changing voluntary contractions. *J. Physiol. (Lond.)* **230**, 371–390.
402. Mizuno N. and Sauerland E. K. (1970) Trigeminal proprioceptive projections to the hypoglossal nucleus and the cervical ventral gray column. *J. Comp. Neurol.* **139**, 215–226.
403. Mizuno N., Sauerland E. K. and Clemente C. D. (1968) Projections from the orbital gyrus in the cat. I: to brain stem structures. *J. Comp. Neurol.* **133**, 436–476.
404. Møller E. (1966) The chewing apparatus: an electromyograph study of the action of the muscles of mastication and its correlation to facial morphology. *Acta Physiol. Scand.* **69**, Suppl. 280.
405. Møller E. (1976) Human muscle patterns. In: Sessle B. J. and Hannam A. G. (ed.), *Mastication and Swallowing: Biological and Clinical Correlates.* Toronto, Univ. of Toronto Press. (In press.)
406. Møller E., Sheikholeslam A. and Lous I. (1971) Deliberate relaxation of the temporal and masseter muscles in subjects with functional disorders of the chewing apparatus. *Scand. J. Dent. Res.* **79**, 478–482.
407. Møller E. and Troelstrup B. (1975) Functional and morphological asymmetry in children with unilateral cross-bite. *J. Dent. Res.* **54**, Special Issue A, L 45. (Abstract.)
408. Montet P. (1959) *L'Egypte et la Bible.* Neuchatel, Delachaux et Niestle.
409. Morimoto T. and Kawamura Y. (1971) Discharge patterns of hypoglossal afferents in a cat. *Brain Res.* **35**, 539–542.
410. Morimoto T. and Kawamura Y. (1972) Inhibitory postsynaptic potentials of hypoglossal motoneurons of the cat. *Exp. Neurol.* **37**, 188–198.
411. Morimoto T., Takata M. and Kawamura Y. (1972) Inhibition of hypoglossal motoneurons by a masseteric nerve volley. *Brain Res.* **43**, 285–288.
412. Morin G., Gaillard F., Liege B. and Galand G. (1974) Réponses proprioceptives unitaires à l'abaissement de la mâchoire enregistrées dans le tronc cérébral de la grenouille. *J. Physiol. (Paris)* **68**, 121–144.
413. Moss F. B. (1968) The relationship between the dimensions of the fibres and the number of nuclei during normal growth of skeletal muscle in the domestic fowl. *J. Anat.* **122**, 555–564.
414. Moss F. B. and Le Blond C. P. (1971) Satellite cells as the source of nuclei in muscles of growing rats. *Anat. Rec.* **170**, 421–436.
415. Moss M. L. (1968) The primacy of functional matrices in oro-facial growth. *Dent. Pract. Dent. Rec.* **19**, 65–73.

416. Moyers R. E. (1973) *Handbook of Orthodontics,* 3rd ed. Chicago, Year Book Medical Publishers.
417. Muhl Z. F. and Grimm A. F. (1974) Adaptability of rabbit digastric muscle to an abrupt change in length: a radiographic study. *Arch. Oral Biol.* **19,** 829–834.
418. Muir A. R. (1961) In:Boyd J. D. (ed.), *Electron Microscopy for Anatomists.* London, Arnold, pp. 267–277.
419. Munro R. R. and Griffin C. J. (1970) Analysis of the electromyography of the masseter muscle and the anterior part of the temporalis muscle in the open-close-clench cycle in man. *Arch. Oral Biol.* **15,** 827–844.
420. Murphy T. R. (1959) Compensatory mechanisms in facial height adjustment to functional tooth attrition. *Aust. Dent. J.* **4,** 312–323.
421. Murphy T. R. (1967) Shortening/inhibition of prime movers: a safety factor in mastication. *Br. Dent. J.* **123,** 578–584.
422. Murphy W. H. (1970) Oxytetracycline microfluorescent comparison of orthodontic retraction into recent and healed extraction sites. *Am. J. Orthod.* **58,** 215–239.
423. Nachlas M. M., Tsou K. C., de Souza E., Cheng C. S. and Seligman A. M. (1957) Cytochemical demonstration of succinic dehydrogenase by the use of a new *p*-nitrophenyl substituted ditetrazole. *J. Histochem. Cytochem.* **5,** 420–436.
424. Nakamura Y., Goldberg L. J. and Clemente C. D. (1967) Nature of suppression of the masseteric monosynaptic reflex induced by stimulation of the orbital gyrus in the cat. *Brain Res.* **6,** 184–198.
425. Nakamura Y., Goldberg L. J., Mizuno N. and Clemente C. D. (1970) Masseteric reflex inhibition induced by afferent impulses in the hypoglossal nerve. *Brain Res.* **18,** 241–255.
426. Nakamura Y., Mori S. and Nagashima H. (1973) Origin and central pathways of crossed inhibitory effects of afferents from the masseteric muscle on the masseteric motoneuron of the cat. *Brain Res.* **57,** 29–42.
427. Nakamura Y., Nagashima H. and Mori S. (1973) Bilateral effects of the afferent impulses from the masseteric muscle on the trigeminal motoneuron of the cat. *Brain Res.* **57,** 15–27.
428. Nakamura Y. and Wu C. Y. (1970) Presynaptic inhibition of jaw-opening reflex by high threshold afferents from the masseter muscle of the cat. *Brain Res.* **23,** 193–211.
429. Nandy K. (1968) Histochemical study of chromatolytic neurons. *Arch. Neu. l.* **18,** 425–434.
430. Nelson P. G. (1966) Interaction between spinal motoneurons of the cat. *J. Neurophysiol.* **29,** 275–287.
431. Ness A. R. (1954) The mechanoreceptors of the rabbit mandibular incisor. *J. Physiol. (Lond.)* **126,** 475–493.
432. Noble W. H. and Martin L. P. (1973) Tooth mobility changes in response to occlusal interference. *J. Prosthet. Dent.* **30,** 412–419.
433. Nordstrom S. H., Bishop M. and Yemm R. (1974) The effect of jaw opening on the sarcomere length of the masseter and temporal muscles of the rat. *Arch. Oral Biol.* **19,** 151–155.
434. Nordstrom S. H. and Yemm R. (1972) Sarcomere length in the masseter muscle of the rat. *Arch. Oral Biol.* **17,** 895–902.
435. Nordstrom S. H. and Yemm R. (1974) The relationship between jaw position and isometric active tension produced by direct stimulation of the rat masseter muscle. *Arch. Oral Biol.* **19,** 353–359.
436. Obwegeser H. (1957) The surgical correction of mandibular prognathism and retrognathia. *J. Oral Surg.* **10,** 677–689.
437. Olson C. B. and Swett C. P. jun. (1966) A functional and histochemical characterisation of motor units in a heterogeneous muscle (flexor digitorum longus) of the cat. *J. Comp. Neurol.* **128,** 475–498.
438. Oosterhuis H. and Bethlem J. (1973) Neurogenic muscle involvement in myasthenia gravis. *J. Neurol. Neurosurg. Psychiatry* **36,** 244–253.
439. O'Rourke J. T. (1949) Significance of tests of biting strength. *J. Am. Dent. Assoc.* **38,** 627–633.
440. Öwall B. and Elmqvist D. (1975) Motor pauses in EMG activity during chewing and biting. *Odontol. Revy* **26,** 17–38.
441. Öwall B. and Møller E. (1974) Oral tactile sensibility during biting and chewing. *Odontol. Revy* **25,** 327–346.

442. Padykula H. A. and Herman E. (1955) The specificity of the histochemical method for adenosine triphosphate. *J. Histochem. Cytochem.* **3,** 170–195.
443. Pameijer J. H. N., Brion M., Glickman I. and Roeber F. W. (1970) Intraoral occlusal telemetry. Part IV. Tooth contact during swallowing. *J. Prosthet. Dent.* **24,** 396–400.
444. Pameijer J. H. N., Brion M., Glickman I. and Roeber F. W. (1970) Intraoral occlusal telemetry. Part V. Effect of occlusal adjustment upon tooth contacts during chewing and swallowing. *J. Prosthet. Dent.* **24,** 429–497.
445. Pameijer J. H. N., Glickman I. and Roeber F. (1968) Intraoral occlusal telemetry. Part II. Registration of tooth contacts in chewing and swallowing. *J. Prosthet. Dent.* **19,** 151–159.
446. Pameijer J. H. N., Glickman I. and Roeber F. W. (1969) Intraoral occlusal telemetry. Part III. Tooth contacts in chewing, swallowing and bruxism. *J. Periodontol.* **40,** 253–258.
447. Parker G. R. (1972) Transseptal fibres and relapse following bodily retraction of teeth: A histologic study. *Am. J. Orthod.* **61,** 331–344.
448. Paukul E. (1904) Die Zuckungsformen von Kanischemuskeln verschiedener Farbe and Struktur. *Arch. Anat. Physiol.* 100–120.
449. Penfield W. and Jasper H. (1954) *Epilepsy and the Functional Anatomy of the Human Brain.* London, Churchill, p. 107.
450. Perkel D. H., Gerstein G. L. and Moore G. P. (1967) Neuronal spike trains and stochastic point processes. I. The single spike train. *Biophys. J.* **7,** 371–418.
451. Perry H. T., Lammie G. A., Main J. and Teuscher G. W. (1960) Occlusion in a stress situation. *J. Am. Dent. Assoc.* **60,** 626–633.
452. Pétrovic M. A., Oudet C. and Gasson N. (1973) Effets des appareils de propulsion et de rétropulsion mandibulaire sur le nombre des sarcomères en serie du muscle ptérygoidien externe et sur la croissance du cartilage condylien du jeune rat. *L'Orthodontie Française* **44,** 191–210.
453. Pfaffmann C. (1939) Afferent impulses from the teeth due to pressure and noxious stimulation. *J. Physiol. (Lond.)* **97,** 207–219.
454. Piatt J. (1945) Origin of the mesencephalic V root cells in Amblystoma. *J. Comp. Neurol.* **82,** 35–54.
455. Picton D. C. A. (1963) Vertical movement of cheek teeth during biting. *Arch. Oral Biol.* **8,** 109–118.
456. Picton D. C. A. and Moss J. P. (1973) The part played by the transseptal fibre system in experimental approximal drift of cheek teeth of monkeys (*Macaca irus*). *Arch. Oral Biol.* **18,** 669–680.
457. Pitts J. D. (1970) Abstracts of Ciba Foundation Symposia Growth Control in Cell Cultures. **16,** 89–105.
458. Pollen D. S. and Ajmone-Marsan C. (1965) Cortical inhibitory postsynaptic potentials and strychninization. *J. Neurophysiol.* **28,** 342–358.
459. Porter R. (1965) Synaptic potentials in hypoglossal motoneurones. *J. Physiol. (Lond.)* **180,** 209–244.
460. Porter R. (1967) Cortical action on hypoglossal motoneurones in cats: a proposed role for a common internuncial cell. *J. Physiol. (Lond.)* **193,** 295–308.
461. Posselt U. and Franzén G. (1960) Registration of the condyle path inclination by intra-oral wax records: variations in three instruments. *J. Prosthet. Dent.* **10,** 441–454.
462. Prieskel H. W. (1965) Some observations on the postural position of the mandible. *J. Prosthet. Dent.* **15,** 625–633.
463. Provost W. A. and Towle H. J. (1972) Determination of the physiologic rest position by electronic measurement. *J. Prosthet. Dent.* **27,** 377–380.
464. Ralston H. J. and Libet B. (1953) The question of tonus in skeletal muscle. *Am. J. Phys. Med.* **32,** 85–92.
465. Ramfjord S. P. and Ash M. (1971) *Occlusion*, 2nd ed. Philadelphia, Saunders.
466. Ramfjord S. P., Kerr D. A. and Ash M. (1966) World Workshop in Periodontics. USA. University of Michigan.
467. Ramfjord S. P. and Kohler C. A. (1959) Periodontal reaction to functional occlusal stress. *J. Periodontol.* **30,** 95–112.
468. Ramon y Cajal S. (1904) Asociación del método del nitrato de plata cou el embrionario. *Trab. del Lab. de Invest. biol.* tIII.
469. Ramon y Cajal S. (1909) *Histologie du système nerveux de l'homme et des vertébrés.* Paris, Maloine.

470. Ramon y Cajal S. (1952) *Histologie du système nerveux de l'homme et des vertébrés.* Madrid, Montana.
471. Ranvier L. (1873) Propriétés et structures différentes des muscles rouge et des muscles blanc chez les lapins et chez les raies. *Compt. Rend.* 77, 1030–1043.
472. Ranvier L. (1874) De quelques faits relatif à l'histologie et à la physiologie des muscles striaés. *Arch. Physiol. Norm. Pathol. (Paris)* **1**, 5–15.
473. Rasmussen G. L. (1946) The olivary peduncle and other fiber projections of the superior olivary complex. *J. Comp. Neurol.* **84**, 141–219.
474. Rasmussen G. L. (1953) Further observations of the efferent cochlear bundle. *J. Comp. Neurol.* **99**, 61–74.
475. Rateitschak K. H. (1970) Reaction of periodontal tissues to artificial forces. In: Eastoe J. E., Picton D. C. A. and Alexander A. G. (ed.), *Prevention of Periodontal Disease. A Symposium.* London, Kimpton, pp. 157–162.
476. Rea T. J. (1972) Human variance in adjusting a fully adjusted articulator by means of a pantograph. M.S. Thesis, University of Michigan.
477. Reimer L. (1973) Physical limits in STEM of thick specimen. Proceedings of the 5th Annual SEM Symposium. 1972.
478. Reinking R. M., Stauffer E. K., Stuart D. G., Taylor A. and Watt D. G. D. (1975) The inhibitory effects of muscle spindle primary afferents investigated by the afferent triggered averaging method. *J. Physiol. (Lond.)* **246**, 20–22 *P.*
479. Reiss R. F. (1962) Theory and simulation of rhythmic behaviour due to reciprocal inhibition in small nerve nets. American Federation of Information Processing Societies. Proceedings 1962 Spring Joint Computer Conference **21**, 171–194.
480. Revel J. P. and Karnovsky M. J. (1967) Hexagonal array of subunits in intercellular junctions of the mouse heart and liver. *J. Cell Biol.* **33**, C7–C12.
481. Ringqvist M. (1971) Histochemical fibre types and fibre sizes in human masticatory muscles. *Scand. J. Dent. Res.* **79**, 366–368.
482. Ringqvist M. (1973) Histochemical enzyme profiles of fibres in human masseter muscles with special regard to fibres with intermediate myofibrillar ATPase reaction. *J. Neurol. Sci.* **18**, 133–141.
483. Ringqvist M. (1973) Fibre sizes of human masseter muscle in relation to bite force. *J. Neurol. Sci.* **19**, 297–305.
484. Ringqvist M. (1974) A histochemical study of temporal muscle fibres in denture wearers and subjects with natural dentition. *Scand. J. Dent. Res.* **82**, 28–39.
485. Ringqvist M. (1976) Size distribution of histochemical fibre types in masseter muscle of adults with different states of occlusion. *J. Neurol. Sci.* (In press.)
486. Rioch J. M. (1934) The neural mechanism of mastication. *Am. J. Physiol.* **108**, 168–176.
487. Ruffer A. (1919) *Food in Egypt.* Cairo, L'Institut d'Egypte.
488. Ruffer M. A. (1921) *Studies in the Palaeopathology of Egypt.* Chicago, Univ. of Chicago Press.
489. Rushworth G., Lishman W. A., Hughes J. T. and Oppenheimer D. R. (1961) Intense rigidity of the arms due to isolation of motoneurones by a spinal tumour, *J. Neurol. Neurosurg. Psychiatry* **24**, 132–142.
490. Ruska H. and Edwards G. A. (1957) A new cytoplasmic pattern in striated muscle fibres and its possible relation to growth. *Growth* **21**, 73–88.
491. Sakada S. (1971) Response of Golgi-Mazzoni corpuscles in the cat periostea to mechanical stimuli. In: Dubner R. and Kawamura Y. (ed.), *Oral-facial Sensory and Motor Mechanisms.* New York, Appleton-Century-Crofts, pp. 105–122.
492. Sakada S. and Kamio E. (1970) Fibre diameters and responses of single units in the periodontal nerve of the cat mandibular canine. *Bull. Tokyo Dent. Coll.* **11**, 223–234.
493. Sakada S. and Kamio E. (1971) Receptive fields and directional sensitivity of single sensory units innervating the periodontal ligaments of the cat mandibular teeth. *Bull. Tokyo Dent. Coll.* **12**, 25–43.
494. Sakada S. and Maeda K. (1967) Correlation between histological structure and response to pressure of mechanoreceptors in the cat mandibular periosteum. *Bull. Tokyo Dent. Coll.* **8**, 181–196.
495. Sato M. and Austin G. (1961) Intracellular potentials of mammalian dorsal root ganglion cells. *J. Neurophysiol.* **24**, 569–582.
496. Sauerland E. K. and Mizuno N. (1969) Cortically induced presynaptic inhibition of trigeminal proprioceptive afferents. *Brain Res.* **13**, 556–568.
497. Scalzi H. A. and Price H. M. (1972) Electron microscopic observations of the sensory

region of the mammalian muscle spindle. In: Banker B. Q. et al. (ed.), *Research in Muscle Development and the Muscle Spindle.* Excerpta Medica, Amsterdam, pp. 254–263.
498. Scapino R. P. (1965) The third joint of the canine jaw. *J. Morphol.* **116,** 23–50.
499. Scarpelli D. G., Hess R. and Pearse A. G. E. (1958) The cytochemical localisation of oxidative enzymes. *J. Biophys. Biochem. Cytol.* **4,** 747–751.
500. Schaerer P., Legault J. V. and Zander H. A. (1966) Mastication under anesthesia. *Helv. Odontol. Acta* **10,** 130–134.
501. Schaerer P., Stallard R. and Zander H. A. (1967) Occlusal interferences and mastication: an electromyographic study. *J. Prosthet. Dent.* **17,** 438–449.
502. Scheinin J. J. (1930) Typing of the cells of the mesencephalic nucleus of the trigeminal nerve in the dog, based on Nissl granule arrangement. *J. Comp. Neurol.* **50,** 109–131.
503. Schmalburch H. Z. (1968) Noniusperioden und Langenwachstum der quergestreiften Muskelfaser. *Z. Mikrosk. Anat. Forsch.* **79,** 493–507.
504. Schmidt J. R. and Harrison J. D. (1972) A method for simultaneous electromyographic and tooth-contact recording. *J. Prosthet. Dent.* **24,** 387–395.
505. Schmidt R. E. (1964) The pharmacology of presynaptic inhibition. In: Eccles J. C. and Schadé J. P. (ed.), *Physiology of Spinal Neurons. Prog. Brain Res.* Vol. 12. Elsevier, Amsterdam, pp. 119–131.
506. Schmitt A., Yu S. K. J. and Sessle B. J. (1973) Excitatory and inhibitory influences from laryngeal and orofacial areas on tongue position in the cat. *Arch. Oral Biol.* **18,** 1121–1130.
507. Schneider A. J. (1928) The histology of the radix mesencephalica n. trigemini in the dog. *Anat. Rec.* **38,** 321–339.
508. Schoen R. (1931) Untersuchungen über Zungen- und Kieferreflexe. I. Mitteilung: der Kieferzungenreflex und andere proprioceptieve Reflexe der Zunge und der Kiefermuskulatur. *Arch. Exp. Path. Pharm.* **160,** 29–48.
509. Schuchardt K. (1961) Experience with the surgical treatment of some deformities of the jaw. Int. Soc. Plastic Surg. Transactions of 2nd Congress London 1959 (Wallace A. B. ed.) Edinburgh, Livingstone, p. 73.
510. Schwartz I. L. and Penefsky Z. J. (1973) Excitation and contraction of muscle. In: Brobeck J. R. (ed.), *Best & Taylor's Physiological Basis of Medical Practice,* 9th ed. Ch. 4. Baltimore, Williams & Wilkins, pp. 61–113.
511. Schweitzer J. M. (1961) Masticatory function in man. *J. Prosthet. Dent.* **11,** 625–647.
512. Schwindling R. and Stark W. (1968) Untersuchung über die Ruheschwebe des Unterkiefers auf electronischen Wege. *Stoma, Heidelb.* **21,** 15–24.
513. Sears T. A. (1973) Servo control of intercostal muscles. In: Desmedt J. E. (ed.), *New Developments in Electromyography and Clinical Neurophysiology.* Vol. 3. Basle, Karger, pp. 404–417.
514. Sessle B. J. (1973) Presynaptic excitability changes induced in single laryngeal primary afferent fibres. *Brain Res.* **53,** 333–342.
515. Sessle B. J. (1976) How are mastication and swallowing programmed and regulated? In: Sessle B. J. and Hannam A. G. (ed.), *Mastication and Swallowing: Biological and Clinical Correlates.* Toronto, Univ. of Toronto Press. (In press.)
516. Sessle B. J. and Greenwood L. F. (1975) Effects of trigeminal tractotomy and of carbamazepine on single trigeminal sensory neurones in cats. *J. Dent. Res.* **54,** (Special Issue B) B201–B206.
517. Sessle B. J. and Kenny D. J. (1973) Control of tongue and facial motility: neural mechanisms that may contribute to movements such as swallowing and sucking. In: Bosma J. F. (ed.), *Fourth Symposium on Oral Sensation and Perception.* Bethesda, D.H.E.W., pp. 222–231.
518. Sessle B. J. and Schmitt A. (1972) Effects of controlled tooth stimulation on jaw muscle activity in man. *Arch. Oral Biol.* **17,** 1597–1607.
519. Shafiq S. A., Gorycki M. A. and Mauro A. (1968) Mitosis during post-natal growth in skeletal and cardiac muscles of the rat. *J. Anat.* **103,** 135–141.
520. Shanahan T. E. J. and Leff A. (1966) Physiologic and mechanical concepts of occlusion. *J. Prosthet. Dent.* **16,** 62–72.
521. Sharkey S. W. (1969) Gnathosonics in orthodontics. 1969 *Trans. Europ. Orthodont. Soc.*
522. Sheppard I. M. (1959) The closing masticatory strokes. *J. Prosthet. Dent.* **9,** 946–951.
523. Sheppard I. M. (1965) The effect of extreme vertical overlap on masticatory strokes. *J. Prosthet. Dent.* **15,** 1035–1042.

524. Sherrington C. S. (1917) Reflexes elicitable in the cat from pinna, vibrissae and jaws. *J. Physiol. (Lond.)* **51,** 404–431.
525. Shwaluk S. (1971) Initiation of reflex activity from the temporomandibular joint of the cat. *J. Dent. Res.* **50,** 1642–1646.
526. Silverman S. I. (1961) *Oral Physiology.* St. Louis, C. V. Mosby Co.
527. Skoglund S. (1973) Joint receptors and kinaesthesis. In: Iggo A. (ed.), *Handbook of Sensory Physiology.* Vol. II, Somatosensory System. Berlin, Springer-Verlag.
528. Smith C. A. and Rasmussen G. L. (1965) Degeneration in the efferent nerve endings in cochlea after axonal section. *J. Cell Biol.* **26,** 63–77.
529. Smith R. D. (1969) Location of the neurons innervating tendon spindles of masticatory muscles. *Exp. Neurol.* **25,** 646–654.
530. Smith R. D. and Marcarian H. Q. (1967) The neuromuscular spindles of the lateral pterygoid muscle. *Anat. Anz.* **120,** 47–53.
531. Smith R. D., Marcarian H. Q. and Niemer W. T. (1967) Bilateral relationships of the trigeminal mesencephalic nuclei and mastication. *J. Comp. Neurol.* **131,** 79–92.
532. Smith R. D., Marcarian H. Q. and Niemer W. T. (1968) Direct projections from the masseteric nerve to the mesencephalic nucleus. *J. Comp. Neurol.* **133,** 495–502.
533. Smith S. E. (1965) Kinetics of rat brain amino acid transport *in vitro. J. Physiol. (Lond.)* **181,** 62 *P.*
534. Söderholm U. (1965) Histochemical localization of esterases, phosphatases and tetrazolium reductases in the motor neurons of the spinal cord of the rat and the effect of nerve division. *Acta Physiol. Scand.* **65,** Suppl. 256.
535. Soltis J. E., Nakfoor P. R. and Bowman D. C. (1971) Changes in ability of patients to differentiate intensity of forces applied to maxillary central incisors during orthodontic treatment. *J. Dent. Res.* **50,** 590–596.
536. Sotelo C., Llinás R. and Baker R. (1974) Structural study of inferior olivary nucleus of the cat: morphological correlates of electrotonic coupling. *J. Neurophysiol.* **37,** 541–559.
537. Spira M. E. and Bennett M. V. L. (1972) Synaptic control of electrotonic coupling between neurons. *Brain Res.* **37,** 294–300.
538. Stefanis C. and Jasper H. (1965) Strychnine reversal of inhibitory potentials in pyramidal tract neurones. *Int. J. Neuropharmacol.* **4,** 125–138.
539. Stein J. M. and Padykula H. A. (1962) Histochemical classification of muscle fibres. *Exp. Neurol.* **26,** 424–432.
540. Stein R. B. (1974) Peripheral control of movement. *Physiol. Rev.* **54,** 215–243.
541. Stein R. B., French A. S., Mannard A. and Yemm R. (1972) New methods for analysing motor function in man and animals. *Brain Res.* **40,** 187–192.
542. Steiner T. S., Thexton A. J. and Weber W. V. (1973) The electromyogram as a measure of reflex response in experimental animals. *J. Appl. Physiol.* **35,** 762–769.
543. Stephens J. A., Gerlach R. L., Reinking R. M. and Stuart A. B. (1973) Fatiguability of medial gastrocnemius motor units in the cat. In: Stein R. B., Pearson K. S., Smith R. S. and Redford J. B. (ed.), *Control of Posture and Locomotion.* New York, Plenum, pp. 179–185.
544. Stewart D. (1927) Some aspects of the innervation of the teeth. *Proc. R. Soc. Med.* **20,** 1675–1685.
545. Stewart W. R. and King R. B. (1963) Fiber projections from the nucleus caudalis in the spinal trigeminal nucleus. *J. Comp. Neurol.* **121,** 271–286.
546. Stones H. H. (1938) An experimental investigation into the association of traumatic occlusion with periodontal disease. *Proc. R. Soc. Med.* **31,** 479–495.
547. Storey A. T. (1973) Reflex functions of the temporomandibular joint. *J. Prosthet. Dent.* **30,** 830–837.
548. Stuart D. G., Mosher C. G. and Gerlach R. L. (1972) Properties and central connections of Golgi tendon organs with special reference to locomotion. In: Barker B. Q. et al. (ed.), *Research in Muscle Development and the Muscle Spindle.* Amsterdam, Excerpta Medica, pp. 437–462.
549. Sumi T. (1970) Activity in single hypoglossal fibers during cortically-induced swallowing and chewing in rabbits. *Pfluegers Arch.* **314,** 329–346.
550. Sumino R. (1971) Central neural pathways involved in the jaw-opening reflex in the cat. In: Dubner R. and Kawamura Y. (ed.), *Oral-facial Sensory and Motor Mechanisms.* New York, Appleton-Century-Crofts, pp. 315–331.
551. Sumino R. and Nakamura Y. (1974) Synaptic potentials of hypoglossal motoneurons

and a common inhibitory interneuron in the trigemino-hypoglossal reflex. *Brain Res.* **73,** 439–454.
552. Sussman H. I. and Simring M. (1972) The distal wedge operation in periodontal therapy: a two-year evaluation. *J. Oral Med.* **27,** 106–109.
553. Svaetichin G. (1951) Analysis of action potentials recorded from single spinal ganglion cells. *Acta Physiol. Scand.* **24,** 23–57.
554. Svanberg G. and Lindhe J. (1973) Experimental tooth hypermobility in the dog. A methodological study. *Odontol. Revy.* **24,** 269–282.
555. Svanberg G. and Lindhe J. (1974) Vascular reactions in the periodontal ligament incident to trauma from occlusion. *J. Clin. Periodontol.* **1,** 58–69.
556. Szentagothai J. (1948) Anatomical considerations of monosynaptic reflex arcs. *J. Neurophysiol.* **11,** 445–453.
557. Szentagothai J. (1949) Functional representation in the motor trigeminal nucleus. *J. Comp. Neurol.* **90,** 111–120.
558. Tabary J. C., Tabary C., Tardieu C., Tardieu G. and Goldspink G. (1972) Physiological and structural changes in the cat's soleus muscle due to immobilization at different lengths by plaster casts. *J. Physiol. (Lond.)* **224,** 231–244.
559. Takata M. and Kawamura Y. (1970) Studies on summation of IPSP of masseter motoneuron. *J. Osaka Univ. Dent. Sch.* **10,** 81–97.
560. Tallgren A. (1957) Changes in adult face height due to ageing, wear and loss of teeth, and prosthetic treatment. *Acta Odontol. Scand.* **15,** Suppl. 24.
561. Tallgren A. (1966) The reduction in face height of edentulous and partially edentulous subjects during long term wear. *Acta Odontol. Scand.* **24,** 195–239.
562. Tardieu C., Tabary J. C. and Tardieu G. (1968) Etude mécanique et électromyographique de réponses à différentes pertubations du maintien postural. *J. Physiol. (Paris)* **60,** 243–259.
563. Tardieu C., Tardieu G., Gagnard L. and Tabary C. (1969) Les rétractions musculaires Etude expérimentale. Conséquences thérapeutiques. *Revue Pratn.* **19,** 1535–1543.
564. Tarkhan A. A. (1934) The innervation of extrinsic ocular muscles. *J. Anat.* **68,** 293–313.
565. Taylor A. (1972) Muscle receptors in the voluntary control of movement. *Paraplegia* **9,** 167–172.
566. Taylor A. and Cody F. W. J. (1974) Jaw muscle spindle activity in the cat during normal movements of eating and drinking. *Brain Res.* **71,** 523–530.
567. Taylor A., Cody F. W. J. and Bosley M. A. (1973) Histochemical and mechanical properties of the jaw muscles of the cat. *Exp. Neurol.* **38,** 99–109.
568. Taylor A. and Davey M. R. (1968) Behaviour of jaw muscle stretch receptors during active and passive movements in the cat. *Nature (Lond.)* **220,** 301–302.
569. Thelander H. E. (1924) The course and distribution of the radix mesencephalica trigemini in the cat. *J. Comp. Neurol.* **37,** 207–220.
570. Thexton A. J. (1969) Reflex control of jaw movement in the cat. Ph.D. Thesis. University of London.
571. Thexton A. J. (1969) Characteristics of reflex jaw opening in the cat. *J. Physiol. (Lond.)* **201,** 67–68 *P.*
572. Thexton A. J. (1971) Experimentally elicited cyclic jaw movement. *J. Dent. Res.* **50,** 696–697. (Abstract.)
573. Thexton A. J. (1971) A cyclical palato-lingual reflex. *J. Dent. Res.* **50,** 1192. (Abstract.)
574. Thexton A. J. (1972) Inhibition and facilitation of temporalis responses. *J. Dent. Res.* **51,** 1266. (Abstract.)
575. Thexton A. J. (1973) Oral reflexes elicited by mechanical stimulation of palatal mucosa in the cat. *Arch. Oral Biol.* **18,** 971–980.
576. Thexton A. J. (1974) Jaw opening and jaw closing reflexes in the cat. *Brain Res.* **66,** 425–433.
577. Thexton A. J. (1974) Some aspects of neurophysiology of dental interest II. Oral reflexes and neural oscillators. *J. Dent.* **2,** 131–137.
578. Thexton A. J. and Hiiemae K. (1974) Does the jaw opening reflex act as an effective protective response ? *J. Dent. Res.* **53,** 1067 (Abstract).
579. Thexton A. J. and Hiiemae K. (1975) The twitch tension characteristics of opossum jaw musculature. *Arch. Oral Biol.* **20,** 743–748.
580. Thexton A. J. and Hiiemae K. (1975) Masticatory electromyographic activity as a function of food consistency. *J. Dent Res.* **54,** (Special Issue A) L95, (Abstract).

581. Thilander B. (1961) Innervation of the temporo-mandibular joint capsule in man. *Trans. R. Sch. Dent. Stockholm* 7, 1–67.
582. Thompson J. R. (1954) Concepts regarding function of the stomatognathic system. *J. Am. Dent. Assoc.* **48,** 626–637.
583. Thompson J. R. and Brodie A. G. (1942) Factors in the position of the mandible. *J. Am. Dent. Assoc.* **29,** 925–941.
584. Thompson H. (1959) Mandibular joint pain. *Br. Dent. J.* **107,** 243–251.
585. Thompson H. (1971) Mandibular dysfunction syndrome. *Br. Dent. J.* **130,** 187–193.
586. Thomson J. C. and Macdonald N. S. (1969) Monitoring mandibular posture. *J. Biomech.* **2,** 319–323.
587. Trapozzano V. R. and Lazzari J. R. (1967) The physiology of the terminal rotational position of the condyles in the temporomandibular joint. *J. Prosthet. Dent.* **17,** 122–133.
588. Troelstrup B. and Møller E. (1970) Electromyography of the temporalis and masseter muscles in children with unilateral cross-bite. *Scand. J. Dent. Res.* **78,** 425–430.
589. Troest T. (1964) Diagnosing minute deflective occlusal contacts. *J. Prosthet. Dent.* **14,** 71–73.
590. Tryde G., Frydenberg O. and Brill N. (1962) An assessment of the tactile sensibility in human teeth. *Acta Odontol. Scand.* **20,** 233–256.
591. Tueller V. M. (1969) The relationship between the vertical dimension of occlusion and forces generated by closing muscles of mastication. *J. Prosthet. Dent.* **22,** 284–288.
592. Uchizono K. (1975) Characteristics of excitatory and inhibitory synapses in the central nervous system of the cat. *Nature (Lond.)* **207,** 642–643.
593. Vallbo Å. B. (1974) Afferent discharge from human muscle spindles in non-contracting muscles. Steady state impulse frequency as a function of joint angle. *Acta Physiol. Scand.* **90,** 303–318.
594. Vallbo Å. B. (1974) Human muscle spindle discharge during isometric voluntary contractions. Amplitude relations between spindle frequency and torque. *Acta Physiol. Scand.* **90,** 319–336.
595. Valverde F. (1962) Reticular formation of the albino rat's brain stem. Cytoarchitecture and corticofugal connections. *J. Comp. Neurol.* **119,** 25–54.
596. Valverde F. (1966) The pyramidal tract in rodents. A study of its relations with the posterior column nuclei, dorso-lateral reticular formation of the medulla oblongata and cervical spinal cord. *Z. Zellforsch. Mikrosk. Anat.* **71,** 298–363.
597. Vedral D. F. and Matzke H. A. (1967) Topographical localization of the muscles of mastication in the motor nucleus of the trigeminal nerve in the cat. *J. Hirnforsch.* **9,** 565–569.
598. Voss H. (1935) Ein besonders reichliches Vorkommen von Muskelspindeln in der tiefen Portion des M. masseter des Menschen und der Arthropoiden. *Anat. Anz.* **81,** 290–292.
599. Voss H. (1956) Zahl und Anordnung der Muskelspindeln in den oberen Zungenbeinmuskeln im M. trapezius und M. latissimus dorsi. *Anat. Anz.* **103,** 443–446.
600. de Vree F. L. and Gans C. (1973) Masticatory responses of pygmy goats (*Capra Hircus*) to different foods. *Am. Zoo.* **13,** 1342–1343. (Abstract.)
601. Wachstein M. and Meisel E. (1955) The distribution of histochemically demonstrable succinic dehydrogenase and of mitochondria in tongue and skeletal muscles. *J.Biophys. Biochem. Cytol.* **1,** 483–487.
602. Waerhaug J. (1955) Pathogenesis of pocket formation in traumatic occlusion. *J. Periodontol.* **26,** 107–118.
603. Wagner A. G. (1971) Comparison of four methods to determine rest position of the mandible. *J. Prosthet. Dent.* **25,** 506–514.
604. Walker L. B. and Rajagopal M. D. (1959) Neuromuscular spindles in the human tongue. *Anat. Rec.* **133,** 438. (Abstract.)
605. Wallenberg A. (1904) Neue Untersuchungen über der Hirnstamm der Taube. III Die cerebrale Trigeminuswurzel. *Anat. Anz.* **25,** 526–528.
606. Warwick R. (1964) Oculomotor organization. In: Bender M. B. (ed.), *The Oculomotor System.* New York, Harper & Row. Chapter 7, pp. 173–201.
607. Watt D. M. (1965) Report on a method of direct occlusal analysis. *Dent. Pract. Dent. Rec.* **15,** 416–420.
608. Watt D. M. (1966) Gnathosonics—a study of sounds produced by the masticatory mechanism. *J. Prosthet. Dent.* **16,** 73–82.

609. Watt D. M. (1966) Clinical applications of gnathosonics. *J. Prosthet. Dent.* **16,** 83–95.
610. Watt D. M. (1967) A gnathosonic study of tooth impact. *Dent. Pract. Dent. Rec.* **17,** 317–324.
611. Watt D. M. (1967) The stereostethoscope—an instrument for clinical gnathosonics. *J. Prosthet. Dent.* **18,** 458–464.
612. Watt D. M. (1968) Gnathosonics in occlusal evaluation. *J. Prosthet. Dent.* **19,** 133–143.
613. Watt D. M. (1968) A study of the reproducibility of articulator settings from graphic records of mandibular movement. *Dent. Pract. Dent. Rec.* **19,** 119–122.
614. Watt D. M. (1969) Recording the sounds of tooth contact: a diagnostic technique for evaluation of occlusal disturbances. *Int. Dent. J.* **19,** 221–238.
615. Watt D. M. (1970) The diagnosis and treatment of gnathic dysfunctions. *J. R. Coll. Surg. Edinb.* **15,** 121–128.
616. Watt D. M. (1970) Use of sound in oral diagnosis. *Proc. R. Soc. Med.* **63,** 793–797.
617. Watt D. M. (1970) Classification of occlusion. *Dent. Pract. Dent. Rec.* **20,** 305–308.
618. Weddell G., Harpman J. A., Lambley D. G. and Young L. (1940) The innervation of the musculature of the tongue. *J. Anat.* **74,** 255–267.
619. Weijs W. A. and Dantuma R. (1975) Electromyography and mechanics of mastication in the albino rat. *J. Morphol.* (In press.)
620. Weinberg E. (1928) The mesencephalic root of the fifth nerve. A comparative anatomical study. *J. Comp. Neurol.* **46,** 249–405.
621. Wentz F. M., Jarabak J. and Orban B. (1958) Experimental occlusal trauma initiating cuspal interferences. *J. Periodontol.* **29,** 117–127.
622. Werner G. and Mountcastle V. B. (1963) The variability of central neural activity in a sensory system and its implications for the central reflections of sensory events. *J. Neurophysiol.* **26,** 958–977.
623. Wilkie J. K. (1964) Preliminary observations on pressor sensory thresholds of anterior teeth. *J. Dent. Res.* **43,** 962. (Abstract.)
624. Williams P. E. and Goldspink G. (1971) Longitudinal growth of striated muscle fibres. *J. Cell Sci.* **9,** 751–767.
625. Wills D. J., Picton D. C. A. and Davies W. I. R. (1974) The effect of the rate of application of force on the intrusion of central incisors in adult monkeys. *J. Dent. Res.* **53,** 1054. (Abstract.)
626. Wilson D. M. and Waldron I. (1968) Models for the generation of the motor output pattern in flying locusts. *Proc. I.E.E.* **56,** 1058–1064.
627. Wilson D. M. and Wyman R. J. (1965) Motor output patterns during random and rhythmic stimulation of locust thoracic ganglia. *Biophys. J.* **5,** 121–143.
628. Wilson S. A. K. (1940) *Neurology.* London, Arnold. p. 1667.
629. Wilson V. J. and Kato M. (1965) Excitation of extensor motoneurons by group II afferent fibers in ipsilateral muscle nerves. *J. Neurophysiol.* **28,** 545–554.
630. Winlock H. E. (1941) Materials used at the embalming of King Tutankhamun, Paper No. 10, New York, Metropolitan Museum of Art.
631. Wold J. E. and Brodal A. (1974) The cortical projection of the orbital and proreate gyri to the sensory trigeminal nuclei in the cat. An experimental anatomical study. *Brain Res.* **65,** 381–395.
632. Woodburne R. T. (1936) A phylogenetic consideration of the primary and secondary centres and connections of the trigeminal complex in a series of vertebrates. *J. Comp. Neurol.* **65,** 403–501.
633. Woollard H. (1930) The innervation of the ocular muscles. *J. Anat.* **65,** 215–223.
634. Woolsey C. N., Górska T., Wetzel A., Erickson T. C., Earls F. J. and Allman J. M. (1972) Complete unilateral section of the pyramidal tract at the medullary level in *Macaca Mulatta. Brain Res.* **40,** 119–123.
635. Wyke B. D. (1967) The neurology of joints. *Ann R. Coll. Surg. Engl.* **41,** 25–50.
636. Wyke B. D. (1974) Neuromuscular mechanisms influencing mandibular posture: a neurologist's review of current concepts. *J. Dent.* **2,** 111–120.
637. Yellin H. and Guth L. (1970) The histochemical classification of muscle fibres. *Exp. Neurol.* **26,** 424–432.
638. Yemm R. (1969) Variations in the electrical activity of the human masseter muscle occurring in association with emotional stress. *Arch. Oral Biol.* **14,** 873–878.
639. Yemm R. (1972) The response of the masseter and temporal muscles following electrical stimulation of oral mucous membrane in man. *Arch. Oral Biol.* **17,** 23–33.

640. Yemm R. (1972) Reflex jaw opening following electrical stimulation of oral mucous membrane in man. *Arch. Oral Biol.* **17,** 513–523.
641. Yemm R. and Berry D. C. (1969) Passive control in mandibular rest position. *J. Prosthet. Dent.* **22,** 30–36.
642. Yemm R. and Nordstrom S. H. (1974) Forces developed by tissue elasticity as a determinant of mandibular resting posture in the rat. *Arch. Oral Biol.* **19,** 347–351.
643. Yu S-K. J., Schmitt A. and Sessle B. J. (1973) Inhibitory effects on jaw muscle activity of innocuous and noxious stimulation of facial and intraoral sites in man. *Arch. Oral Biol.* **18,** 861–870.
644. Zapata P. and Torrealba G. (1971) Mechanosensory units in the hypoglossal nerve of the cat. *Brain Res.* **32,** 349–367.

Additional References

645. Achari N. K. and Thexton A. J. (1972) Diencephalic influences on the jaw opening reflex in the cat. *Arch. Oral Biol.* **17,** 1073–1080.
646. Angel R. W., Garland H. and Moore W. (1973) The unloading reflex during blockade of antagonist muscle nerves. *Electroencephalogr. Clin. Neurophysiol.* **34,** 303–307.
647. Baker G. I. and Laskin D. M. (1969) Histochemical characterisation of the muscles of mastication. *J. Dent. Res.* **48,** 97–104.
648. Beaudreau D. E., Daugherty W. F. jun. and Masland W. S. (1969) Two types of pause in masticatory muscles. *Am. J. Physiol.* **216,** 16–21.
649. Blair G. A. S. and Gordon D. S. (1973) Trigeminal neuralgia and dental malocclusions. *Br. Med. J.* **4,** 38–40.
650. Buchtal F., Dahl K. and Rosenfalck P. (1973) Rise time of the spike potential in fast and slowly contracting muscle of man. *Acta Physiol. Scand.* **87,** 261–269.
651. Burke R. E., Rymer W. Z. and Walsh J. V. (1973) Organisation of synaptic input to defined types of motor units in cat medial gastrocnemius muscle. *Proc. Soc. Neuroscience* **45,** 5.
652. Cooper S. (1966) Muscle spindles and motor units. In: Andrew B. L. (ed.), *Control and Innervation of Skeletal Muscle.* Edinburgh, Livingstone.
653. Foster G. E. (1973) Mesencephalic trigeminal nucleus of rat. *J. Anat.* **114,** 293. (Abstract.)
654. Hagbarth K-E. and Vallbo Å. B. (1969) Single unit recordings from muscle nerves in human subjects. *Acta Physiol. Scand.* **76,** 321–334.
655. Henneman E. (1968) Peripheral mechanisms involved in the control of muscle. In: Mountcastle, V. B. (ed.), *Medical Physiology,* 12th ed., St. Louis, C. V. Mosby. pp. 1697–1716.
656. Hiiemae K. (1971) The structure and function of the jaw muscles in the rat; II. Fibre type and composition. *Zool. J. Linn.* **50,** 101–109.
657. Hiiemae K. and Thexton A. J. (1974) Twitch tension characteristics of opossum jaw musculature. *J. Dent. Res.* **53,** 1067. (Abstract.)
658. Hubbard J. E. and Di Carlo V. (1973) Fluorescence histochemistry of monoamine-containing cell bodies in the brain stem of the squirrel monkey (*Saimiri sciureus*). I. The locus caeruleus. *J. Comp. Neurol.* **147,** 533–566.
659. Ingervall B. (1974) Relation between height of the articular tubercle of the temporo-mandibular joint and facial morphology. *Angle Orthod.* **44,** 15–24.
660. Jankowska E., Jukes M. G. M., Lund S. and Lundberg A. (1967) The effect of DOPA on the spinal cord. 6. Half-centre organization of interneurones transmitting effects from the flexor reflex afferents. *Acta Physiol. Scand.* **70,** 389–402.
661. Lund J. P. and Lamarre Y. (1974) Activity of neurons in the lower precentral cortex during voluntary and rhythmical jaw movements in the monkey. *Exp. Brain Res.* **19,** 282–299.
662. Melzack, R. (1973) *The Puzzle of Pain.* Harmondsworth, Penguin.
663. Millar J. (1973) Joint afferent fibres responding to muscle stretch, vibration and contraction. *Brain Res.* **63,** 380–383.

664. Møller E. (1974) Action of the muscles of mastication. In: Kawamura Y. (ed.) *Frontiers of Oral Physiology* Vol. I, Basel, Karger, pp. 121–158.
665. Montet P. (1946) *La Vie Quotidienne en Egypte au Temps des Ramses.* Paris, Libraire Hachette.
666. Nakamura M., Takatori S., Nozaki S. and Kikuchi M. (1975) Monosynaptic reciprocal control of trigeminal motoneurons from the medial bulbar reticular formation. *Brain Res.* **89**, 144–148.
667. Olsen C. B., Carpenter D. O. and Henneman E. (1968) Orderly recruitment of muscle action potentials. *Arch. Neurol. Psychiat. (Lond.)* **19**, 591–597.
668. Sauerland E. K., Nakamura Y., and Clemente C. D. (1967) The role of the lower brain stem in cortically induced inhibition of somatic reflexes in the cat. *Brain Res.* **6**, 164–180.
669. Scher E. A. (1958) The locating peg tray. *Int. Dent. J.* **8**, 40–42.
670. Scher E. A. (1965) The alginate resting record. *Proc. Br. Soc. Study Prosthet. Dent.* pp. 31–32.
671. Spillane J. D. and Wells C. E. C. (1964) The neurology of Jennerian vaccination. *Brain* **87**, 1–44.
672. Tanaka R. (1974) Reciprocal Ia inhibition during voluntary movements in man. *Exp. Brain Res.* **21**, 529–540.
673. Touloumis C., Patry Y., Lund J. P., Richmond F. and Lamarre Y. (1975) Golgi tendon organs in masseter and temporalis muscles of the kitten. *J. Dent. Res.* **54** (Special Issue A), 109 (Abstract.)